Problems and Solutions in Optics and Photonics

ABOUT THE AUTHORS

Ajoy Ghatak obtained his MSc from Delhi University and PhD from Cornell University. He has recently retired from IIT Delhi as Professor of Physics. He received the 2008 SPIE Educator Award in recognition of *"his unparalleled global contributions to the field of fiber optics research, and his tireless dedication to optics education worldwide..."* and the 2003 Optical Society of America Esther Hoffman Beller Award in recognition of his *"outstanding contributions to optical science and engineering education"*. He is also a recipient of the CSIR SS Bhatnagar Award, 16th Khwarizmi International Award, the International Commission for Optics Galileo Galilei Award and the UGC Meghnad Saha Award for his research work. He has authored/ co-authored several books in the areas of *Optics*, *Fiber Optics*, *Quantum Mechanics* and *Special Theory of Relativity*. His latest book is entitled *Albert Einstein*: *A Glimpse of his Life*, *Philosophy and Science*. He received DSc (Honoris Causa) from University of Burdwan in 2007.

K Thyagarajan obtained his MSc from Delhi University and PhD from IIT Delhi and is currently Professor of Physics at IIT Delhi. He has held Visiting positions at various research laboratories and Universities in France, USA, Canada, Hong Kong and Japan. He has published more than 140 research papers in international journals, has filed five patent applications and is the co-author (with Professor A K Ghatak) of seven books, In 1998, he was the co-recipient of the "Fiber Optic Person of the Year 1997" Award by Lucent Technologies–Finolex and Voice and Data, India. In 2003, he was decorated with the title of *"Officier dans l'ordre des Palmes Académiques"* by the French Government, in 2005, he was elected *Fellow of the Optical Society of America* and in 2008, he was elected *Fellow of the Indian National Academy of Engineers*.

Problems and Solutions in Optics and Photonics

Ajoy Ghatak and K Thyagarajan
Department of Physics
IIT Delhi

Tata McGraw Hill Education Private Limited
NEW DELHI

McGraw-Hill Offices
New Delhi New York St Louis San Francisco Auckland Bogotá Caracas
Kuala Lumpur Lisbon London Madrid Mexico City Milan Montreal
San Juan Santiago Singapore Sydney Tokyo Toronto

Tata McGraw-Hill

Published by the Tata McGraw Hill Education Private Limited,
7 West Patel Nagar, New Delhi 110 008.

Problems and Solutions in Optics and Photonics

ISBN (13): 978-1-25-900316-5
ISBN (10): 1-25-900316-7

Vice President and Managing Director—McGraw-Hill Education, Asia-Pacific Region: *Ajay Shukla*

Head—Higher Education Publishing and Marketing: *Vibha Mahajan*
Publishing Manager—SEM & Tech. Ed.: *Shalini Jha*
Editorial Executive: *Smruti Snigdha*
Development Editor: ***Renu Upadhyay***
Copy Editor: *Preyoshi Kundu*

Sr Production Manager: ***Satinder S Baveja***
Asst Manager Production: ***Anjali Razdan***

Marketing Manager—SEM & Tech. Ed.: *Biju Ganesan*
Product Specialist—SEM & Tech. Ed.: *Sachin Tripathi*

General Manager—Production: ***Rajender P Ghansela***
Asst General Manager—Production: ***B L Dogra***

Typeset at Tej Composers, WZ-391, Madipur, New Delhi 110063, and printed at A.P. Offset Pvt. Ltd., 25/487, Zulfe Bengal, Dilshad Garden, New Delhi - 110095.

Cover Printer: A.P. Offset

AZLACRACRYZXL

The McGraw-Hill Companies

Contents

Preface

Ever since the first fabrication of the laser in 1960, there has been a tremendous growth in the fields of optics and photonics with extremely important applications in many diverse areas from optical amplifiers to laser physics, fiber optics to optical communications, optical data processing to holography, etc. These applications have as background the phenomena of wave propagation, interference, diffraction and polarisation. Although there are many text books in this area, we feel that the concepts get better understood if the students work out a large number of problems. This book is a collection of problems (and their solutions) starting from basic phenomena in optics to their applications and we hope that this book will help students to get a better understanding of Optics and Photonics. The book is divided into many chapters covering various aspects of optics and photonics and each chapter has a number of problems followed by their solutions, placed at the end of the chapter. This will enable the student to refer to the solutions when necessary and at the same provide an opportunity to solve some or all of these problems by himself or herself before consulting the solutions.

For over 40 years, we have been teaching courses related to optics and photonics at IIT Delhi – these courses have been taught both at the undergraduate as well as at the postgraduate level. A large number of problems have emerged from the teaching of these courses.

The topics covered in the present edition have been included, keeping in mind the needs of students pursuing B Sc and M Sc Physics, apart from courses like B Tech.

Our Objective

- To make learning and understanding smoother for the students.
- Identify important topics of the subject and the basic concepts underlying these.
- Arranging them chapterwise giving the fundamental formulae, which are ultimately the key to solving problems.

Features of the Book

1. *A Quick Review* – Important formulae and topics discussed in every chapter.
2. Includes latest applications in the field.
3. *Questions and Answers* – Important concepts further enumerated in question and answer format.
4. *Problems based on principles, concepts and numericals* – Over 400 numericals covering a wide range of difficulty provided with final answers.

5. *Objective Type Questions/ Multiple Choice Questions* for better understanding of the concepts and principles.

Chapter Organisation

The book is divided into 24 chapters. Each chapter begins with a quick review on the important concepts and principles, helping students understand the chapter better.

Chapter 1 deals with Matrix method in paraxial optics. **Chapter 2** elucidates Fermat's principle, Snell's law and the ray equation. **Chapters 3**, **4**, and **5** will make students understand Optical instruments, Aberrations, and Huygens' principle and applications, respectively.

The concept of interference has been divided into two parts-division of wave front has been dealt in **Chapter 6** whereas Division of amplitude has been discussed in **Chapter 7**. **Chapter 8** explains Multiple beam interferometry. **Chapters 9** and **10** are based on Fraunhofer diffraction.

Chapters 11, **12**, **13**, **14**, and **15** deal with Fresnel diffraction, Fourier optics and holography, Polarisation I: Basics and double refraction, Polarisation II: Jones vectors and Jones matrices, and Wave equation and its solutions.

Topics like Group velocity and Pulse dispersion, Basic laser physics, Basic concepts of Fiber optics and ray optics consideration in multimode fibers, Basic waveguide theory and Concept of modes, and Single mode fibers in fiber optics have been explained in **Chapters 16**, **17**, **18**, **19**, and **20**.

The last four chapters in this book i.e., **Chapters 21 to 24** help in understanding concepts of integrated optics, electro-optic effect, acousto-optic effect, and nonlinear optics.

In addition, the book also contains a number of **multiple choice questions and references along with suggested readings** at the end of the book.

Acknowledgements

Dr Ghatak is very grateful to Department of Science and Technology, Government of India for supporting this endeavour and providing financial assistance under their USERS program.

At IIT Delhi, we were very fortunate to have the opportunity to interact with many outstanding colleagues and students which helped us in creating a large number of problems that we have put together in this book. In particular, we would like to thank Professor Ishwar Goyal, Professor B D Gupta, Professor Arun Kumar, Professor Bishnu Pal, Professor Anurag Sharma, Professor M R Shenoy, Professor Kehar Singh and Professor Ravi Varshney for many stimulating discussions.

Dr Ghatak and Dr Thyagarajan would like to express their gratitude and thank the following reviewers who took out time to review the book.

Faculty Reviewers

Kallol Bhattacharya	Dept of Applied Optics and Photonics, Calcutta University, Kolkata
Neeraj Pandey	Instruments Research and Development Establishment, DRDO, Dehradun

S K Saha Indian Institute of Astrophysics, Bengaluru

Student Reviewer

Ravi Sreedhar Loyola College, Chennai

Sincere thanks to the editorial and publishing professionals of Tata McGraw-Hill for their keen interest and support in bringing out this edition in its present form.

AJOY GHATAK
K THYAGARAJAN

Feedback

We have tried to work out each and every problem very carefully; nevertheless, if there are any errors (or any suggestions), we will be very grateful if they are pointed out to us. Also we would greatly appreciate suggestions for introducing new problems (and solutions) in the book. We will acknowledge the same when we introduce them in future editions. Our email addresses are *ajoykghatak@gmail.com* and *ktrajan@gmail.com*.

Publishers' Note

Do you have a feature request? A suggestion? We are always open to new ideas (the best ideas come from you!). You may send your comments to ***tmh.sciencemathsfeedback@gmail.com*** (don't forget to mention the title and author name in the subject line).

Piracy-related issues may also be reported.

Matrix Method in Paraxial Optics

1

A Quick Review

In the paraxial approximation, rays remain close to the optical axis and are assumed to make small angles with the axis. Such a ray is initially specified by a 2 × 1 matrix with elements λ_1 and x_1

$$\begin{pmatrix} \lambda_1 \\ x_1 \end{pmatrix}$$

where x_1 represents the distance from the axis and the parameter λ is defined by the following equation:

$$\lambda = n \sin \alpha \tag{1}$$

which represents the product of the refractive index and the sine of the angle that the ray makes with the z axis (see Fig. 1.1).

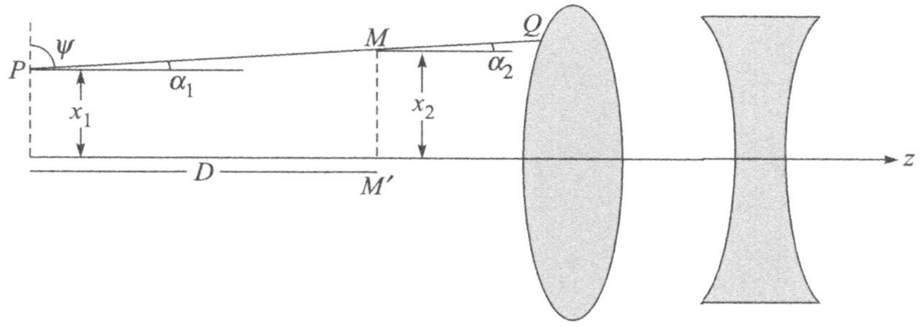

Fig. 1.1 In a homogeneous medium, the ray travels in a straight line.

1.1 EFFECT OF TRANSLATION

If a ray is initially specified by a 2 × 1 matrix with elements λ_1 and x_1, then after propagating through a distance D in a homogeneous medium of refractive index n_1, the final ray is given by

$$\begin{pmatrix} \lambda_2 \\ x_2 \end{pmatrix} = T \begin{pmatrix} \lambda_1 \\ x_1 \end{pmatrix} \tag{2}$$

where, $$T = \begin{pmatrix} 1 & 0 \\ D/n_1 & 1 \end{pmatrix} \tag{3}$$

The matrix T is known as the translation matrix and represents the effect of translation through a distance D in a homogeneous medium of refractive index n_1.

1.2 EFFECT OF REFRACTION

Consider a ray incident on a spherical surface (of radius R) separating two media of refractive indices n_1 and n_2 (see Fig. 1.2). If (λ', x') and (λ'', x'') represent the coordinates of the ray at A' (just before refraction), and at A'' (just after refraction), then

$$\begin{pmatrix} \lambda'' \\ x'' \end{pmatrix} = \mathfrak{R} \begin{pmatrix} \lambda' \\ x' \end{pmatrix} \tag{4}$$

where, $$\mathfrak{R} = \begin{pmatrix} 1 & -P \\ 0 & 1 \end{pmatrix} \tag{5}$$

and $$P = \frac{n_2 - n_1}{R} \tag{6}$$

is known as the *power* of the refracting surface. The 2 × 2 matrix $\mathfrak{R}$ characterises refraction through the spherical surface. Note that

$$\det \mathfrak{R} = \det T = 1 \tag{7}$$

We will be using the analytical geometry sign convention so that the coordinates on the left of the point P are negative and coordinates on the right of P are positive (see Figs. 1.2 and 1.3).

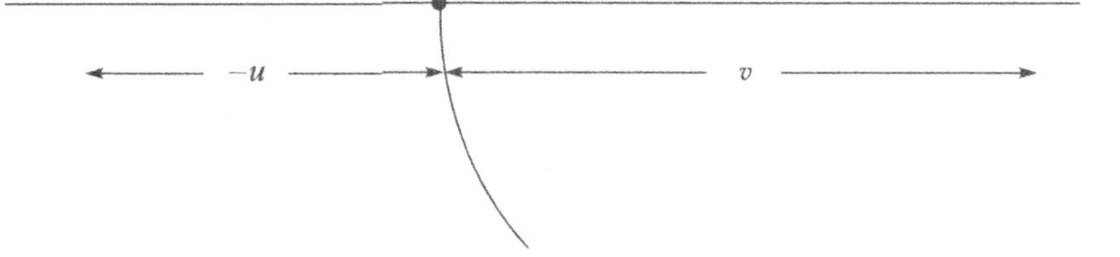

Fig. 1.2 Imaging by a spherical refracting surface separating two media of refractive indices n_1 and n_2.

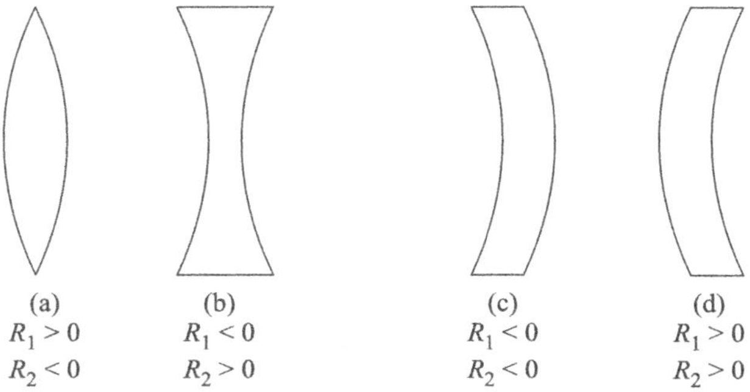

Fig. 1.3 Signs of R_1 and R_2 for different types of lenses.

In general, an optical system made up of a series of lenses can be characterised by the refraction and translation matrices. If a ray is specified by (λ', x') when it enters an optical system and is specified by (λ'', x'') when it leaves the system (points P and Q in Fig. 1.4), then we can, in general, write

$$\begin{pmatrix} \lambda'' \\ x'' \end{pmatrix} = S \begin{pmatrix} \lambda' \\ x' \end{pmatrix} \tag{8}$$

where the matrix
$$S = \begin{pmatrix} b & -a \\ -d & c \end{pmatrix} \tag{9}$$

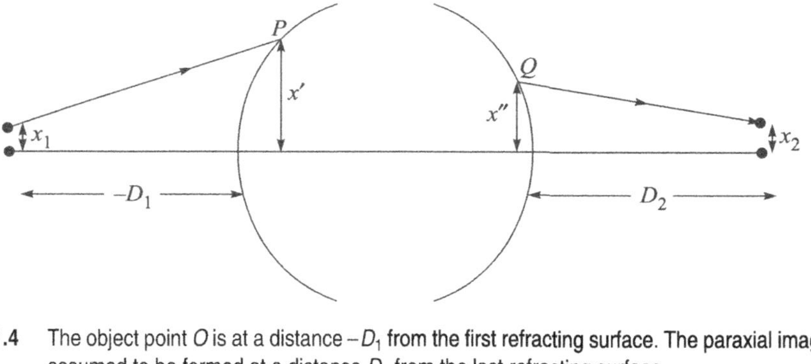

Fig. 1.4 The object point O is at a distance $-D_1$ from the first refracting surface. The paraxial image is assumed to be formed at a distance D_2 from the last refracting surface.

is called the *system matrix* and is determined solely by the optical system. We must note that:

The quantities b and c are dimensionless. The quantities a and P have the dimension of inverse length, and the quantity d has the dimension of length. In general, the units will not be given; however, it will be implied that a and P are in cm^{-1} and d is in cm.

We consider an object point O is at a distance $-D_1$ from the first refracting surface. The paraxial image is assumed to be formed at a distance D_2 from the last refracting surface (see Fig. 1.4). Thus,

$$\begin{pmatrix} \lambda_2 \\ x_2 \end{pmatrix} = \begin{pmatrix} 1 & 0 \\ D_2 & 1 \end{pmatrix} \begin{pmatrix} b & -a \\ -d & c \end{pmatrix} \begin{pmatrix} 1 & 0 \\ -D_1 & 1 \end{pmatrix} \begin{pmatrix} \lambda_1 \\ x_1 \end{pmatrix} \tag{10}$$

$$= \begin{pmatrix} b + aD_1 & -a \\ bD_2 + aD_1D_2 - cD_1 - d & c - aD_2 \end{pmatrix} \begin{pmatrix} \lambda_1 \\ x_1 \end{pmatrix} \tag{11}$$

Thus,
$$x_2 = (bD_2 + aD_1D_2 - cD_1 - d)\lambda_1 + (c - aD_2)x_1 \tag{12}$$

For a ray emanating from the axial object point (i.e., for $x_1 = 0$) the image plane is determined by the condition $x_2 = 0$. Thus, for the image plane we must have

$$bD_2 + aD_1D_2 - cD_1 - d = 0 \tag{13}$$

or,
$$\frac{b}{D_1} + a - \frac{c}{D_2} - \frac{d}{D_1D_2} = 0 \tag{14}$$

which would give us the relationship between the distances D_1 and D_2. When $D_1 = \infty$, $D_2 = \frac{c}{a}$. Corresponding to the image plane, we have

$$\begin{pmatrix} \lambda_2 \\ x_2 \end{pmatrix} = \begin{pmatrix} b + aD_1 & -a \\ 0 & c - aD_2 \end{pmatrix} \begin{pmatrix} \lambda_1 \\ x_1 \end{pmatrix} \tag{15}$$

For $x_2 \neq 0$, we obtain

$$x_2 = (c - aD_2)x_1$$

Consequently, the magnification of the system $M(= x_2/x_1)$ would be given by

$$M = \frac{x_2}{x_1} = c - aD_2 \tag{16}$$

Further, since

$$\begin{vmatrix} b + aD_1 & -a \\ 0 & c - aD_2 \end{vmatrix} = 1$$

we obtain
$$b + aD_1 = \frac{1}{c - aD_2} = \frac{1}{M} \tag{17}$$

1.3 UNIT PLANES

The unit planes are two planes, one each in the object and the image space, between which the magnification M is unity; i.e., any paraxial ray emanating from the unit plane in the object space will emerge at the same height from the unit plane in the image space. Thus, if d_{u1} and d_{u2} represent the distances of the unit planes from the refracting surfaces (see Fig. 1.5)[1] we obtain from Eq. (17)

$$b + ad_{u1} = \frac{1}{c - ad_{u2}} = 1 \tag{18}$$

or,
$$d_{u1} = \frac{1 - b}{a} \tag{19}$$

[1] Obviously, if we consider U_1 as an object plane, then U_2 is the corresponding image plane.

$$d_{u2} = \frac{c-1}{a} \tag{20}$$

Hence, the unit planes are determined completely by the elements of the system matrix S. It will be convenient to measure distances from the unit planes. Thus, if u is the distance of the object plane from the first unit plane and v is the distance of the corresponding image plane from the second unit plane (see Fig. 1.5), we obtain

$$D_1 = u + d_{u1} = u + \frac{1-b}{a} \tag{21}$$

and

$$D_2 = v + d_{u2} = v + \frac{c-1}{a} \tag{22}$$

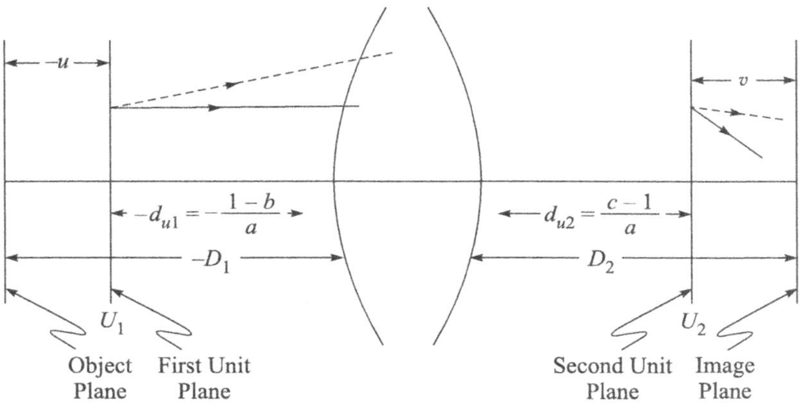

Fig. 1.5 U_1 and U_2 are the two unit planes. A ray emanating at any height from the first unit plane will cross the second unit plane at the same height.

Now, from Eq. (13) we have

$$D_2 = \frac{d + cD_1}{b + aD_1} \tag{23}$$

Substituting for D_1 and D_2 from Eq. (21) and (22), we get

$$v + \frac{c-1}{a} = \frac{d + cu + c(1-b)/a}{b + au + (1-b)}$$

or,

$$v = \frac{ad - bc + c(au+1) - (c-1)(1+au)}{a(1+au)}$$

$$= \frac{au}{a(1+au)} \tag{24}$$

where we have used the condition that

$$\det S = bc - ad = 1 \tag{25}$$

On simplification, we obtain

$$\frac{1}{v} - \frac{1}{u} = a \tag{26}$$

Thus, $1/a$ represents the focal length of the system if the distances are measured from the two unit planes.

PROBLEMS

1.1 (a) Consider imaging by a spherical surface (of radius of curvature R) separating two media of refractive indices n_1 and n_2. If (λ_1, x_1) and (λ_2, x_2) represent the coordinates of the ray at O, and at I (see Fig. 1.2), show that

$$\begin{pmatrix} \lambda_2 \\ x_2 \end{pmatrix} = \begin{pmatrix} 1+\dfrac{Pu}{n_1} & -P \\ \dfrac{v}{n_2}\left(1+\dfrac{Pu}{n_1}\right)-\dfrac{u}{n_1} & \left(1-\dfrac{vP}{n_2}\right) \end{pmatrix} \begin{pmatrix} \lambda_1 \\ x_1 \end{pmatrix} \tag{27}$$

(b) Using the above result derive the following equation determining the paraxial image

$$\frac{n_2}{v} - \frac{n_1}{u} = P = \frac{n_2 - n_1}{R} \tag{28}$$

1.2 Obtain the system matrix for a thick lens, and derive the thick lens and the thin lens formulae.

1.3 In continuation of the previous problem, determine the positions of the unit planes for a thick double convex lens with $|R_1| = |R_2|$.

1.4 (a) Obtain the elements of the system matrix of a combination of two thin lenses of focal lengths f_1 and f_2 separated by a distance t.

(b) Consider a system of two thin convex lenses of focal lengths 10 cm and 30 cm separated by a distance of 20 cm in air (see Fig. 1.6). Determine the system matrix elements and the positions of the unit planes. Assume a parallel beam of light incident from the left. Determine the positions of the unit planes and determine the image point. Using the unit planes draw the ray diagram.

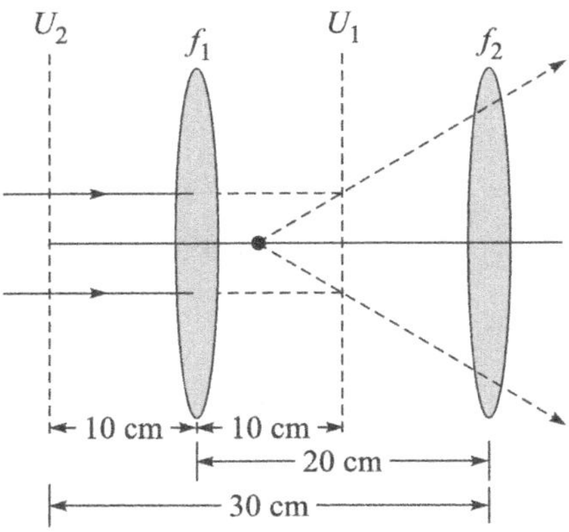

Fig. 1.6

1.5 Consider a thick biconvex lens whose magnitudes of the radii of curvature of the first and second surfaces are 45 cm and 30 cm respectively. The thickness of the lens is 5 cm and the refractive index of the material of the lens is 1.5. Determine the elements of system matrix and positions of the unit planes and determine the image point of an object at a distance of 90 cm from the first surface.

1.6 Consider a hemisphere of radius 20 cm and refractive index 1.5. If H_1 and H_2 denote the positions of the first and second principal points, then show that $AH_1 = 13.3$ cm and that H_2 lies on the second surface as shown in Fig. 1.7. Further, show that the focal length is 40 cm.

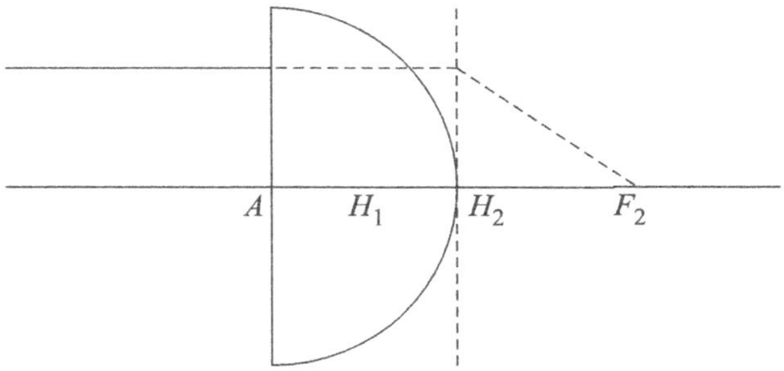

Fig. 1.7

1.7 Consider a thick lens of the form shown in Fig. 1.8; the radii of curvature of the first and second surfaces are −10 cm and +20 cm respectively and the thickness of the lens is 1.0 cm. The refractive index of the material of the lens is 1.5. Determine the positions of the principal planes.

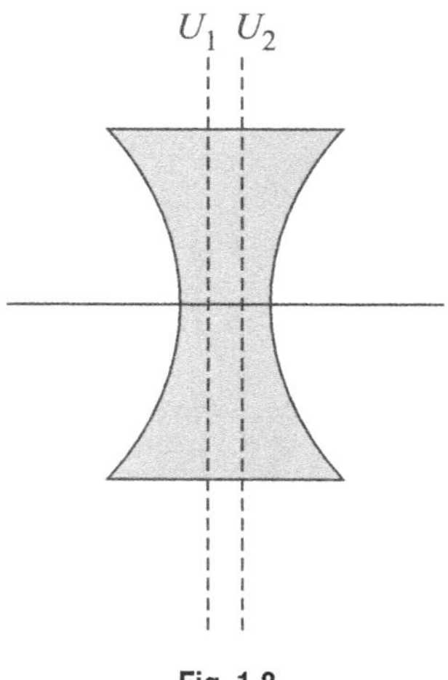

Fig. 1.8

1.8 Consider a sphere of radius 20 cm of refractive index 1.6. Find the positions of the paraxial focal point and the unit planes.

1.9 Consider a lens combination consisting of a convex lens (of focal length +15 cm) and a concave lens (of focal length −20 cm) separated by 25 cm. Determine the system matrix elements and the positions of the unit planes. For an object (of height 1 cm) placed at a distance of 27.5 cm from the convex lens, determine the size and position of the image.

1.10 Consider a system of two thin lenses as shown in Fig. 1.9. For a 1 cm tall object at a distance of 40 cm from the convex lens, calculate the position and size of the image.

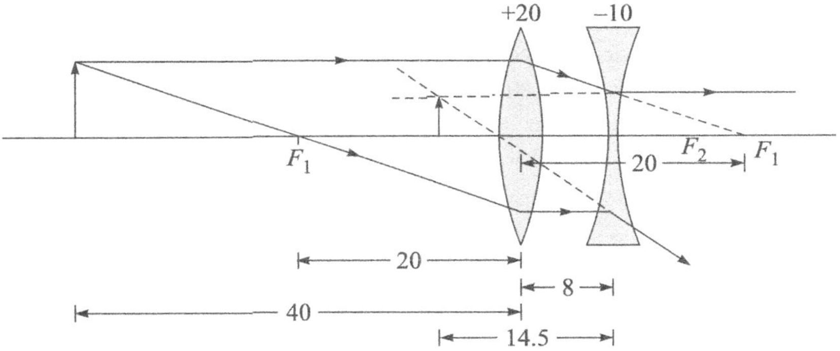

Fig. 1.9 A combination of two thin lenses.

SOLUTIONS

1.1 (a) Let (λ_1, x_1), (λ', x'), (λ'', x'') and (λ_2, x_2) represent the coordinates of the ray at O, A' (just before refraction), A'' (just after refraction), and I respectively (see Fig. 1.2). Thus,

$$\begin{pmatrix}\lambda' \\ x'\end{pmatrix} = \begin{pmatrix}1 & 0 \\ -u/n_1 & 1\end{pmatrix}\begin{pmatrix}\lambda_1 \\ x_1\end{pmatrix}$$

$$\begin{pmatrix}\lambda'' \\ x''\end{pmatrix} = \begin{pmatrix}1 & -P \\ 0 & 1\end{pmatrix}\begin{pmatrix}\lambda' \\ x'\end{pmatrix}$$

$$\begin{pmatrix}\lambda_2 \\ x_2\end{pmatrix} = \begin{pmatrix}1 & 0 \\ v/n_2 & 1\end{pmatrix}\begin{pmatrix}\lambda'' \\ x''\end{pmatrix}$$

or,

$$\begin{pmatrix}\lambda_2 \\ x_2\end{pmatrix} = \begin{pmatrix}1 & 0 \\ v/n_2 & 1\end{pmatrix}\begin{pmatrix}1 & -P \\ 0 & 1\end{pmatrix}\begin{pmatrix}1 & 0 \\ -u/n_1 & 1\end{pmatrix}\begin{pmatrix}\lambda_1 \\ x_1\end{pmatrix}$$

Simple manipulations give

$$\begin{pmatrix} \lambda_2 \\ x_2 \end{pmatrix} = \begin{pmatrix} 1+\dfrac{Pu}{n_1} & -P \\ \dfrac{v}{n_2}\left(1+\dfrac{Pu}{n_1}\right)-\dfrac{u}{n_1} & \left(1-\dfrac{vP}{n_2}\right) \end{pmatrix} \begin{pmatrix} \lambda_1 \\ x_1 \end{pmatrix} \tag{29}$$

(b) Using the above equation we obtain

$$x_2 = \left[\frac{v}{n_2}\left(1+\frac{Pu}{n_1}\right)-\frac{u}{n_1}\right]\lambda_1 + \left(1-\frac{vP}{n_2}\right)x_1 \tag{30}$$

For a ray emanating from an axial object point (i.e., for $x_1 = 0$) the image plane is determined by the condition $x_2 = 0$. Thus in the above equation, the coefficient of λ_1 should vanish and therefore

$$\frac{u}{n_1} = \frac{v}{n_2}\left(1+\frac{Pu}{n_1}\right)$$

or,
$$\frac{n_2}{v} - \frac{n_1}{u} = P = \frac{n_2 - n_1}{R} \tag{31}$$

which is the equation determining the paraxial image. Hence, on the image plane

$$\begin{pmatrix} \lambda_2 \\ x_2 \end{pmatrix} = \begin{pmatrix} 1+\dfrac{Pu}{n_1} & -P \\ 0 & 1-\dfrac{vP}{n_2} \end{pmatrix} \begin{pmatrix} \lambda_1 \\ x_1 \end{pmatrix} \tag{32}$$

giving
$$x_2 = \left(1-\frac{vP}{n_2}\right)x_1 \tag{33}$$

Thus, the magnification is given by

$$m = \frac{x_2}{x_1} = 1 - \frac{vP}{n_2} \tag{34}$$

which on using Eq. (31) gives

$$m = \frac{n_1 v}{n_2 u} \tag{35}$$

1.2 Let us consider a lens of thickness t and made of a material of relative refractive index n (see Fig. 1.10). Let R_1 and R_2 be the radii of curvature of the two surfaces. The ray is assumed to strike the first surface of the lens at P and emerge from point Q; let the coordinates of the ray at P and Q be

$$\begin{pmatrix} \lambda_1 \\ x_1 \end{pmatrix} \quad \text{and} \quad \begin{pmatrix} \lambda_2 \\ x_2 \end{pmatrix} \tag{36}$$

where λ_1 and λ_2 are the optical direction cosines of the ray at P and Q; x_1 and x_2 are the distances of points P and Q from the axis (see Fig. 1.10). The ray, in propagating from P to Q, undergoes two refractions [one at the first surface (whose radius of curvature is R_1) and the other at the second surface (whose radius of curvature is R_2)] and a translation through a distance[2] t in a medium of refractive index n. Thus

$$\begin{pmatrix} \lambda_2 \\ x_2 \end{pmatrix} = \begin{pmatrix} 1 & -P_2 \\ 0 & 1 \end{pmatrix}\begin{pmatrix} 1 & 0 \\ t/n & 1 \end{pmatrix}\begin{pmatrix} 1 & -P_1 \\ 0 & 1 \end{pmatrix}\begin{pmatrix} \lambda_1 \\ x_1 \end{pmatrix} \tag{37}$$

where, $$P_1 = \frac{n-1}{R_1} \quad \text{and} \quad P_2 = \frac{1-n}{R_2} = -\frac{n-1}{R_2} \tag{38}$$

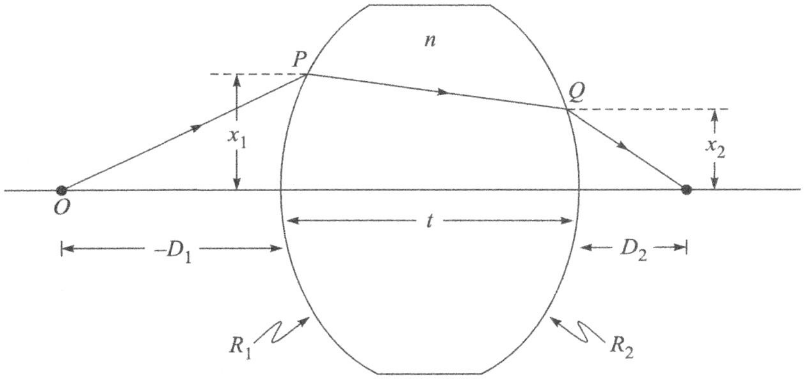

Fig. 1.10 A paraxial ray passing through a thick lens of thickness *t*.

represent the powers of the two refracting surfaces. Thus, our system matrix is given by

$$S = \begin{pmatrix} b & -a \\ -d & c \end{pmatrix} = \begin{pmatrix} 1 & -P_2 \\ 0 & 1 \end{pmatrix}\begin{pmatrix} 1 & 0 \\ t/n & 1 \end{pmatrix}\begin{pmatrix} 1 & -P_1 \\ 0 & 1 \end{pmatrix}$$

$$= \begin{pmatrix} 1 - \dfrac{P_2 t}{n} & -P_1 - P_2\left(1 - \dfrac{t}{n}P_1\right) \\ \dfrac{t}{n} & 1 - \dfrac{t}{n}P_1 \end{pmatrix} \tag{39}$$

Thus for a thick lens, the parameters of the system matrix are

$$a = P_1 + P_2\left(1 - \frac{t}{n}P_1\right); \quad b = 1 - \frac{P_2 t}{n}; \quad c = 1 - \frac{t}{n}P_1; \quad d = -\frac{t}{n} \tag{40}$$

To calculate the focal length, we note from Eq. (26) that

$$\frac{1}{f} = a = P_1 + P_2\left(1 - \frac{t}{n}P_1\right)$$

2. Notice that since we are dealing with paraxial rays, the distance between P and Q is approximately t.

Thus, $$\frac{1}{f} = (n-1)\left(\frac{1}{R_1} - \frac{1}{R_2}\right) + \frac{(n-1)^2 t}{nR_1R_2} \tag{41}$$

For a thin lens, $t \to 0$ and the system matrix takes the following form:

$$S = \begin{pmatrix} 1 & -P_1 - P_2 \\ 0 & 1 \end{pmatrix} \tag{42}$$

Thus for a thin lens,

$$a = P_1 + P_2 \qquad b = 1 \qquad c = 1 \qquad d = 0 \tag{43}$$

Substituting the above values of a, b, c, and d in Eq. (13), we obtain

$$D_2 + (P_1 + P_2) D_1 D_2 - D_1 = 0$$

or, $$\frac{1}{D_2} - \frac{1}{D_1} = P_1 + P_2$$

$$= (n-1)\left(\frac{1}{R_1} - \frac{1}{R_2}\right) \tag{44}$$

or, $$\frac{1}{D_2} - \frac{1}{D_1} = \frac{1}{f} \tag{45}$$

where, $$f = \frac{1}{P_1 + P_2} = \left[(n-1)\left(\frac{1}{R_1} - \frac{1}{R_2}\right)\right]^{-1} \tag{46}$$

represents the focal length of the lens. Equation (45) is the well-known thin lens formula. Thus for a thin lens, the system matrix takes the following form:

$$S = \begin{pmatrix} 1 & -\frac{1}{f} \\ 0 & 1 \end{pmatrix} \tag{47}$$

1.3 For a thick lens one obtains [using Eqs. (19), (20) and (40)]

$$d_{u1} = \frac{P_2 t}{n} \frac{1}{P_1 + P_2[1 - (t/n)P_1]} \tag{48}$$

and $$d_{u2} = -\frac{t}{n} \frac{P_1}{P_1 + P_2[1 - (t/n)P_1]} \tag{49}$$

For a thick double convex lens with $|R_1| = |R_2|$

$$P_1 = P_2 = \frac{n-1}{R} \tag{50}$$

where, $R = |R_1| = |R_2|$. Thus,

$$d_{u1} = \frac{t}{n} \frac{1}{2 - \frac{t}{n}\frac{n-1}{R}} \approx \frac{t}{2n} \tag{51}$$

and $$d_{u2} = -\frac{t}{n} \frac{1}{2 - \frac{t}{n}\frac{n-1}{R}} \approx -\frac{t}{2n} \tag{52}$$

where we have assumed $t << R$ which is indeed the case for most thick lenses. The positions of the unit planes are shown in Fig. 1.11.

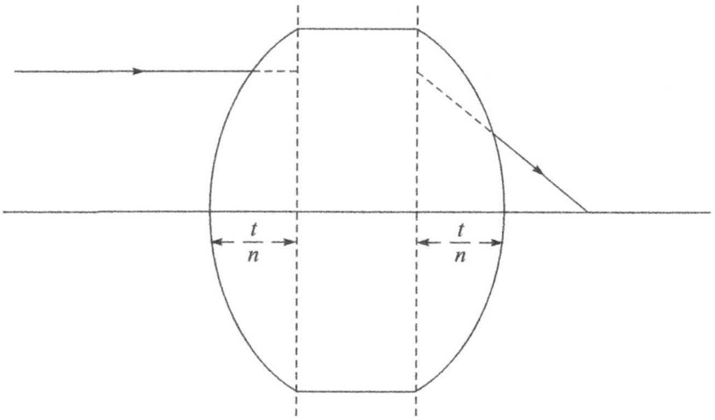

Fig. 1.11 Unit planes of a thick symmetric biconvex lens.

1.4 (a) The system matrix is given by (see Fig. 1.6)

$$S = \begin{pmatrix} 1 & -\frac{1}{f_2} \\ 0 & 1 \end{pmatrix} \begin{pmatrix} 1 & 0 \\ t & 1 \end{pmatrix} \begin{pmatrix} 1 & -\frac{1}{f_1} \\ 0 & 1 \end{pmatrix}$$

$$= \begin{pmatrix} \left(1-\frac{t}{f_2}\right) & -\left(\frac{1}{f_1}+\frac{1}{f_2}-\frac{t}{f_1 f_2}\right) \\ t & \left(1-\frac{t}{f_1}\right) \end{pmatrix} \quad (53)$$

Thus, the elements of the system matrix are

$$a = \frac{1}{f_1}+\frac{1}{f_2}-\frac{t}{f_1 f_2}; \quad b = 1-\frac{t}{f_2}; \quad c = 1-\frac{t}{f_1}; \quad d = -t \quad (54)$$

(b) Since $f_1 = 10$ cm, $f_2 = 30$ cm, $t = 20$ cm

$$= \begin{pmatrix} \left(1-\frac{20}{30}\right) & -\left(\frac{1}{10}+\frac{1}{30}-\frac{20}{300}\right) \\ 20 & \left(1-\frac{20}{10}\right) \end{pmatrix}$$

$$= \begin{pmatrix} \frac{1}{3} & -\frac{2}{30} \\ 20 & -1 \end{pmatrix}$$

Thus $a = \frac{1}{15}$, $b = \frac{1}{3}$, $c = -1$, $d = -20$; $d_{u1} = \frac{1-b}{a} = 10$ cm; $d_{u2} = \frac{c-1}{a} = -30$ cm and the first convex lens is in the middle of the two unit planes (see Fig. 1.6). $f = \frac{1}{a} = 15$ cm.

Now, $$\frac{1}{v} - \frac{1}{u} = \frac{1}{f} = a$$

For $u = -\infty$, $v = f = 15$ cm (distance measured from the second unit plane). The final image is virtual and is 15 cm away (on the left) from the second lens.

1.5 $R_1 = 45$ cm, $R_2 = -30$ cm, $t = 5$ cm, $n = 1.5$

$$P_1 = \frac{n-1}{R_1} = \frac{1}{90} \quad \text{and} \quad P_2 = \frac{1-n}{R_2} = \frac{1}{60}$$

represents the power of the two refracting surfaces. Thus,

$$a = P_1 + P_2\left(1 - \frac{t}{n}P_1\right) = \frac{11}{405} \approx 0.02716; \quad b = 1 - \frac{P_2 t}{n} = \frac{17}{18} \approx 0.9444$$

$$c = 1 - \frac{t}{n}P_1 = \frac{26}{27} \approx 0.9630; \quad d = -\frac{t}{n} = -\frac{10}{3}$$

$$d_{u1} = \frac{1-b}{a} = \frac{1}{18} \times \frac{405}{11} \approx 2.0455 \text{ cm;}$$

$$d_{u2} = \frac{c-1}{a} = -\frac{405}{27 \times 11} \approx -1.3636 \text{ cm}$$

Now, $\frac{1}{v} - \frac{1}{u} = \frac{1}{f} = a$; $u = -90 - 2.0455 = -92.0455$ cm. Thus

$$\frac{1}{v} = 0.02716 - \frac{1}{92.0455} \Rightarrow v = 61.37 \text{ cm} \quad \text{(from the second unit plane)}$$

Thus, the image will be at a distance of 61.37 – 1.3636 ≈ 60 cm from the second surface (see Fig. 1.12).

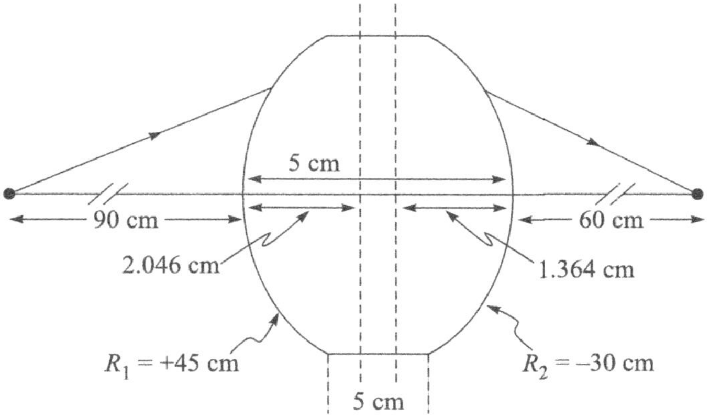

Fig. 1.12 Figure for Problem 1.5.

1.6 $R_1 = \infty, \quad R_2 = -20$ cm, $\quad t = 20$ cm, $\quad n = 1.5$

$$P_1 = \frac{n-1}{R_1} = 0 \quad \text{and} \quad P_2 = \frac{1-n}{R_2} = \frac{1}{40} = 0.025$$

$$a = P_1 + P_2\left(1 - \frac{t}{n}P_1\right) = P_2 = 0.025; \quad b = 1 - \frac{P_2 t}{n} = \frac{2}{3}$$

$$c = 1 - \frac{t}{n}P_1 = 1; \quad d = -\frac{t}{n} = -\frac{200}{15}$$

$$d_{n1} = d_{u1} = \frac{1-b}{a} = \frac{40}{3} \approx 13.33 \text{ cm}; \quad d_{n2} = d_{u2} = \frac{c-1}{a} = 0$$

Thus the first nodal (and unit) plane is at a distance of +13.33 cm from the plane surface and the second nodal (and unit) plane lies on the second surface. Further, for $u = -\infty$, the equation

$$\frac{1}{v} - \frac{1}{u} = a \text{ becomes } v = \frac{1}{a} = 40 \text{ cm}$$

which is the focal length.

1.7 $R_1 = -10$ cm, $\quad R_2 = +20$ cm, $\quad t = 1.0$ cm, $\quad n = 1.5$

$$P_1 = \frac{n-1}{R_1} = -\frac{1}{20} = -0.05; \quad P_2 = -\frac{n-1}{R_2} = -\frac{1}{40} = -0.025$$

$$a = P_1 + P_2\left(1 - \frac{t}{n}P_1\right) = -0.07583; \quad b = 1 - \frac{P_2 t}{n} = 1.0167$$

$$c = 1 - \frac{t}{n}P_1 = 1.0333; \quad d = -\frac{t}{n} = -\frac{1}{1.5} = -0.6667$$

$$d_{u1} = \frac{1-b}{a} \approx +0.220 \text{ cm}; \quad d_{u2} = \frac{c-1}{a} \approx -0.440 \text{ cm}$$

1.8 We can easily calculate $P_1 = 0.03$, $P_2 = 0.03$ and since $t = 40$ cm and $n = 1.6$, we get

$$b = 0.25, \; a = 0.0375, \; d = -25 \quad \text{and} \quad c = 0.25.$$

Alternatively, one can multiply the matrices to obtain

Second surface to image	Refraction at second surface	Transmission through glass	Refraction at the first surface
$\begin{pmatrix} 1 & 0 \\ v & 1 \end{pmatrix}$	$\begin{pmatrix} 1 & (1-1.6)/20 \\ 0 & 1 \end{pmatrix}$	$\begin{pmatrix} 1 & 0 \\ 40/1.6 & 1 \end{pmatrix}$	$\begin{pmatrix} 1 & -(1.6-1)/20 \\ 0 & 1 \end{pmatrix}$

$$= \begin{pmatrix} 1 & 0 \\ v & 1 \end{pmatrix} \begin{pmatrix} 0.25 & -0.0375 \\ 25 & 0.25 \end{pmatrix}$$

$$= \begin{pmatrix} 0.25 & -0.0375 \\ 25 + 0.25v & 0.25 - 0.0375v \end{pmatrix}$$

Thus at the image plane, the ray coordinates are

$$\begin{pmatrix} \lambda_2 \\ x_2 \end{pmatrix} = \begin{pmatrix} 0.25 & -0.0375 \\ 25 + 0.25v & 0.25 - 0.0375v \end{pmatrix} \begin{pmatrix} \lambda_1 \\ x_1 \end{pmatrix}$$

This gives us

$$x_2 = (25 + 0.25v)\lambda_1 + (0.25 - 0.0375v)x_1$$

To determine the focal distance v, consider a ray incident parallel to the axis for which $\lambda_1 = 0$. The focal plane would be that plane for which x_2 is also zero. This gives us

$$0.0375v = 0.25 \quad \text{or} \quad v = 6.7 \text{ cm}$$

(see Fig. 1.13). The system matrix elements are

$$a = \frac{1}{f} = 0.0375 \text{ cm}^{-1} \quad \Rightarrow \quad f \approx 26.7 \text{ cm}$$

$$b = 0.25 \qquad c = 0.25 \qquad d = -25 \text{ cm}$$

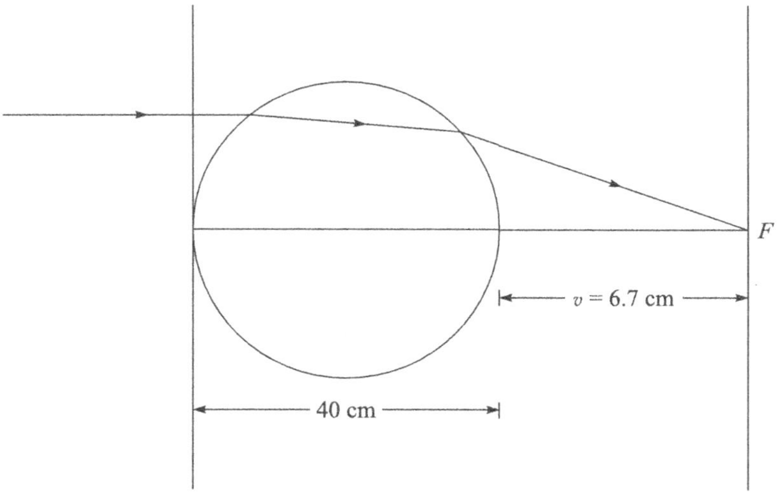

Fig. 1.13 Imaging by a sphere of radius 20 cm and refractive index 1.6

The unit planes are given by

$$d_{u1} = \frac{1-b}{a} = 20 \text{ cm}$$

and

$$d_{u2} = \frac{c-1}{a} = -20 \text{ cm}$$

Thus, both the unit planes pass through the center of the sphere. In this example, we cannot use the approximation $t << R$.

1.9 $f_1 = +15$ cm $\quad f_2 = -20$ cm $\quad t_1 = 25$ cm

Thus, we readily get [see Solution to Problem 1.4(a)]

$$a = \frac{1}{10} = \frac{1}{f} \qquad b = \frac{45}{20} \qquad c = -\frac{2}{3} \qquad d = -25$$

and $$d_{u1} = \frac{1-b}{a} = -12.5 \text{ cm} \qquad d_{u2} = \frac{c-1}{a} = -\frac{50}{3} \text{ cm}$$

Thus, the distance of the object from the first unit plane is given by

$$u = -27.5 - (-12.5) = -15 \text{ cm} \quad \text{(see Fig. 1.14).}$$

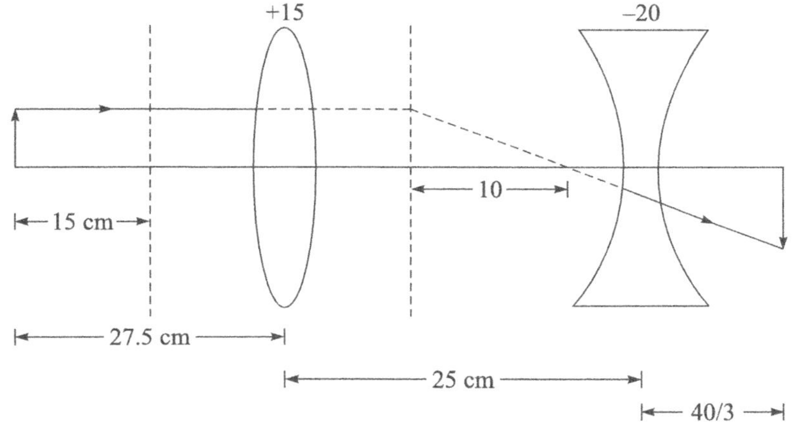

Fig. 1.14 Figure for Problem 1.9.

Since $f = +10$ cm, we get

$$v = 30 \text{ cm}$$

which represents the distance of the image plane from the second unit plane. Thus, the image is at a distance of $30 - 50/3 = 40/3$ cm from the concave lens. The magnification is given by

$$M = \frac{v}{u} = -2$$

1.10 Let v be the distance of the image plane from the concave lens. Thus the matrix, which when operated on the object column matrix gives the image column matrix, is given by

Concave lens to image	Concave lens	Convex lens to concave lens	Convex lens	Object to convex lens
$\begin{pmatrix} 1 & 0 \\ v & 1 \end{pmatrix}$	$\begin{pmatrix} 1 & +1/10 \\ 0 & 1 \end{pmatrix}$	$\begin{pmatrix} 1 & 0 \\ 8 & 1 \end{pmatrix}$	$\begin{pmatrix} 1 & -1/20 \\ 0 & 1 \end{pmatrix}$	$\begin{pmatrix} 1 & 0 \\ 40 & 1 \end{pmatrix}$

$$= \begin{pmatrix} 1 & 0 \\ v & 1 \end{pmatrix} \begin{pmatrix} 2.2 & 0.01 \\ +32 & 0.6 \end{pmatrix}$$

$$= \begin{pmatrix} 2.2 & 0.01 \\ 2.2v + 32 & 0.6 + 0.01v \end{pmatrix}$$

The image plane would correspond to

$$32 + 2.2v = 0$$

or, $$v \approx -14.5 \text{ cm}$$

i.e., it is at a distance of 14.5 cm to the left of the concave lens. If we compare this with Eq. (45), we obtain

$$M = 0.6 + 0.01v = 0.6 - 0.01\left(\frac{32}{2.2}\right) = +\frac{1}{2.2}$$

Fermat's Principle, Snell's Law and Ray Equation

2

A Quick Review

2.1 FERMAT'S PRINCIPLE

Fermat's principle states that the actual ray path between two points is the one for which the optical path length is stationary with respect to variations of the path.

$$\delta \int_{A \to B} n\,ds = 0 \tag{1}$$

Using Fermat's principle one can derive Snell's law of refraction (see Problem 2.1):

$$n_1 \sin \phi_1 = n_2 \sin \phi_2 \tag{2}$$

where ϕ_1 is the angle of incidence and ϕ_2 the angle of refraction (see Fig. 2.1).

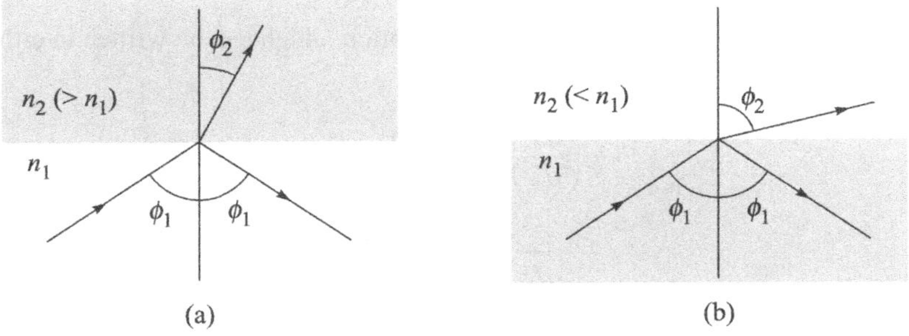

Fig. 2.1 (a) For a ray incident on a denser medium, the ray bends towards the normal and the angle of refraction is less than the angle of incidence. (b) For a ray incident on a rarer medium, the ray bends away from the normal and the angle of refraction is greater than the angle of incidence. In each refraction, the Snell's law $n_1 \sin \phi_1 = n_2 \sin \phi_2$ is obeyed.

We assume that the refractive index depends only on the x-coordinate. Such an inhomogeneous medium can be thought of as a limiting case of a medium consisting of a continuous set of thin slices of media of different refractive indices – [see Fig. 2.2(a)]. At each interface, the light ray satisfies Snell's law and one obtains

$$n_1 \sin \phi_1 = n_2 \sin \phi_2 = n_3 \sin \phi_3 = \ldots \tag{3}$$

Thus, in the limiting case of a continuous variation of refractive index [see Fig. 2.2(b)], the product

$$n(x)\sin\phi(x) = n(x)\cos\theta(x) = n_1\cos\theta_1 = \tilde{\beta} \tag{4}$$

is an invariant of the ray path; we will denote this invariant by $\tilde{\beta}$. In the above equation $\theta(x)$ is the angle that the ray makes with the z-axis. The value of this invariant may be determined from the fact that if the ray initially makes an angle θ_1 (with the z-axis) at a point where the refractive index is n_1, then the value of $\tilde{\beta}$ is $n_1 \cos\theta_1$.

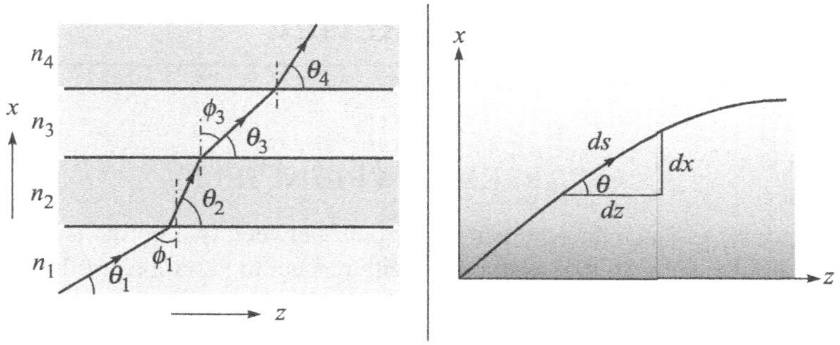

Fig. 2.2 (a) In a layered structure, the ray bends in such a way that the product $n_1 \sin\phi_1 = n_2 \sin\phi_2 = n_3 \sin\phi_3 = \ldots$ remains constant. (b) For a medium with continuously varying refractive index, the ray path bends in such a way that the product $n(x)\sin\phi(x)$ [$= n(x)\cos\theta(x)$] remains constant.

2.2 RAY EQUATIONS IN INHOMOGENOUS MEDIA

Equation (4) can be used to derive the ray equation which can be written in either of the following forms (see Problem 2.4):

$$\left(\frac{dx}{dz}\right)^2 = \frac{n^2(x)}{\tilde{\beta}^2} - 1 \tag{5}$$

or,

$$\frac{d^2x}{dz^2} = \frac{1}{2\tilde{\beta}^2}\frac{dn^2}{dx} \tag{6}$$

The above equations represent rigorously correct ray equations when the refractive index depends only on the x-coordinate.

PROBLEMS

2.1 Obtain the laws of refraction (i.e., Snell's law) from Fermat's principle.

2.2 Consider a spherical refracting surface SPM separating two media of refractive indices n_1 and n_2 (see Fig. 2.3). The point C represents the center of the spherical surface SPM. Consider two points O and Q such that the points O, C and Q are in a straight line. Calculate the optical path length OSQ

in terms of the distances x, y, r and the angle θ (see Fig. 2.3). Use Fermat's principle to find the ray connecting the two points O and Q. Also, assuming the angle θ to be small, determine the paraxial image of the point O.

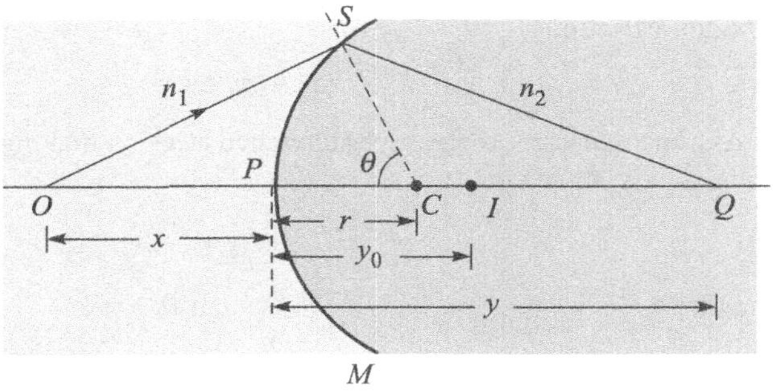

Fig. 2.3 *SPM* is a spherical refracting surface separating two media of refractive indices n_1 and n_2. *C* represents the center of the spherical surface.

2.3 Consider a set of rays, parallel to the axis, and incident on a paraboloidal reflector (see Fig. 2.4). Show, by using Fermat's principle, that all rays will pass through the focus of the paraboloid; a paraboloid is obtained by rotating a parabola about its axis. This is the reason why paraboloidal reflectors are used to focus parallel rays from a distant source, like in radio astronomy.

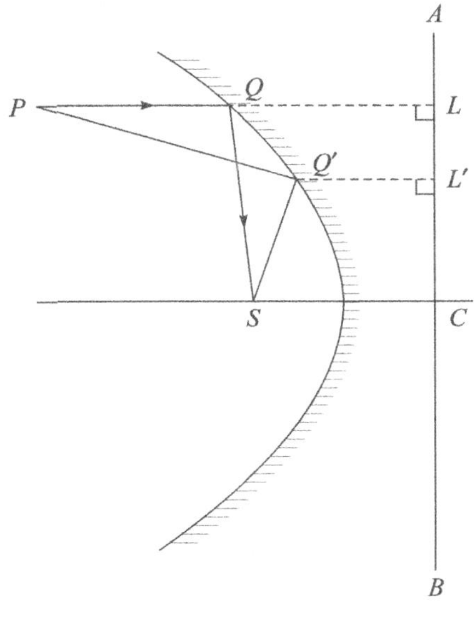

Fig. 2.4 All rays parallel to the axis of a paraboloidal reflector pass through the focus after reflection (the line *ACB* is the directrix). It is for this reason that antennas (for collecting electromagnetic waves) or solar collectors are often paraboloidal in shape.

2.4 Assume the refractive index to depend only on the x-coordinate. Use Snell's law [Eq. (4)] to derive the ray equation [Eqs (5) and (6)].

2.5 Solve the ray equation in a homogeneous medium for which $n(x)$ is a constant.

2.6 (a) Obtain the ray paths in a medium characterised by the following refractive index variation

$$n(x) = n_0 + kx \tag{7}$$

Assume that at $z = 0$, the ray is launched at $x = x_1$ making an angle θ_1, with the z-axis; thus

$$x(z = 0) = x_1$$

and

$$\left.\frac{dx}{dz}\right|_{z=0} = \tan\theta_1$$

(b) Assume that $k \approx 1.234 \times 10^{-5}\ \text{m}^{-1}$ and rays are launched at $x = x_1 = 1.5$ m, where $n(x_1) = 1.00026$. Plot the ray paths when the angle that the ray makes with the horizontal axis are $+0.2°$, $0°$, $-0.2°$, $-0.28°$, $-0.3486°$ and $-0.5°$.

2.7 (a) Consider an optical waveguide characterised by the refractive index distribution is usually written in the form:

$$n^2(x) = n_1^2\left[1 - 2\Delta\left(\frac{x}{a}\right)\right]^2, \quad |x| < a \quad \text{CORE}$$

$$= n_2^2 = n_1^2(1 - 2\Delta), \quad |x| > a \quad \text{CLADDING} \tag{8}$$

The region $|x| < a$ is known as the core of the waveguide and the region $|x| > a$ is known as the cladding. Assuming that at $z = 0$, $x = 0$, show that for $n_2 < \tilde{\beta} < n_1$, the ray paths are sinusoidal.

(b) Derive an expression for the periodical length z_p of the sinusoidal path.

(c) Assume $n_1 = 1.5$, $\Delta = 0.01$ and $a = 20$ μm; if θ_1 is the angle that the ray makes with the z axis at $x = 0$, calculate the values of z_p corresponding to $\theta_1 = 4°$, $8.13°$ and $20°$.

2.8 In continuation of the previous problem, for $n_2 < \tilde{\beta} < n_1$, the ray path (inside the core) of a parabolic index waveguide is given by

$$x = x_0 \sin \Gamma z \tag{9}$$

where,

$$x_0 = \frac{1}{\gamma}\sqrt{n_1^2 - \tilde{\beta}^2} \tag{10}$$

$$\gamma = n_1 \frac{\sqrt{2\Delta}}{a} \tag{11}$$

and

$$\Gamma = \frac{\gamma}{\tilde{\beta}} \tag{12}$$

Calculate the time taken by a ray to traverse a certain length through the parabolic index waveguide; such a calculation is of considerable importance in fiber optic communication systems (see Sec. Eq. (17) of Chapter 18).

2.9 Discuss the ray paths in a medium characterized by the following refractive index variation

$$n^2(x) = n_1^2 \qquad x < 0$$
$$= n_1^2 - gx \qquad x > 0 \tag{13}$$

Obtain the ray path for a ray incident on the origin ($x = 0, z = 0$) making an angle θ_1 with the z-axis.

2.10 Consider a refractive index variation which saturates to a constant value as $x \to \infty$:

$$n^2(x) = n_0^2 + n_2^2(1 - e^{-\alpha x}); \quad x > 0 \tag{14}$$

with $$n_0 = 1.000233, n_2 = 0.45836 \quad \text{and} \quad \alpha = 2.303 \text{ m}^{-1} \tag{15}$$

Calculate the angle at which the ray should be launched at $x = 0.43$ m, so that it becomes horizontal at $x = 0.2$ m.

SOLUTIONS

2.1 To obtain the laws of refraction, let PQ be a surface separating two media of refractive indices n_1 and n_2 as shown in Fig. 2.5. Let a ray starting from the point A, intersect the interface at R and proceed to B along RB. Clearly, for minimum optical path length, the incident ray, the refracted ray and the normal to the interface must all lie in the same plane. To determine that point R for which the optical path length from A to B is a minimum, we drop perpendiculars AM and BN from A and B respectively on the interface PQ. Let $AM = h_1$, $BN = h_2$ and $MR = x$. Then since A and B are fixed, $RN = L - x$, where $MN = L$ is a fixed quantity. The optical path length from A to B, by definition, is

$$L_{op} = n_1 AR + n_2 RB = n_1\sqrt{x^2 + h_1^2} + n_2\sqrt{(L - x)^2 + h_2^2} \tag{16}$$

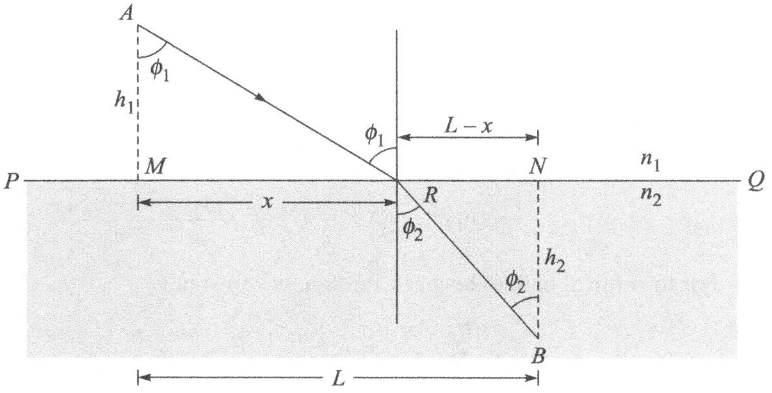

Fig. 2.5 *A* and *B* are two points in media of refractive indices n_1 and n_2. The ray path connecting *A* and *B* will be such that $n_1 \sin \phi_1 = n_2 \sin \phi_2$.

To minimise this, we must have

$$\frac{dL_{\text{op}}}{dx} = 0 \tag{17}$$

i.e.,

$$\frac{n_1 x}{\sqrt{x^2 + h^2}} - \frac{n_2 (L - x)}{\sqrt{(L - x)^2 + h^2}} = 0 \tag{18}$$

Further, as can be seen from Fig. 2.5

$$\sin \phi_1 = \frac{x}{\sqrt{x^2 + h_1^2}}$$

and

$$\sin \phi_2 = \frac{(L - x)}{\sqrt{(L - x)^2 + h_2^2}}$$

Thus, Eq. (18) becomes

$$n_1 \sin \phi_1 = n_2 \sin \phi_2 \tag{19}$$

which is the Snell's law of refraction.

2.2 From the triangle *SOC* (see Fig. 2.3) we have

$$OS = [(x + r)^2 + r^2 - 2(x + r) r \cos \theta]^{1/2}$$

$$\approx \left[x^2 + 2rx + 2r^2 - 2(xr + r^2)\left(1 - \frac{\theta^2}{2}\right)\right]^{1/2}$$

$$\approx x\left[1 + \frac{rx + r^2}{x^2}\theta^2\right]^{1/2} \approx x + \frac{1}{2}r^2\left(\frac{1}{r} + \frac{1}{x}\right)\theta^2$$

where we have assumed θ (measured in radians) to be small so that we may use the expression

$$\cos \theta \approx 1 - \frac{\theta^2}{2}$$

and also make a binomial expansion. Similarly, by considering the triangle *SCQ* we would have

$$SQ \approx y - \frac{1}{2}r^2\left(\frac{1}{r} - \frac{1}{y}\right)\theta^2$$

Thus, the optical path length *OSQ* is given by

$$L_{\text{op}} = n_1\, OS + n_2\, SQ$$

$$\approx (n_1 x + n_2 y) + \frac{1}{2}r^2\left[\frac{n_1}{x} + \frac{n_2}{y} - \frac{n_2 - n_1}{r}\right]\theta^2 \tag{20}$$

For the optical path to be an extremum, we must have

$$\frac{dL_{\text{op}}}{d\theta} = 0 = r^2\left[\frac{n_1}{x} + \frac{n_2}{y} - \frac{n_2 - n_1}{r}\right]\theta \tag{21}$$

Thus, unless the quantity inside the square brackets is zero we must have $\theta = 0$ implying that the *only* ray connecting the points O and Q will be the straight line path OPQ which also follows from Snell's law because the ray OP hits the spherical surface normally and should proceed undeviated.

On the other hand, if the value of y was such that the quantity inside the square brackets was zero, i.e., if y was equal to y_0 such that

$$\frac{n_2}{y_0} + \frac{n_1}{x} = \frac{n_2 - n_1}{r} \tag{22}$$

then $dL_{op}/d\theta$ would vanish for *all* values of θ; of course, θ is assumed to be small – which is the paraxial approximation. Now, if the point I corresponds to $PI = y_0$ (see Fig. 2.3) then *all* paths like OSI are allowed ray paths implying that *all* (paraxial) rays emanating from O will pass through I and I will therefore represent the paraxial image point. Obviously, all rays like OSI (which start from O and pass through I) take the *same* amount of time in reaching the point I.

We should mention that Eq. (22) is a particular form of the equation determining the paraxial image point

$$\frac{n_2}{v} - \frac{n_1}{u} = \frac{n_2 - n_1}{R} \tag{23}$$

with the sign convention that all distances measured to the right of the point P are positive and those to its left negative. Thus $u = -x$, $v = +y$ and $r = +R$.

In order to determine whether the ray path OPQ corresponds to minimum time or maximum time or stationary, we must determine the sign of $d^2L_{op}/d\theta^2$ which is given by

$$\frac{d^2 L_{op}}{d\theta^2} = r^2\left[\frac{n_1}{x} + \frac{n_2}{y} - \frac{n_2 - n_1}{r}\right] = r^2 n_2\left[\frac{1}{y} - \frac{1}{y_0}\right]$$

Obviously, if $y > y_0$ (i.e., the point Q is on the right of the paraxial image point I) $d^2L_{op}/d\theta^2$ is negative and the ray path OPQ corresponds to *maximum* time in comparison with nearby paths and conversely. On the other hand, if $y = y_0$, $d^2L_{op}/d\theta^2$ will vanish implying that the extremum corresponds to stationarity.

2.3 Consider a ray PQ, parallel to the axis of the parabola, incident at the point Q (see Fig. 2.4). In order to find the reflected ray, one has to draw a normal at the point Q and then draw the reflected ray. It can be shown from geometrical considerations that the reflected ray QS will always pass through the focus S. However, this procedure will be quite cumbersome and as we will show below, the use of Fermat's principle leads us to the desired results immediately.

In order to use Fermat's principle we try to find out the ray connecting the focus S and an arbitrary point P (see Fig. 2.4). Let the ray path be $PQ'S$. According to Fermat's principle the ray path will correspond to a minimum value of $PQ' + Q'S$. From the point Q' we drop a perpendicular $Q'L'$, on the directrix AB. From the definition of the parabola it follows that $Q'L' = Q'S$.

Thus,

$$PQ' + Q'S = PQ' + Q'L'$$

Let L be the foot of the perpendicular drawn from the point P on AB. Then, for $PQ' + Q'L'$ to be a minimum, the point Q should lie on the straight line PL, and thus the actual ray which connects the points P and S will be $PQ + QS$ where PQ is parallel to the axis. Therefore, all rays parallel to the axis will pass through S and conversely, all rays emanating from the point S will become parallel to the axis after suffering a reflection.

2.4 If ds represents the infinitesimal arc length along the curve, then

$$(ds)^2 = (dx)^2 + (dz)^2 \tag{24}$$

or,

$$\left(\frac{ds}{dz}\right)^2 = \left(\frac{dx}{dz}\right)^2 + 1 \tag{25}$$

Now, if we refer to Fig. 2.2(b), we find that

$$\frac{dz}{ds} = \cos\theta = \frac{\tilde{\beta}}{n(x)} \tag{26}$$

Thus Eq. (25) becomes

$$\left(\frac{dx}{dz}\right)^2 = \frac{n^2(x)}{\tilde{\beta}^2} - 1 \tag{27}$$

For a given $n(x)$ variation, Eq. (27) can be integrated to give the ray path $x(z)$; however, it is often more convenient to put Eq. (27) in a slightly different form by differentiating it with respect to z:

$$2\frac{dx}{dz}\frac{d^2x}{dz^2} = \frac{1}{\tilde{\beta}^2}\frac{dn^2}{dx}\frac{dx}{dz}$$

or,

$$\frac{d^2x}{dz^2} = \frac{1}{2\tilde{\beta}^2}\frac{dn^2}{dx} \tag{28}$$

Equations (27) and (28) represent rigorously correct ray equations when the refractive index depends only on the x-coordinate and we may use either of them to determine the ray paths.

2.5 In a homogeneous medium for which $n(x)$ is a constant. In such a case, the *RHS* of Eq. (28) is zero and one obtains

$$\frac{d^2x}{dz^2} = 0$$

Integrating the above equation twice with respect to z, we obtain

$$x = Az + B$$

which is the equation of a straight line, as it ought to be in a homogeneous medium.

2.6 Consider the refractive index variation

$$n(x) = n_0 + kx \tag{29}$$

For the above profile, the ray equation [Eq. (28)] takes the form

$$\frac{d^2x}{dz^2} = \frac{1}{2\tilde{\beta}^2}\frac{dn^2}{dx} = \frac{k}{\tilde{\beta}^2}[n_0 + kx]$$

or,

$$\frac{d^2X}{dz^2} = \kappa^2 X(z) \tag{30}$$

where,

$$X \equiv x + \frac{n_0}{k} \quad \text{and} \quad \kappa = \frac{k}{\tilde{\beta}} \tag{31}$$

Thus, the ray path is given by

$$x(z) = -\frac{n_0}{k} + C_1 e^{\kappa z} + C_2 e^{-\kappa z} \tag{32}$$

where the constants C_1 and C_2 are to be determined from initial conditions. We assume that at $z = 0$, the ray is launched at $x = x_1$ making an angle θ_1 with the z-axis; thus

$$x(z=0) = x_1$$

and

$$\left.\frac{dx}{dz}\right|_{z=0} = \tan\theta_1$$

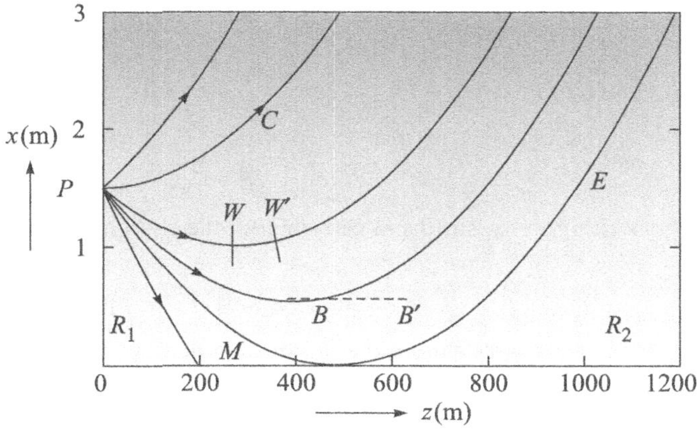

Fig. 2.6 Ray paths in a medium characterised by a linear variation of refractive index [see Eq. (7)] with parameters as given in Problem 6(b). The object point is at a height of 1.5 m and the curves correspond to +0.2°, 0°, −0.2°, −0.28°, −0.3486° and −0.5°. The shading shows that the refractive index increases with x.

Elementary manipulations would give us

$$C_1 = \frac{1}{2}\left[x_1 + \frac{1}{k}(n_0 + n_1\sin\theta_1)\right] \tag{33}$$

and

$$C_2 = \frac{1}{2}\left[x_1 + \frac{1}{k}(n_0 - n_1\sin\theta_1)\right] \tag{34}$$

where $n_1 = n_0 + kx_1$ represents the refractive index at $x = x_1$ and we have used the fact that

$$\tilde{\beta} = n_1 \cos \theta_1 \tag{35}$$

Now, $k \approx 1.234 \times 10^{-5}$ m^{-1} and $n(x_1) = 1.00026$ where $x_1 = 1.5$ m; thus $n_0 = 1.0002415$. Figure 2.6 shows the ray path as given by Eq. (32) for $\theta_1 = +0.2°$, $0°$, $-0.2°$, $-0.28°$, $-0.3486°$ and $-0.5°$.

2.7 (a) In the core of the waveguide, we write the refractive variation as

$$n^2(x) = n_1^2 - \gamma^2 x^2 \tag{36}$$

where, $$\gamma = n_1 \frac{\sqrt{2\Delta}}{a}$$

We will use Eq. (27) to determine the ray paths. Equation (27) can be written as

$$\int \frac{dx}{\sqrt{n^2(x) - \tilde{\beta}^2}} = \pm \frac{1}{\tilde{\beta}} \int dz \tag{37}$$

Substituting for $n^2(x)$, we get

$$\int \frac{dx}{\sqrt{x_0^2 - x^2}} = \pm \Gamma \int dz \tag{38}$$

where, $$x_0 = \frac{1}{\gamma} \sqrt{n_1^2 - \tilde{\beta}^2} \tag{39}$$

and $$\Gamma = \frac{\gamma}{\tilde{\beta}} \tag{40}$$

Writing $x = x_0 \sin \theta$ and carrying out the straightforward integration, we get

$$x = \pm x_0 \sin [\Gamma(z - z_0)] \tag{41}$$

We can always choose the origin such that $z_0 = 0$ so that the general ray path would be given by

$$x = \pm x_0 \sin \Gamma z \tag{42}$$

(b) For $n_1 = 1.5$, $\Delta = 0.01$, $a = 20$ µm, we get $n_2 = 1.485$ and $\gamma = 1.0607 \times 10^4$ m^{-1}. Obviously, rays will be guided in the core if $n_2 < \tilde{\beta} < n_1$. When $\tilde{\beta} = n_2$, the ray path will become horizontal at the core-cladding interface. For $\tilde{\beta} < n_2$, the ray will be incident at the core-cladding interface at an angle greater than the critical angle and the ray will be refracted away. Thus, we may write

$$\begin{aligned} n_2 < \tilde{\beta} < n_1 \quad &\Rightarrow \quad \text{Guided rays} \\ \tilde{\beta} < n_2 \quad &\Rightarrow \quad \text{Refracting rays} \end{aligned} \tag{43}$$

In Fig. 2.7, the ray paths shown correspond to $\theta_1 = 4°$, $8.13°$ and $20°$; the corresponding values of $\tilde{\beta}$ are approximately 1.496 ($> n_2$), 1.485 ($= n_2$) and 1.410 ($< n_2$)—the last ray undergoes refraction at the core-cladding interface. It may be readily seen that the periodical length z_p of the sinusoidal path is given by

$$z_p = \frac{2\pi}{\Gamma} = \frac{2\pi a \cos\theta_1}{\sqrt{2\Delta}} \tag{44}$$

Thus for the two rays shown in Fig. 2.7 (with $\theta_1 = 4°$ and $8.13°$) the values of z_p would be

0.8864 mm and 0.8796 mm

respectively. Indeed, in the paraxial approximation, $\cos\theta_1 \approx 1$ and all rays have the same periodic length.

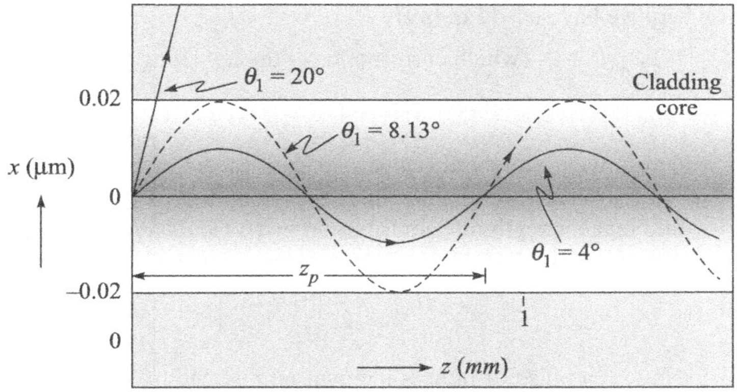

Fig. 2.7 Typical ray paths in a parabolic index medium for parameters as given in Problem 7(c) for $\theta_1 = 4°$, $8.13°$ and $20°$.

2.8 The ray path (inside the core) is given by

$$x = x_0 \sin\Gamma z \tag{45}$$

where x_0 and Γ have been defined earlier. Let $d\tau$ represent the time taken by a ray to traverse the are length ds:

$$d\tau = \frac{ds}{c/n(x)} \tag{46}$$

Since, $$n(x)\frac{dz}{ds} = \tilde{\beta}$$

[see Eq. (26)] we may write Eq. (46) as

$$d\tau = \frac{1}{c\tilde{\beta}} n^2(x)\, dz$$

$$= \frac{1}{c\tilde{\beta}} [n_1^2 - \gamma^2 x^2]\, dz$$

or, $$d\tau = \frac{1}{c\tilde{\beta}} [n_1^2 - \gamma^2 x_0^2 \sin^2 \Gamma z]\, dz \tag{47}$$

where in the last step we have used Eq. (45). Thus, if $\tau(z)$ represents the time taken by the ray to traverse a distance z along the waveguide then

$$\tau(z) = \frac{n_1^2}{c\tilde{\beta}}\int_0^z dz - \frac{\gamma^2 x_0^2}{c\tilde{\beta}}\int_0^z \frac{1-\cos(2\Gamma z)}{2}dz$$

$$= \frac{1}{c\tilde{\beta}}\left[n_1^2 - \frac{1}{2}\gamma^2 x_0^2\right]z + \frac{\gamma^2 x_0^2}{2c\tilde{\beta}}\frac{1}{2\Gamma}\sin 2\Gamma z$$

or,

$$\tau(z) = \frac{1}{2c\tilde{\beta}}[n_1^2 + \tilde{\beta}^2]z + \frac{(n_1^2 - \tilde{\beta}^2)}{4c\gamma}\sin 2\Gamma z \tag{48}$$

where we have used Eq. (39).

When $\tilde{\beta} = n_1$ (which corresponds to the ray along the z-axis)

$$\tau(z) = \frac{z}{c/n_1} \tag{49}$$

which is what we should have expected as the ray will *always* travel with speed c/n_1. For large values of z, the second term on the RHS of Eq. (48) would make a negligible contribution to $\tau(z)$ and we may write

$$\tau(z) \approx \frac{1}{2c}\left[\tilde{\beta} + \frac{n_1^2}{\tilde{\beta}}\right]z \tag{50}$$

Now, if a pulse of light is incident on one end of the waveguide, it would in general excite all rays and since different rays take different amounts of time, the pulse will get temporally broadened. Thus, for a parabolic index waveguide, this broadening will be given by

$$\Delta\tau = \tau(\tilde{\beta} = n_2) - \tau(\tilde{\beta} = n_1)$$

or,

$$\Delta\tau = \frac{z}{2c}\frac{(n_1 - n_2)^2}{n_2} \approx \frac{zn_2}{2c}\Delta^2 \tag{51}$$

where in the last step we have assumed

$$\Delta \equiv \frac{n_1^2 - n_2^2}{2n_1^2} \approx \frac{n_1 - n_2}{n_2} \tag{52}$$

For $n_1 \approx 1.5$ and $\Delta \approx 0.01$, we get

$$\Delta\tau \approx 0.25 \text{ ns/km} \tag{53}$$

2.9 In the region $x > 0$, $n^2(x)$ decreases linearly with x and Eq. (23) takes the form

$$\frac{d^2x}{dz^2} = -\frac{g}{2\tilde{\beta}^2}$$

The general solution of which is given by

$$x(z) = -\frac{g}{4\tilde{\beta}^2}z^2 + K_1 z + K_2 \tag{54}$$

Consider a ray incident on the origin ($x = 0$, $z = 0$) as shown in Fig. 2.8. Thus

$$K_2 = 0 \quad \text{and} \quad \tilde{\beta} = n_1 \cos \theta_1 \tag{55}$$

Further,
$$\left.\frac{dx}{dz}\right|_{z=0} = K_1 = \tan \theta_1 \tag{56}$$

Thus, the ray path will be given by

$$\left.\begin{aligned} x(z) &= (\tan \theta_1) z && z < 0 \\ &= -\frac{gz}{4\tilde{\beta}^2}(z - z_0) && 0 < z < z_0 \\ &= -\frac{gz_0}{4\tilde{\beta}^2}(z - z_0) && z > z_0 \end{aligned}\right\} \tag{57}$$

where,
$$z_0 = \frac{2n_1^2}{g} \sin 2\theta_1$$

Thus in the region $0 < z < z_0$, the ray path is a parabola. Typical ray paths are shown in Fig. 2.8, the calculations corresponds to

$$n_1 = 1.5, g = 0.1 \text{ m}^{-1}$$

and different rays corresponds to

$$\theta_1 = \frac{\pi}{9}, \frac{\pi}{6}, \frac{\pi}{4} \quad \text{and} \quad \frac{\pi}{3}$$

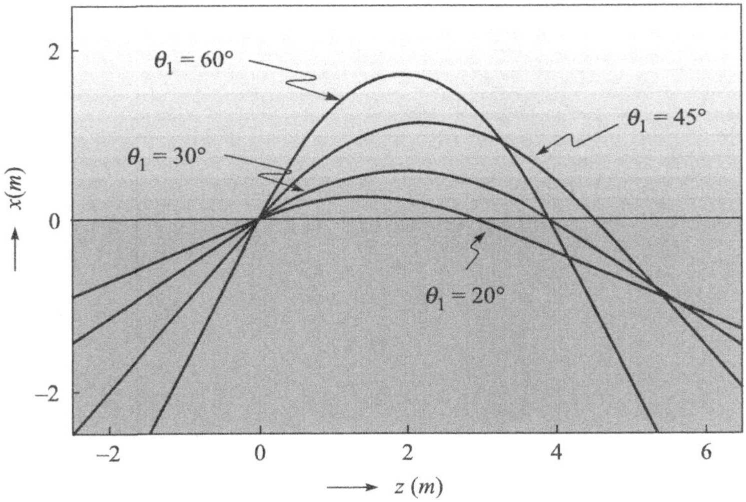

Fig. 2.8 Parabolic ray paths (corresponding to θ_1 = 20°, 30°, 45° and 60°) in a medium characterised by refractive index variation as given in Problem 2.9. The ray paths in the region $x < 0$ are straight lines.

2.10 Now, at $x = 0.43$ m, $n(x) = n_1 \approx 1.0642$ and at $x = 0.2$ m, $n(x) \approx 1.03827$. Thus, if θ_1 represents the angle that the ray makes with the z-axis at the launching point, then

$$n_1 \cos \theta_1 = 1.03827 \times \cos 0$$

implying

$$\theta_1 \approx 13°$$

Further, since the ray becomes horizontal at $x = 0.2$ m, the value of the invariant is given by $\tilde{\beta} = 1.03827$.

Optical Instruments

3

A Quick Review

3.1 THE EYE

The eye can be considered an optical instrument which forms images of external objects on the retina. The interior part of the eye is a liquid having a refractive index of 1.33 (equal to that of water). The eye lens is a double convex lens having a refractive index of about 1.4. The ciliary muscles permit the change of the power of the eye lens which allows us to accommodate and focus objects at different distances. In fact since the cornea is a curved surface it also acts like a lens and has a power of about 43 Diopters. The eye lens has a nominal power of 17 Diopters. The distance from the cornea to the retina is about 24 mm.

3.2 MAGNIFYING GLASS

The magnifying power of a magnifying glass is defined by

$$M = \frac{\text{angle subtended by the virtual image formed by the lens}}{\text{angle subtended by the object when located at a distance 25 cm from the eye}} \quad (1)$$

Visual acuity (VA) is defined and measured in terms of the smallest angular size of the character that can be recognized by the eye. The smallest recognizable letter should subtend an angle of 5 minutes of arc from the eye and each element of the letter should subtend an angle of 1 minute of arc. The normal VA is 1 minute of arc which is approximately equal to 2.9×10^{-4} radians.

3.3 COMPOUND MICROSCOPE

Figure 3.1 shows a compound microscope consisting of an objective lens of focal length f_o and eye piece of focal length f_e. A real image of the object placed very close to the focus of the objective is formed by the objective and the eye piece provides for angular magnification.

The lateral magnification provided by the objective lens is

$$M_o = -\frac{h_i}{h_o} = -\frac{L}{f_o} \quad (2)$$

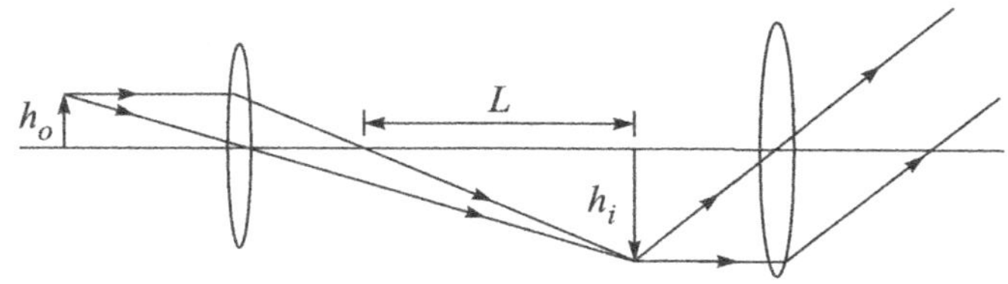

Fig. 3.1 A compound microscope consisting of an objective and an eye piece. The objective forms a real image and the eye piece magnifies the image so that the eye can resolve the details.

The quantity L represents the distance between the back focus and the position of the image and is often referred to as the tube length. In standard microscopes L = 160 mm.

Similarly the angular magnification of the eye piece is given by

$$M_e = \frac{254}{f_e} \tag{3}$$

where we have used the fact that the least distance of distinct vision is 254 mm and the focal length is in mm.

Thus, the overall magnification of the compound microscope is given by

$$M = M_o M_e = -\frac{160}{f_o} \times \frac{254}{f_e} \tag{4}$$

3.4 TELESCOPE

Figure 3.2 shows a telescope in which parallel rays from a far off object are focused by an objective of focal length f_o and the image formed by the objective is then magnified by the eye piece of focal length f_e. The angular magnification of the telescope is given by

$$M = -\frac{f_o}{f_e} \tag{5}$$

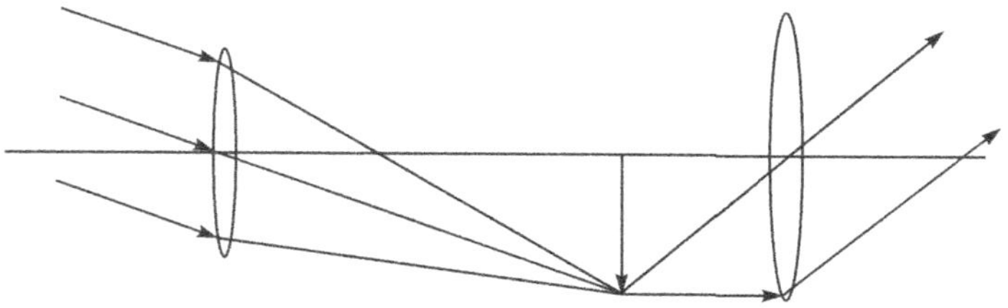

Fig. 3.2 A telescope consisting of an objective lens and an eye piece. Parallel rays from a distant object get focused on the focal plane of the objective and the eye piece provides for magnification of the image.

PROBLEMS

3.1 Consider a compound microscope formed using an objective of focal length 20 mm and an eye piece of focal length 30 mm. (a) What would be the magnification of the microscope assuming a tube length of 160 mm? (b) What will be the corresponding distance of the object from the objective?

3.2 Consider a telescope formed using an objective of focal length 20 cm and eye piece of focal length 2 cm. (a) What should be the distance between the objective and the eye piece? (b) What will be the angular magnification of the telescope?

3.3 Consider a camera with a lens of focal length 150 mm. (a) For $f\#$ numbers of $f/16$ and $f/8$, what should be the diameters of the aperture openings? (b) If the exposure time required for a picture taken with $f/8$ is 1 s, then what exposure time would be appropriate for the same picture taken with $f/16$? The quantity $f\#$ is the ratio of the focal length to the diameter of the lens and is referred to as the F-number.

3.4 Consider a magnifying glass of focal length 5 cm. An object is placed in front of the lens so that the magnified virtual image is formed at the distance of 25 cm from the lens. What will be the corresponding magnification?

3.5 Consider a compound microscope made up of an objective lens of focal length 3 cm and an eye piece lens of focal length 5 cm. The distance between the objective and the eye piece is 20 cm. Obtain the magnification of the microscope assuming that the final image is formed at a distance of 25.4 cm from the eye.

3.6 Consider a telescope with an objective lens of focal length 25 cm and an eye piece with a focal length of 5 cm. If the telescope is used to view an object kept at a distance of 1 m, obtain the angular magnification of the telescope.

3.7 Consider a microscope with an objective having a focal length of 3 mm and a tube length of 160 mm. If the eyepiece has a magnification of 10, obtain the overall magnification of the microscope.

3.8 Consider a microscope with objective and eye piece of focal lengths 6 cm and the distance between the objective and the eye piece of 206 mm. If the final image formed by the eye piece is assumed to be at infinity to allow for relaxed viewing, obtain the overall magnification of the microscope. Obtain also the distance of the object from the objective of the microscope.

3.9 An astronomical telescope is to be constructed using an objective and an eye lens to achieve a magnification of 20 and having a distance between the objective and the eye lens of 254 mm. (a) What should be the focal lengths of the objective and the eye lens? (b) If the diffraction limited resolution of the image formed by the telescope is to be resolvable by the eye, what should be the diameter of the objective? Assume an eye pupil diameter of 4 mm.

3.10 A person wishes to read a board with letters of height 1 cm placed at a distance of 1000 m. What telescope magnification would be required?

SOLUTIONS

3.1 (a) Using Eq. (4) we obtain for the magnification of the microscope as

$$M = -\frac{160}{20} \times \frac{254}{30} \approx 65$$

(b) For the given set of parameters, the distance of the image from the objective would be $L + f_o = 180$ mm. Using the standard lens formula we obtain for the object distance from the objective to be 22.5 mm.

3.2 (a) The distance between the objective and the eye piece should be equal to the sum of the focal lengths of objective and the eye piece and hence should be 22 cm.

(b) The angular magnification of the telescope would be 10.

3.3 (a) Since $f\#$ is the ratio of focal length to diameter, the diameter of the aperture for $f/16$ and $f/8$ would be 9.375 mm and 18.75 mm respectively.

(b) Since the exposure depends on the area of the aperture, with the halving of diameter when changing from $f/8$ to $f/16$, the required exposure time would increase by a factor of 4. Thus, the required exposure time for $f/16$ would be 4s.

3.4 For the virtual image to be formed at a distance of 25 cm from the lens, the object must be placed at a distance u given by

$$\frac{1}{u} = \frac{1}{5} + \frac{1}{25}$$

giving $u \approx 4.17$ mm. Hence, the magnification is given by

$$M = \frac{25}{4.17} \approx 6$$

3.5 For the virtual image to be formed at a distance of 25.4 cm from the eye piece, the image formed by the objective should be at a distance u_e from the eye piece, where

$$\frac{1}{u_e} = \frac{1}{f_e} + \frac{1}{25.4}$$

This gives us $u_e \approx 4.2$ cm. Since the distance between the objective and the eye piece is 20 cm, the image formed by the objective lies at a distance of $20 - 4.2 = 15.8$ cm from the objective. For this, the distance u_o of the object from the objective should be

$$\frac{1}{u_o} = \frac{1}{f_o} - \frac{1}{15.8}$$

which gives us $u_e \approx 3.7$ cm. Hence, the magnification of the objective is $M_o = 15.8/3.7 \approx 4.3$. The magnification of the eye piece is $M_e = 25.4/4.2 \approx 6$. Hence, the overall magnification of the microscope is $M = M_o \times M_e \approx 25.8$.

3.6 From the lens formula we find that the image formed by the objective will lie at a distance of 33.33 cm from the objective. For the eye piece to form the image at a distance of 25.4 cm, the distance of the image formed by the objective from the eye piece must be 4.2 cm. Thus, the overall magnification will be $M = \frac{33.33}{100} \times \frac{25.4}{4.2} \approx 2.$

If the height of the object placed at a distance of 1 m from the objective is 1 cm, then the angle subtended by the object at the eye would be 0.01 radians. The final image subtends an angle of 2/25.4 = 0.078 radians. Hence, the angular magnification of the telescope is 7.8.

3.7 For a tube length of 160 mm and an objective focal length of 3 mm, the overall magnification is 160/3 × 10 = 533.3.

3.8 If the final image is at infinity, then the real image formed by the objective should be at the front focus of the eye piece, i.e., at a distance of 6 mm from the eye piece. Thus, the angular magnification of the eye piece is 250/6 = 41.67. Since the distance of the image from the objective lens is 200 mm and the focal length of the objective is 6 mm, the distance of the object from the objective is 6.19 mm. Thus, the magnification of the objective would be 200/6.19 = 32.31. Hence, the overall magnification of the microscope is 1346.

3.9 (a) The magnification of the telescope is the ratio of the focal lengths of the objective and the eye lens. Hence,

$$M = -\frac{f_o}{f_e} = -20$$

Also the distance between the objective and the eye lens should be equal to the sum of the focal lengths of the objective and the eye lens. Hence,

$$f_o + f_e = 254 \text{ mm}$$

Solving the above two equations, we obtain f_e = 12.1 mm and f_o = 241.9 mm.

(b) The height of the image (h_i) produced by the objective at the limit of resolution is given by

$$h_i = \frac{1.22\lambda f_o}{d_o}$$

The angle subtended by this image on the eye when it lies at the front focal plane of the eye piece is h_i/f_e. For this to be resolvable by the eye, we must have

$$\frac{h_i}{f_e} = \frac{1.22\lambda}{d_{\text{eye}}}$$

Thus we obtain

$$d_o = \frac{f_o}{f_e} d_{\text{eye}}$$

which gives us d_o = 80 mm.

3.10 The angle subtended by the object at the eye is 10^{-5} radians. This angle should be increased to 5 minutes of arc or 15×10^{-4} radians to be recognizable. Hence the required magnification is 150.

Aberrations

4

A Quick Review

4.1 CHROMATIC ABERRATION

Let us consider a parallel beam of white light incident on a thin convex lens as shown in Fig. 4.1. Since blue light gets refracted more than red light, the point at which the blue light would focus is nearer the lens than the point at which the red light would focus. Thus, the image will appear to be coloured. The difference in the focal lengths corresponding to red and blue colours is approximately given by (see Problem 4.1):

$$f_r - f_b = \left(\frac{n_b - n_r}{n - 1} \right) \tag{1}$$

where n_b and n_r represent the refractive indices for the blue and red colours respectively and $n = (n_b + n_r)/2$. Equation (1) gives rise to what is known as chromatic aberration (see Problems 4.2-4.5).

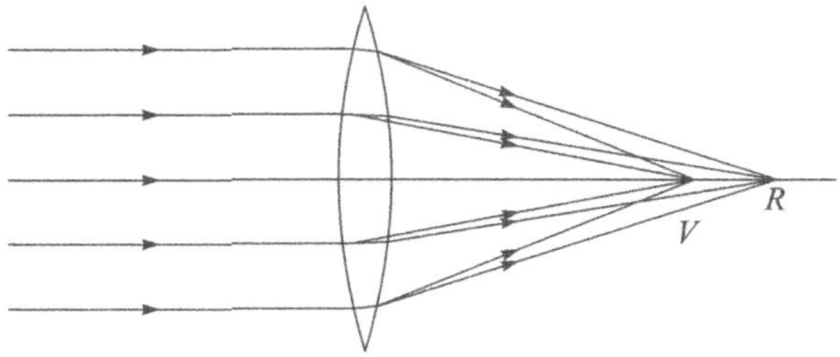

Fig. 4.1 When white light consisting of a continuous range of wavelengths is incident on a lens, then each wavelength refracts by different amounts; this leads to chromatic aberration.

4.2 SPHERICAL ABERRATION

Let a beam of light parallel to the axis be incident on a thin lens [see Fig. 4.2(a)]. The light rays after passing the lens bend towards the axis and cross the axis at some point. If we restrict ourselves to the paraxial region, then we can see that all rays cross the z-axis at the same point which is at a distance f_p from the lens; f_p is known as the paraxial focal length of the lens. If one does not restrict to the paraxial region, then in general, rays which are incident at different heights on the lens, hit the axis at different points. For example, for a convex lens, the marginal rays (which are incident near the periphery of the lens) focus at a point closer than the focal point of paraxial rays [see Fig. 4.2(a)]. Similarly, for a concave lens, rays which are incident farther from the axis appear to be emerging from a point which is nearer to the lens [see Fig. 4.2(b)]. The point at which the paraxial rays strike the axis (F_P) is called the paraxial focus and the point at which the rays near the periphery strike is called the marginal focus (F_M). The distance along the axis between the paraxial image point and the image corresponding to marginal rays (i.e., rays striking the edge of the lens) is termed longitudinal spherical aberration. Similarly, the distance between the paraxial image point and the point at which the marginal ray strikes the paraxial image plane is called the lateral spherical aberration [see Fig. 4.2(a)]. The image on any plane (normal to the z-axis) is a circular patch of light; however, as can be seen from Fig. 4.2(a), on a plane AB the circular patch has the least diameter. This is called the *circle of least confusion.* It may be mentioned that for an object lying on the axis of a cylindrically symmetric system (like a system of coaxial lenses), the image will suffer only from spherical aberration. All other off-axis aberrations like coma, astigmatism, etc., will be absent.

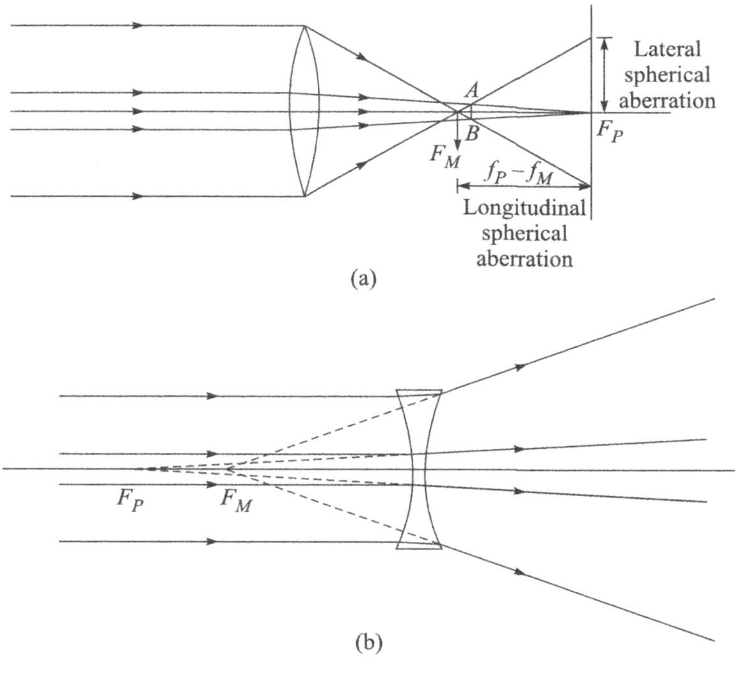

Fig. 4.2 (a) For a converging lens the focal point for marginal rays lies closer to the lens than the focal point for paraxial rays. The distance between the paraxial focal point and the marginal focal point is known as the longitudinal spherical aberration and the radius of the image at the paraxial focal plane is known as the lateral spherical aberration. The combined effect of defocusing and spherical aberration leads to the formation of a circle of least confusion, where the image would have the minimum diameter. (b) The spherical aberration of a diverging lens.

The calculation of the spherical aberration even for a single spherical refracting surface is quite cumbersome (see, e.g., Ref. Sm1), we just give the final results:

$$\Delta z = -\frac{(n_2 - n_1)}{2n_2\left(\frac{1}{z_0} + \frac{n_2 - n_1}{n_1 R}\right)^2}\left(\frac{1}{R} + \frac{1}{z_0}\right)^2 \times \left(-\frac{n_2 + n_1}{n_1 z_0} + \frac{1}{R}\right)h^2 \tag{2}$$

where R represents the radius of curvature of the surface, n_1 and n_2 represent the refractive indices of the media on the left and right of the spherical surface (see Fig. 4.3). For a plane surface $R = \infty$, Eq. (2) reduces to Eq. (24) with $n = n_2/n_1$.

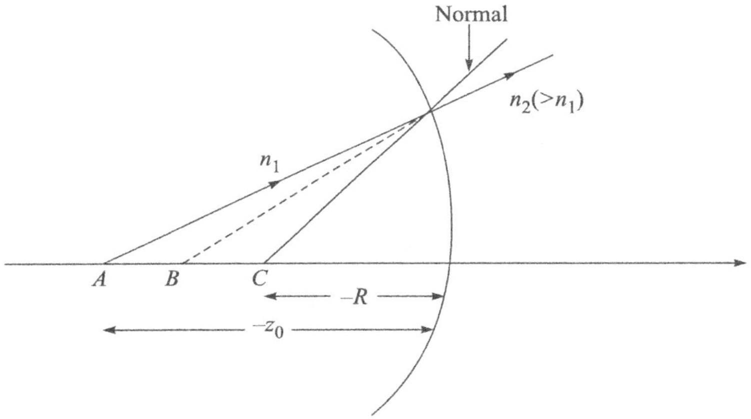

Fig. 4.3 The aplanatic points of a spherical refracting surface.

In a similar manner, for a set of rays incident parallel to the axis, one can show that the coefficient of spherical aberration of a thin lens made of a material of refractive index n and placed in air, with the surfaces having radii of curvatures R_1 and R_2 would be given by

$$A = -\frac{f(n-1)}{2n^2} \times \left[-\left(\frac{1}{R_2} - P\right)^2 \left\{\frac{1}{R_2} - P(n+1)\right\} + \frac{1}{R_1^3}\right] \tag{3}$$

where, $$P = \frac{1}{f} = \left[(n-1)\left(\frac{1}{R_1} - \frac{1}{R_2}\right)\right] \tag{4}$$

represents the power of the lens. The coefficient A is such that when it is multiplied by the cube of the height of the ray at the lens, one obtains the lateral spherical aberration. Thus, the lateral spherical aberration for rays hitting the lens at a height h would be

$$S_{\text{lat}} = Ah^3 = -\frac{f(n-1)h^3}{2n^2} \times \left[-\left(\frac{1}{R_2} - P\right)^2 \left\{\frac{1}{R_2} - P(n+1)\right\} + \frac{1}{R_1^3}\right] \tag{5}$$

The longitudinal spherical aberration (which represents the difference between the marginal focal length and the paraxial focal length) would be given by

$$S_{\text{long}} = Ah^2 f$$
$$= -\frac{(n-1)f^2h^2}{2n^2} \times \left[\frac{1}{R_1^3} - \left(\frac{1}{R_2} - \frac{n+1}{f}\right)\left(\frac{1}{R_2} - \frac{1}{f}\right)^2\right] \tag{6}$$

For a converging lens, S_{long} will always be negative implying that the marginal rays focus closer to the lens. For a thin lens of given power (i.e. of a given focal length), one can define a quantity q, called the shape factor, by the following relation:

$$q = \frac{R_2 + R_1}{R_2 - R_1} \tag{7}$$

where R_1 and R_2 are the radii of curvatures of the two surfaces. For a given focal length of the lens, one can control the spherical aberration by changing the value of q. This procedure is called bending of the lens. Figure 4.4 shows the variation of spherical aberration with q for $n = 1.5$, $f = 40$ cm (i.e., $P = 0.025$ cm^{-1}) and $h = 1$ cm. It can be seen that for values of $q \approx 0.7$, the (magnitude of the) spherical aberration is minimum (but not zero). Thus, by choosing proper values of the radii, the spherical aberration can be minimised. It may be mentioned that the value $q = +1$ implies $R_2 = \infty$ and hence it corresponds to a plano-convex lens with the convex side facing the incident light. On the other hand, for a plano-convex lens with the plane side facing the incident light $R_1 = \infty$ and $q = -1$. Thus, the spherical aberration is dependent on how the deviation is divided between the surfaces.

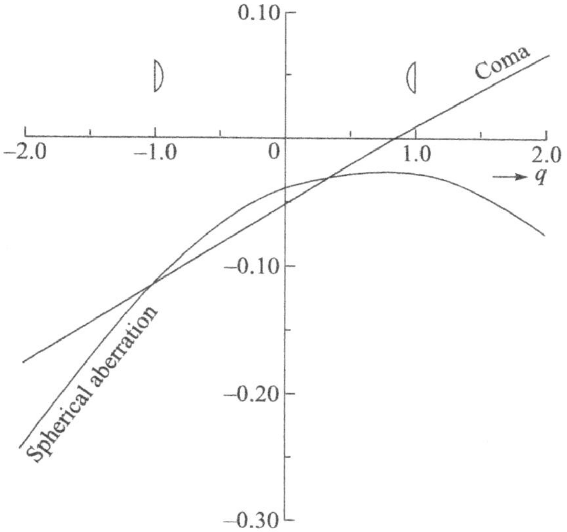

Fig. 4.4 Variation of spherical aberration and coma with the shape factor of a thin lens with $n = 1.5$, $f = 40$ cm and $h = 1$ cm. For calculating the coma we have assumed $\tan\theta = 1$, i.e., rays make an angle of 45° with the axis.

For a parallel beam of rays incident on a lens and inclined at angle with the z-axis (see Fig. 4.5), one can show that the coma in the image is given by (see, e.g., Ref. Gh2):

$$\text{Coma} = \frac{3(n-1)fh^2 \tan\theta}{2} \times \left[\frac{(n-1)(2n+1)}{nR_1R_2} - \frac{(n^2-n-1)}{n^2R_1^2} - \frac{n}{R_2^2}\right] \tag{8}$$

In Fig. 4.4 we have plotted the variation of coma with the shape factor q. It can immediately be seen that for a lens with $q = 0.8$, coma is zero. Also both spherical aberration and coma are close to a minimum for a plano-convex (with the convex side facing the incident light) for which $q = 1.0$. As such, plano-convex lenses are extensively used in eyepieces.

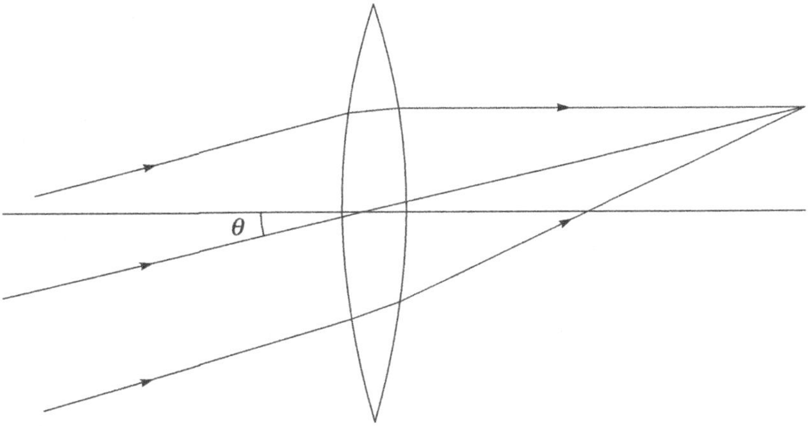

Fig. 4.5 Parallel rays (inclined at an angle θ with the axis) incident on a thin lens.

PROBLEMS

4.1 Show that the chromatic aberration for a thin lens is given by Eq. (1).

4.2 Consider an optical system of two thin lenses made of different materials placed in contact with each other. For example, one of the lenses may be made of crown glass and the other of flint glass. Show that for the lens combination to have the same focal length for the blue and red colours, we must have

$$\frac{\omega}{f} + \frac{\omega'}{f'} = 0 \tag{9}$$

where,

$$\omega = \frac{n_b - n_r}{n-1} \quad \text{and} \quad \omega' = \frac{n'_b - n'_r}{n'-1} \tag{10}$$

are known as the dispersive powers. Since ω and ω' are both positive, f and f' must be of opposite signs.

4.3 An achromatic doublet of focal length 20 cm is to be made by placing a convex lens of borosilicate crown glass in contact with a diverging lens of dense flint glass. Assuming $n_r = 1.51462$, $n_b = 1.52264$, $n_r' = 1.61216$ and $n_b' = 1.62901$, calculate the focal length of each lens; here the unprimed and the primed quantities refer to the borosilicate crown glass and dense flint glass respectively.

4.4 Consider a separated doublet consisting of two thin lenses of focal lengths f and f' and separated by a distance t. Calculate the focal length of the combination and show that the chromatic aberration is very small if the distance between the two lenses is equal to the mean of the focal lengths. (This is indeed the case for the Huygens' eyepiece).

4.5 An achromatic cemented doublet of focal length 25 cm is to be made from a combination of an equiconvex flint glass lens ($n_b = 1.50529$, $n_r = 1.49776$) and a crown glass lens ($n_b = 1.66270$, $n_r = 1.64357$). Calculate the radii of curvatures of the different surfaces and the focal lengths of each of the two lenses.

4.6 Rays parallel to the axis are incident on a spherical refracting surface of radius R separating media of refractive index n_1 and n_2. Assume $n_1 = 1.0$, $n_2 = 1.5$ and $R = 10$ cm; the height x may be assumed to be 0.0001 cm, 1.0 cm, 2.0 cm and 3.0 cm.

Write a small program to obtain the exact point at which the refracted ray will intersect the axis as a function of the height x of the incident ray. Discuss the longitudinal spherical aberration of the image.

4.7 Consider a point object in front of a plane refracting surface. Obtain the paraxial image point and calculate the aberration in the image when we consider rays which hit the refracting surface at a height h.

4.8 Consider a plane glass slab of thickness d made of a material of refractive index n, placed in air. By simple application of Snell's law obtain an expression for the spherical aberration of the slab.

4.9 Consider a spherical refracting surface of radius R. Show that for a point A [see Fig. 4.3] such that

$$z_0 = \frac{n_1 + n_2}{n_1} R \tag{11}$$

the spherical aberration is zero. Notice that both z_0 and R are negative quantities.

4.10 Calculate the longitudinal spherical aberration of a thin plano-convex lens made of a material of refractive index 1.5 and whose curved surface has a radius of curvature of 10 cm, for rays incident at a height of 1 cm. Compare the values of the aberration when the convex side and the plane side face the incident light.

4.11 Consider a lens made up of a material of refractive index 1.5 with a focal length 25 cm. Assuming $h = 0.5$ cm and $\theta = 45°$, obtain the spherical aberration and coma for the lens or various values of the shape factor q and plot the variation in a manner similar to that shown in Fig. 4.4.

SOLUTIONS

4.1 The focal length of a thin lens is given by

$$\frac{1}{f} = (n-1)\left(\frac{1}{R_1} - \frac{1}{R_2}\right) \tag{12}$$

If a change of n by δn (the change of n is due to the change in the wavelength of the light) results in a change of f by δf then we obtain by differentiating the above equation

$$-\frac{\delta f}{f^2} = \delta n\left(\frac{1}{R_1} - \frac{1}{R_2}\right) = \frac{\delta n}{n-1}\frac{1}{f}$$

$$\delta f = -f\frac{\delta n}{n-1} \tag{13}$$

which represents the chromatic aberration of a thin lens. If n_b and n_r represent the refractive indices for the blue and red colours respectively, then

$$f_r - f_b = \left(\frac{n_b - n_r}{n-1}\right) \tag{14}$$

would represent the chromatic aberration.

4.2 We consider an optical system of two thin lenses made of different materials placed in contact with each other. For example, one of the lenses may be made of crown glass and the other of flint glass. We will find the condition for this lens combination to have the same focal length for the blue and red colours. Let n_b, n_y and n_r represent the refractive indices for the material of the first lens corresponding to the blue, yellow and red colours respectively. Similarly, n_b', n_y' and n_r' represent the corresponding refractive indices for the second lens. If f_b and f_b' represent the focal lengths for the first and the second lens corresponding to the blue colour, and if F_b represents the focal length of the combination of the two lenses (placed in contact), then

$$\frac{1}{F_b} = \frac{1}{f_b} + \frac{1}{f_b'} = (n_b - 1)\left(\frac{1}{R_1} - \frac{1}{R_2}\right) + (n_b' - 1)\left(\frac{1}{R_1'} - \frac{1}{R_2'}\right) \tag{15}$$

where R_1 and R_2 represent the radii of curvatures of the first and second surface for the first lens and, as before, the primed quantities refer to the second lens. Thus, we may write

$$\frac{1}{F_b} = \frac{n_b - 1}{n-1}\frac{1}{f} + \frac{n_b' - 1}{n' - 1}\frac{1}{f'} \tag{16}$$

where,
$$\frac{1}{f} \equiv (n-1)\left(\frac{1}{R_1} - \frac{1}{R_2}\right), \tag{17}$$

$$\frac{1}{f'} \equiv (n'-1)\left(\frac{1}{R_1'} - \frac{1}{R_2'}\right) \tag{18}$$

$$n \equiv \frac{n_b + n_r}{2} \approx n_y, \quad n' \equiv \frac{n'_b + n'_r}{2} \approx n'_y \tag{19}$$

f and f' represent the focal lengths of the first and second lens corresponding to a mean colour which is around the yellow region. Similarly, the focal length of the combination corresponding to the red colour would be given by

$$\frac{1}{F_r} = \frac{n_r - 1}{n - 1}\frac{1}{f} + \frac{n'_r - 1}{n' - 1}\frac{1}{f'} \tag{20}$$

For the focal length of the combination to be equal for blue and red colours, we must have

$$\frac{n_b - 1}{n - 1}\frac{1}{f} + \frac{n'_b - 1}{n' - 1}\frac{1}{f'} = \frac{n_r - 1}{n - 1}\frac{1}{f} + \frac{n'_r - 1}{n' - 1}\frac{1}{f'}$$

or

$$\frac{\omega}{f} + \frac{\omega'}{f'} = 0 \tag{21}$$

where,

$$\omega = \frac{n_b - n_r}{n - 1} \quad \text{and} \quad \omega' = \frac{n'_b - n'_r}{n' - 1} \tag{22}$$

are known as the dispersive powers. Since ω and ω' are both positive, f and f' must be of opposite signs for the validity of Eq. (21). A lens combination which satisfies Eq. (21) is known as an achromatic doublet (see Fig. 4.2). It may be mentioned that if the two lenses are made of the same material, then $\omega = \omega'$ and Eq. (9) would imply $f = -f'$; such a combination will have an infinite focal length. Thus, for an achromatic doublet the two lenses must be of different materials.

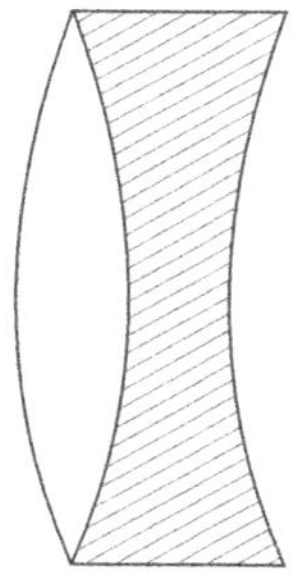

Fig. 4.6 An achromatic doublet.

4.3 $n \approx \dfrac{n_b + n_r}{2} = \dfrac{1.52264 + 1.51462}{2} = 1.51863$

$$n' \approx \frac{n'_b + n'_r}{2} = \frac{1.62901 + 1.61216}{2} = 1.62058$$

Thus,

$$\omega = \frac{1.52264 - 1.51462}{1.51863 - 1} = 0.01546$$

and

$$\omega' = \frac{1.62901 - 1.61216}{1.62058 - 1} = 0.02715$$

Substituting in Eq. (9), we obtain

$$\frac{0.01546}{f} + \frac{0.02715}{f'} = 0$$

or,

$$\frac{f}{f'} = -0.56942$$

Now, for the lens combination to be of focal length 20 cm we must have

$$\frac{1}{f} + \frac{1}{f'} = \frac{1}{20}$$

or,
$$\frac{1}{f}[1 - 0.56942] = \frac{1}{20}$$

or,
$$f = 20 \times 0.43058 = 8.61 \text{ cm}$$

and
$$f' = -\frac{f}{0.56942} = -15.1 \text{ cm}$$

4.4 We consider two thin lenses of focal lengths f and f' separated by a distance t (see Fig. 4.7). The focal length of the combination F, would be

$$\frac{1}{F} = \frac{1}{f} + \frac{1}{f'} - \frac{t}{ff'} \tag{23}$$

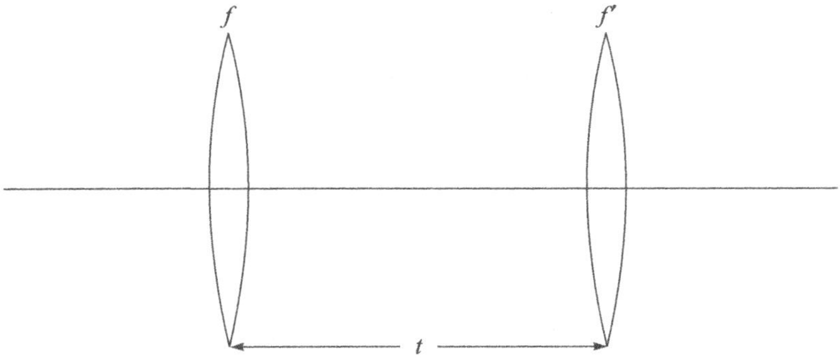

Fig. 4.7 The separated doublet.

The focal length of the first lens would be given by

$$\frac{1}{f} = (n-1)\left(\frac{1}{R_1} - \frac{1}{R_2}\right) \tag{24}$$

with a similar expression for $\frac{1}{f'}$. If Δf and Δn represent the changes in the focal length and in the refractive index due to a change in the wavelength, then by differentiating Eq. (12), we obtain

$$-\frac{\Delta f}{f^2} = \Delta n \left(\frac{1}{R_1} - \frac{1}{R_2}\right) = \frac{\Delta n}{(n-1)f}$$

Thus, differentiating Eq. (23), we obtain

$$-\frac{\Delta F}{F^2} = -\frac{\Delta f}{f^2} - \frac{\Delta f'}{f'^2} + \frac{t}{f}\frac{\Delta f'}{f'^2} + \frac{t}{f'}\frac{\Delta f}{f^2}$$

$$= \frac{\Delta n}{(n-1)f} + \frac{\Delta n'}{(n'-1)f'} - \frac{t}{f}\frac{\Delta n'}{(n'-1)f'} - \frac{t}{f'}\frac{\Delta n}{(n-1)f}$$

$$= \frac{\omega}{f} + \frac{\omega'}{f'} - \frac{t}{ff'}(\omega + \omega') \tag{25}$$

where, as before, ω and ω' represent the dispersive powers. Consequently, for the combination to have the same focal length for blue and red colours we should have

$$\frac{\omega}{f} + \frac{\omega'}{f'} = \frac{t}{ff'}(\omega + \omega')$$

or,

$$t = \frac{\omega f' + \omega' f}{\omega + \omega'} \tag{26}$$

If both the lenses are made of the same material, then $\omega = \omega'$ and the above equation simplifies to

$$t = \frac{f + f'}{2} \tag{27}$$

implying that the chromatic aberration is very small if the distance between the two lenses is equal to the mean of the focal lengths.

4.5 For the equiconvex flint glass lens

$$n_b = 1.50529 \quad \text{and} \quad n_r = 1.49776$$

Thus,

$$n = \frac{n_b + n_r}{2} = 1.501525 \quad \text{and} \quad \omega = \frac{n_b - n_r}{2} \approx 0.00501$$

Similarly, for a crown glass lens

$$n_b' = 1.66270 \quad \text{and} \quad n_r' = 1.64357$$

Thus,

$$n' = \frac{n_b' + n_r'}{2} = 1.654035 \quad \text{and} \quad \omega' = \frac{n_b' - n_r'}{2} \approx 0.01157$$

For the system to be achromatic

$$0 = \frac{\omega}{f} + \frac{\omega'}{f'} = \frac{0.00501}{f} + \frac{0.01157}{f'} \Rightarrow \frac{f}{f'} \approx -0.433$$

Now, for the lens combination to be of focal length 25 cm, we must have

$$\frac{1}{f} + \frac{1}{f'} = \frac{1}{25} \Rightarrow \frac{1}{f}[1 - 0.433] = \frac{1}{25} \Rightarrow f \approx 14.2 \text{ cm}$$

$$\Rightarrow \qquad f' \approx -32.8 \text{ cm}$$

Now $\frac{1}{f} = (n-1)\left(\frac{1}{R_1} - \frac{1}{R_2}\right) = (n-1)\frac{2}{R_1}$, when we have assumed $R_2 = -R_1$.

Thus

$$R_1 = 2(n-1)f = 2 \times 0.5015 \times 14.2 \approx 14.2 \text{ cm}$$

$$R_2 = -R_1 = -14.2 \text{ cm} \Rightarrow R_1' = -14.2 \text{ cm}$$

$$\Rightarrow \qquad \frac{1}{f'} = (n'-1)\left(\frac{1}{R_1'} - \frac{1}{R_2'}\right) \Rightarrow \frac{1}{R_2'} = \frac{1}{R_1'} - \frac{1}{(n'-1)f'}$$

$$\Rightarrow \qquad \frac{1}{R_2'} = -\frac{1}{14.2} + \frac{1}{0.654 \times 32.8} \Rightarrow R_2' \approx -42 \text{ cm}$$

4.6

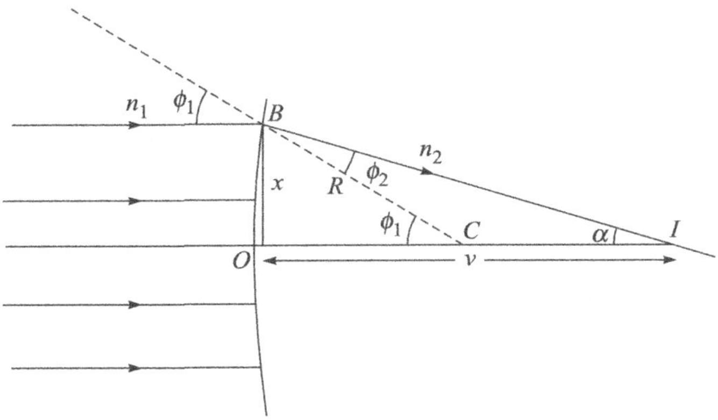

Fig. 4.8 A parallel beam of light is incident on a spherical surface separating two media of refractive indices n_1 and n_2. The point C represents the center of curvature of the refracting surface.

Referring to Fig. 4.8, we may write

$$\sin\phi_1 = \frac{x}{R};$$

$$n_1 \sin\phi_1 = n_2 \sin\phi_2 \Rightarrow \phi_2 = \sin^{-1}\left(\frac{n_1 \sin\phi_1}{n_2}\right)$$

$$\alpha = \phi_1 - \phi_2$$

In ΔBCI: $$\frac{\sin\phi_2}{CI} = \frac{\sin\alpha}{CB} = \frac{\sin\alpha}{R}$$

Thus, $$OI = OC + CI = R + R\frac{\sin\phi_2}{\sin\alpha} = R\left(1 + \frac{\sin\phi_2}{\sin\alpha}\right)$$

The distance OI is represented by v. A simple MATLAB program is given below. We cannot take $x = 0$ – we take it to be a small number. The program corresponds to $n_1 = 1.0$, $n_2 = 1.5$, $R = 10$ cm and x is measured in cm.

```
clear all;
clc;
n1=1.0;
n2=1.5;
r=10.0;
x=0.00001;
phi1=asin(x/r);
phi2=asin(x*n1/(r*n2));
alpha=phi1-phi2;
v=r*(1.+ (sin(phi2)/sin(alpha)))
x=1.0;
phi1=asin(x/r);
phi2=asin(x*n1/(r*n2));
```

```
alpha=phi1-phi2;
v=r*(1.+ (sin(phi2)/sin(alpha)))
x=2.0;
phi1=asin(x/r);
phi2=asin(x*n1/(r*n2));
alpha=phi1-phi2;
v=r*(1.+ (sin(phi2)/sin(alpha)))
x=3.0;
phi1=asin(x/r);
phi2=asin(x*n1/(r*n2));
alpha=phi1-phi2;
v=r*(1.+ (sin(phi2)/sin(alpha)))
```

The answer comes out as v = 30.0000 cm, v = 29.9332 cm, v = 29.7312 cm and v = 29.3891 cm respectively which shows the longitudinal spherical aberration of the image.

The paraxial formula

$$\frac{n_2}{v} - \frac{n_1}{R} = \frac{n_2 - n_1}{R}$$

gives us $$\frac{1.5}{v} - \frac{1.0}{-\infty} = \frac{0.5}{10} \Rightarrow v = 30 \text{ cm}$$

4.7

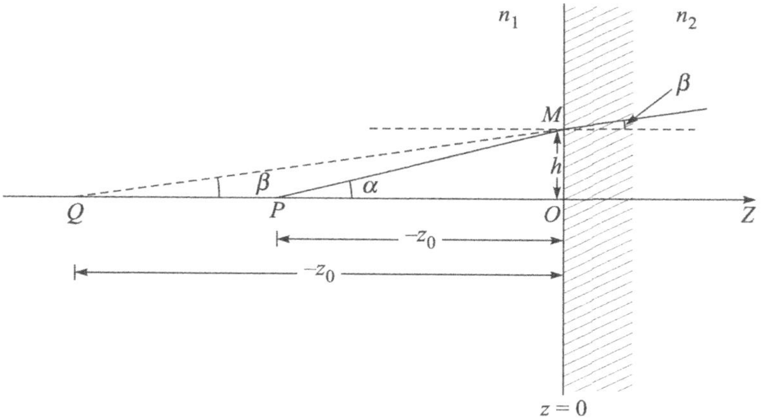

Fig. 4.9 Refraction at a plane surface.

Let the plane of the refracting surface be chosen as the plane $z = 0$. Let P be the object point. The z-axis is chosen to be along the normal (PO) from the point P to the surface. The plane $z = 0$ separates two media of refractive indices n_1 and n_2 (see Fig. 4.9); in the figure we have assumed $n_2 > n_1$. Consider a ray PM incident on the refracting surface (from the object) at a height h as shown in the above figure. The refracted ray appears to emerge from the point Q. We assume the origin to be at the point O. Let the z-coordinates of the points P and Q be z_0 and z_1 respectively. Obviously, both z_0 and z_1 would be negative quantities and the distances OP and OQ would be $-z_0$ and $-z_1$ respectively (see Fig. 4.9). We have to determine z_1 in terms of z_0. From Snell's law we know that

$$\sin \alpha = n \sin \beta \tag{28}$$

where α and β are the angles that the incident and refracted rays make with the z-axis and

$$n = \frac{n_2}{n_1} \tag{29}$$

Now, from Fig. 4.9 we have

$$-z_1 = h \cot \beta = \frac{h}{\sin \beta}\sqrt{1-\sin^2\beta}$$

or,
$$z_1 = -\frac{nh}{\sin \alpha}\sqrt{1-\frac{1}{n^2}\sin^2\alpha} \tag{30}$$

where we have used Eq. (28). Since

$$\sin \alpha = \frac{h}{\sqrt{h^2+z_0^2}} \tag{31}$$

we obtain

$$z_1 = -\frac{nh}{h}(h^2+z_0^2)^{1/2}\left[1-\frac{1}{n^2}\frac{h^2}{(h^2+z_0^2)}\right]^{1/2} \tag{32}$$

or,
$$z_1 = -n|z_0|\left[1+\frac{h^2}{z_0^2}\right]^{1/2}\left[1-\frac{h^2}{n^2 z_0^2}\left(1+\frac{h^2}{z_0^2}\right)^{-1}\right]^{1/2} \tag{33}$$

The value of z_1 given by the above equation is an exact expression in terms of z_0. It can at once be seen that the image distance, z_1, is a complicated function of the height h, at which the ray strikes the refracting surface. In the limit of $h \to 0$, i.e., for paraxial rays, we get

$$z_1 = -n|z_0| \tag{34}$$

which is the expression for the image distance in the paraxial region. To the next order of approximation, assuming $|h/z_0| << 1$, we get

$$z_1 = -n|z_0|\left[1+\frac{h^2}{2z_0^2}\right]\left[1-\frac{h^2}{2n^2 z_0^2}\right]$$

$$\approx -n|z_0|\left[1+\frac{h^2}{2z_0^2 n^2}(n^2-1)\right] \tag{35}$$

Thus the aberration is given by

$$\Delta z = -\frac{h^2}{2n|z_0|}(n^2-1) \tag{36}$$

The above equation gives the longitudinal spherical aberration. The negative sign implies that the nonparaxial rays appear to emanate from a point which is farther away from the paraxial image point.

4.8

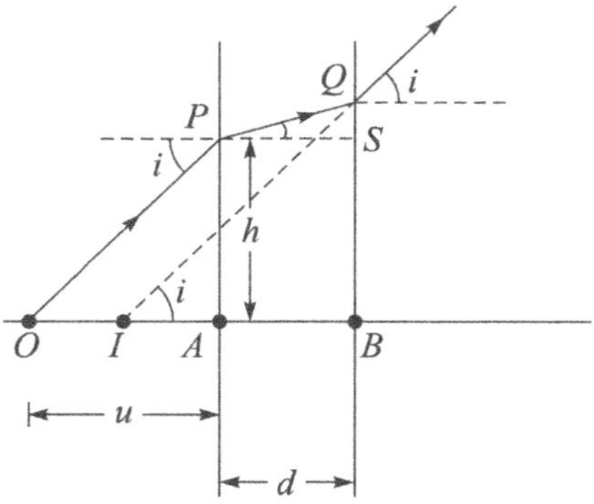

Fig. 4.10 Figure for Problem 4.8

The object is assumed to be at the point O and the ray (undergoing refraction at height h) appears to come from the point I as shown in the figure. Now,

$$QS = d \tan r$$

Thus, $$BQ = BS + SQ = h + d \tan r.$$

Further, $$\sin i = \frac{h}{\sqrt{h^2 + u^2}} \quad \text{and} \quad \sin i = n \sin r$$

$$\Rightarrow \quad \sin r = \frac{h}{\sqrt{n^2 h^2 + n^2 u^2}} \Rightarrow \tan r = \frac{h}{\sqrt{(n^2 - 1) h^2 + n^2 u^2}}$$

Now, $$IB = \frac{BQ}{\tan i} = \frac{h + d \tan r}{h/u}$$

$$= \frac{u}{h}\left[h + \frac{dh}{\sqrt{(n^2 - 1) h^2 + n^2 u^2}}\right]$$

$$= u + \frac{ud}{nu}\left[1 + \frac{(n^2 - 1)h^2}{n^2 u^2}\right]^{-1/2}$$

which is an exact expression. For $h/u << 1$

$$IB \approx u + \frac{d}{n} - \frac{(n^2 - 1) h^2 d}{2n^3 u^2}$$

Thus, the spherical aberration is $\approx -\dfrac{(n^2 - 1) h^2 d}{2n^3 u^2}$.

4.9 For $z_0 = \frac{n_1 + n_2}{n_1} R$, one of the factors in Eq. (2) vanishes and the spherical aberration is zero. Indeed, it can be rigorously shown that all rays emanating from the point A appear to diverge from the point B.

4.10 When the convex side is facing incident light [see Fig. 4.11(a)]

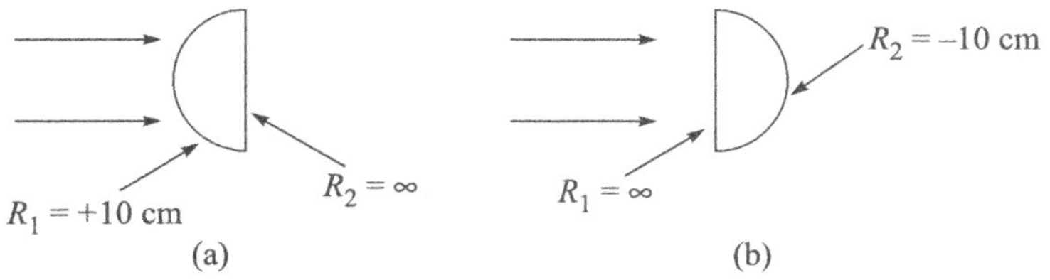

Fig. 4.11

$$R_1 = +10 \text{ cm}, \quad R_2 = \infty, \quad h = 1 \text{ cm}$$

$$\frac{1}{f} = (n-1)\left(\frac{1}{R_1} - \frac{1}{R_2}\right) = \frac{0.5}{10} = \frac{1}{20} \Rightarrow f = 20 \text{ cm}$$

and $$P = \frac{1}{f} = \frac{1}{20}.$$

Thus,

$$A = -\frac{f(n-1)}{2n^2}\left[-\left(\frac{1}{R_2} - P\right)^2 \left\{\frac{1}{R_2} - P(n+1)\right\} + \frac{1}{R_1^3}\right]$$

$$= -\frac{20 \times 0.5}{2 \times 1.5 \times 1.5}\left[-\frac{1}{400}\left\{-\frac{1}{20} \times 2.5\right\} + \frac{1}{10 \times 10 \times 10}\right]$$

$$\approx -2.917 \times 10^{-3} \text{ cm}^{-2}$$

Thus, $S_{\text{long}} = Ah^2 f = -2.917 \times 10^{-3} \times 1 \times 1 \times 20 \approx -0.058$ cm

(a) When the plane side is facing incident light [see Fig. 4.11(b)]

$R_1 = \infty$, $R_2 = -10$ cm, $h = 1$ cm. Obviously, $f = 20$ cm and $P = \frac{1}{20}$ cm^{-1}.

Thus, $$A = -\frac{20 \times 0.5}{2 \times 1.5 \times 1.5}\left[-\left(-\frac{1}{10} - \frac{1}{20}\right)^2 \left\{-\frac{1}{10} - \frac{2.5}{20}\right\} + 0\right]$$

$$= -0.01125 \text{ cm}^{-2}$$

Thus, $S_{\text{long}} = Ah^2 f = -0.225$ cm

4.11 We first note that

$$\frac{1}{f} = (n-1)\left(\frac{1}{R_1} - \frac{1}{R_2}\right) \Rightarrow \frac{1}{R_1} - \frac{1}{R_2} = \frac{1}{(n-1)f}$$

Also, $$q = \frac{R_2 + R_1}{R_2 - R_1} = \frac{\dfrac{1}{R_1} + \dfrac{1}{R_2}}{\dfrac{1}{R_1} - \dfrac{1}{R_2}} \Rightarrow \frac{1}{R_1} + \frac{1}{R_2} = \frac{q}{(n-1)f}$$

Thus, $$\frac{1}{R_1} = \frac{(q+1)}{2(n-1)f} \quad \text{and} \quad \frac{1}{R_2} = \frac{(q-1)}{2(n-1)f}$$

The longitudinal spherical aberration is given by

$$S_{\text{long}} = Ah^2 f$$

$$= -\frac{(n-1)f^2h^2}{2n^2}\left\{\frac{(q+1)^2}{8(n-1)^3f^3}\left[\frac{q-1}{2(n-1)f} - \frac{1}{f}\right]^2 - \left[\frac{q-1}{2(n-1)f} - \frac{(n+1)}{f}\right]\right\}$$

Similarly, the expression for coma is given by

$$\text{Coma} = \frac{3(n-1)B}{2} fh^2 \tan^2\theta$$

where, $$B = \left[\frac{(n-1)(2n+1)}{nR_1R_2} - \frac{n^2-n-1}{n^2R_1^2} - \frac{n}{R_2^2}\right]$$

$$= \left[\frac{(n-1)(2n+1)(q^2-1)}{4n(n-1)^2f^2} - \frac{(n^2-n-1)(q+1)^2}{4n^2(n-1)^2f^2} - \frac{n(q-1)^2}{4(n-1)^2f^2}\right]$$

Huygens' Principle and its Applications

5

A Quick Review

5.1 INTRODUCTION

Huygens' theory is essentially based on a geometrical construction which allows us to determine the shape of the wavefront at any time, if the shape of the wavefront at an earlier time is known. According to Huygens' principle, each point of a wavefront is a source of secondary disturbance and the wavelets emanating from these points spread out in all directions with the speed of the wave. The envelope of these wavelets gives the shape of the new wavefront. In Fig. 5.1, S_1S_2 represents the shape of the wavefront (emanating from the point O) at a particular time which we denote as $t = 0$. The medium is assumed to be homogeneous and isotropic. Let us suppose we want to determine the shape of the wavefront after a time interval of Δt. Then, with each point on the wavefront as center, we draw spheres of radius $v\Delta t$, where v is the speed of the wave in that medium. If we draw a common tangent to all these spheres, then we obtain the envelope which is again a sphere centered at O. Thus, the shape of the wavefront at a later time Δt is the sphere $S_1'S_2'$.

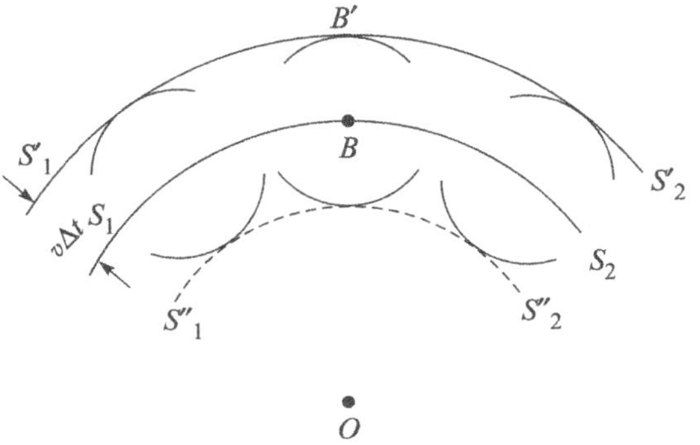

Fig. 5.1 Huygens' construction for the determination of the shape of the wavefront, given the shape of the wavefront at an earlier time. S_1S_2 is a spherical wavefront centered at O at a time, say $t = 0$. $S_1'S_2'$ corresponds to the state of the wavefront at a time Δt, which is again spherical and centered at O. The dashed curve represents the backwave.

Let us consider spherical waves emanating from the point source O and striking the obstacle A (see Fig. 5.2). According to the rectilinear propagation of light (which is also predicted by corpuscular theory) one should obtain a shadow in the region PQ of the screen. This is not rigorously true and one does obtain a finite intensity in the region of the geometrical shadow. However, at the time of Huygens, light was known to travel in straight lines and Huygens explained this by assuming that the secondary wavelets do not have any amplitude at any point not enveloped by the wavefront.

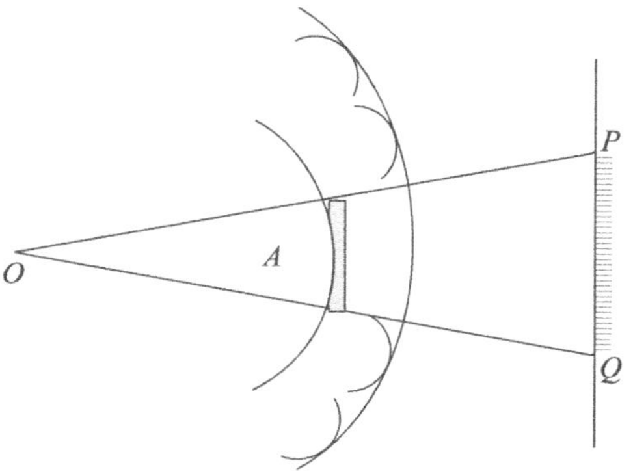

Fig. 5.2 Rectilinear propagation of light. O is a point source emitting spherical waves and A is an obstacle which forms a shadow in the region PQ of the screen.

In order to explain diffraction phenomena, Fresnel modified the principle and postulated that the secondary wavelets mutually interfere. The Huygens' principle along with the fact that the secondary wavelets mutually interfere, is known as the *Huygens–Fresnel principle.* This principle can be used to understand diffraction phenomena from different apertures (see Chapters 9-11).

PROBLEMS

5.1 Use Huygens' principle to obtain Snell's law.

5.2 Obtain the law of reflection using Huygens' principle

5.3 Consider a point source placed in front of a spherical surface of radius of curvature R separating media of refractive indices n_1 and n_2 (see Fig. 5.5). Use Huygens' principle and obtain the relationship between the object distance u and image distance v.

5.4 From the formula for refraction at a single interface, obtain the lens formula.

5.5 Consider a vibrating source moving through a medium with a speed V. Let the speed of propagation of the wave in the medium be v. Show that if $V > v$ then a conical wavefront is set up whose half-angle is given by

$$\theta = \sin^{-1}\left(\frac{v}{V}\right) \tag{1}$$

5.6 Use Huygens' principle to study how a plane wavefront incident along the z-direction on an inhomogenous medium with a refractive index variation given by

$$n^2(x, y) = n_1^2 - \gamma^2 (x^2 + y^2) \tag{2}$$

will get modified.

5.7 Use Huygens' principle to study the reflection of a spherical wave emanating from a point on the axis at a concave mirror of radius of curvature R and obtain the mirror equation

$$\frac{1}{u} + \frac{1}{v} = \frac{2}{R} \tag{3}$$

5.8 Consider a plane wave incident obliquely on the face of a prism. Using Huygens' principle, construct the transmitted wavefront and show that the deviation produced by the prism is given by

$$\delta = i + t - A \tag{4}$$

where A is the angle of the prism, i and t are the angles of incidence and transmittance.

SOLUTIONS

5.1 Let S_1S_2 be a surface separating two media with different speeds of propagation of light v_1 and v_2 as shown in Fig. 5.3. Let A_1B_1 be a plane wavefront incident on the surface at an angle i; A_1B_1 represents the position of the wavefront at an instant $t = 0$.

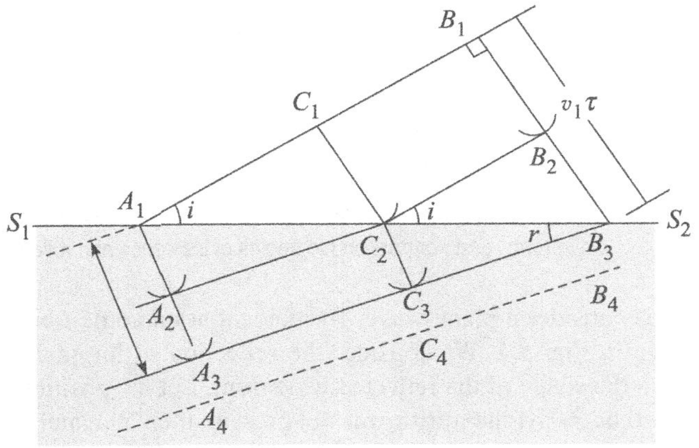

Fig. 5.3 Refraction of a plane wavefront A_1B_1 by a plane interface S_1S_2 separating two media with different velocities of propagation of light v_1 and v_2 ($< v_1$); i and r are the angles of incidence and refraction respectively. $A_2C_2B_2$ corresponds to the shape of the wavefront at an intermediate time τ_1. Notice that $r < i$.

Let τ be the time taken for the wavefront to travel the distance B_1B_3. Then $B_1B_3 = v_1\tau$. In the same time the light would have travelled a distance $A_1A_3 = v_2\tau$ in the second medium. (Note that the lines A_1A_3, B_1B_3, etc. are always normal to the wavefront; these represent rays in isotropic media). It can easily be seen that the incident and refracted rays make angles i and r with the normal. In order to determine the shape of the wavefront at the instant $t = \tau$ we consider an arbitrary point C_1 on the wavefront. Let the time taken for the disturbance to travel the distance C_1C_2 be τ_1. Thus, $C_1C_2 = v_1\tau_1$. From the point C_2 we draw a secondary wavelet of radius $v_2(\tau - \tau_1)$. Similarly from the point A_1 we draw a secondary wavelet of radius $v_2\tau$. The envelope of these secondary wavelets is shown as $A_3C_3B_3$. The shape of the wavefront at the intermediate time τ_1 is shown as $A_2C_2B_2$ and clearly $B_1B_2 = C_1C_2 = v_1\tau_1$ and $A_1A_2 = v_2\tau_1$. In the right-angled triangles $B_2C_2B_3$ and $C_3C_2B_3$, $\angle B_2C_2B_3 = i$ (the angle of incidence) and $\angle C_2B_3C_3 = r$ (the angle of refraction). Clearly,

$$\frac{\sin i}{\sin r} = \frac{B_2B_3/C_2B_3}{C_2C_3/C_2B_3} = \frac{v_1}{v_2} = \frac{n_2}{n_1}$$

or,

$$n_1 \sin i = n_2 \sin r \tag{5}$$

which is known as the Snell's law.

5.2

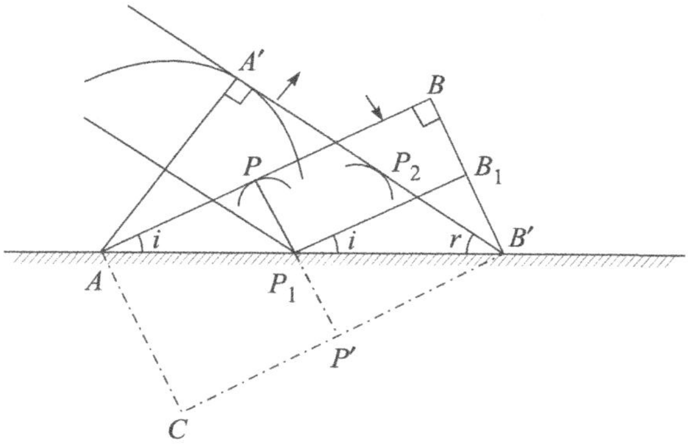

Fig. 5.4 Reflection of a plane wavefront *AB* incident on a plane mirror. *A′B′* is the reflected wavefront; *i* and *r* correspond to angles of incidence and reflection respectively.

Let us consider a plane wave AB incident at an angle i on a plane mirror as shown in Fig. 5.4. We consider the reflection of the plane wave and try to obtain the shape of the reflected wavefront. Let the position of the wavefront at $t = 0$ be AB. If the mirror was not present, then at a later time τ the position of the wavefront would have been CB', where $BB' = PP' = AC = v\tau$ and v is the speed of propagation of the wave. In order to determine the shape of the reflected wavefront at the instant $t = \tau$, we consider an arbitrary point P on the wavefront AB and let τ_1 be the time taken by a disturbance to reach the point P_1 from P. From the point P_1, we draw a sphere of radius $v(\tau - \tau_1)$.

We draw a tangent plane on this sphere from the point B'. Since $BB_1 = PP_1 = v\tau_1$, the distance B_1B' will be equal to P_1P_2 [$= v(\tau - \tau_1)$]. If we consider triangles P_2P_1B' and B_1P_1B' then the side P_1B' is common to both and since $P_1P' = B'B_2$, and since both the triangles are right-angled triangles, $\angle P_2B'P_1 = \angle B_1P_1B'$. The former is the angle of reflection and the latter is the angle of incidence. Thus, we have the law of reflection; when a plane wavefront gets reflected from a plane surface, the angle of reflection is equal to the angle of incidence and the reflected wave is a plane wave.

5.3 Let us consider spherical waves (emanating from the point P) incident on the curved spherical surface SBS'. Let the shape of the wavefront at the time $t = 0$ be ABC [see Fig. 5.5(a)]. In the absence of the spherical surface, the shape of the wavefront at a later time τ would have been $A_1B_1C_1$ where $AA_1 = BB_1 = CC_1 = v_1\tau$. We consider an arbitrary point Q on the wavefront ABC and let τ_1 be the time taken for the disturbance to reach the point Q' (on the surface of the spherical wave); thus $QQ' = v_1\tau_1$. In order to determine the shape of the refracted wavefront at a later time τ, we draw a sphere of radius $v_2(\tau - \tau_1)$ from the point Q'. We may draw similar spheres from other points on the spherical surface; in particular, the radius of the spherical wavefront from the point B, which is equal to BB_2 will be $v_2\tau$. The envelope of these spherical wavelets is shown as $A_1B_2C_1$ which, in general, will not be a sphere. However, a small portion of any curved surface can be considered as a sphere and in this approximation we may consider $A_1B_2C_1$ to be a sphere whose center of curvature is at the point M. The spherical wavefront will, therefore, converge towards the point M and hence the point M represents the real image of the point P. In actual practice the refracted wave will not be a spherical wave and hence it will not converge to a single point; this fact is responsible for the aberrations (see Chapter 4).

We adopt a sign convention in which all distances, measured to the left of the point B, are negative and all distances measured to the right of the point B are positive. Thus,

$$PB = -u \tag{6}$$

where u itself is a negative quantity. Further, since the point M lies on the right of B, we have

$$BM = v \tag{7}$$

and similarly,

$$BO = R \tag{8}$$

where O represents the center of curvature of the spherical surface.

In order to derive a relation between u, v and R we use a theorem in geometry, according to which,

$$(A_1G)^2 = GB \times (2R - GB) \tag{9}$$

where G is the foot of the perpendicular on the axis PM [see Fig. 5.5(b)]. In Fig. 5.5(b) the diameter $B'OB$ intersects the chord A_1GC_1 normally. If $GB << R$, then

$$(A_1G)^2 \approx 2R(GB)$$

Consider the spherical surface SBS' [see Fig. 5.5(a)] whose radius is R. Clearly,

$$(A_1G)^2 = (2R - GB)GB \approx 2R(GB) \tag{10}$$

where we have assumed $GB << R$. Similarly by considering the spherical surface $A_1B_2C_1$ (whose center is at the point M) we obtain

$$(A_1G)^2 \approx 2v(GB_2) \tag{11}$$

where $v = BM \approx B_2M$. In a similar manner,

$$(A_1G)^2 \approx 2(-u)GB_1 \tag{12}$$

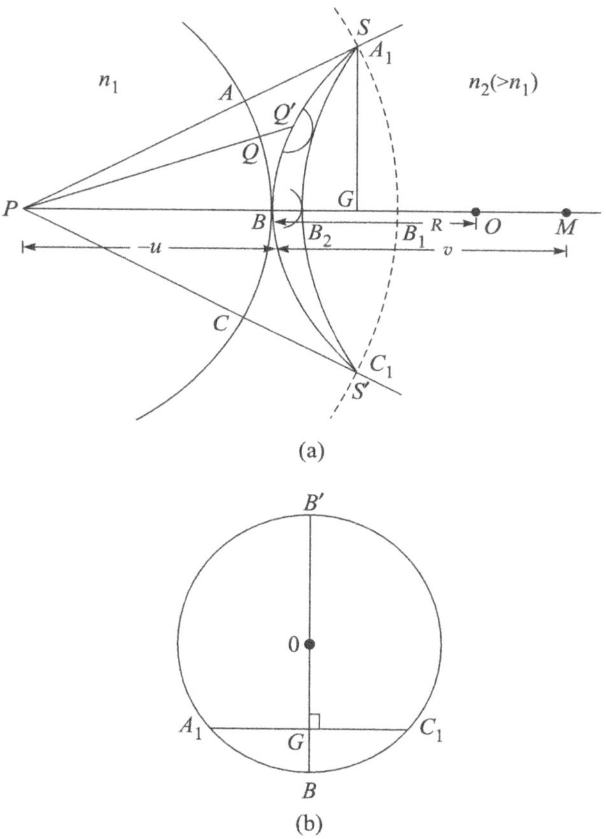

Fig. 5.5 (a) Refraction of a spherical wave ABC (emanating from the point source P) by a convex spherical surface SBS' separating media of refractive indices n_1 and n_2 ($> n_1$). $A_1B_2C_1$ is the refracted wavefront, which is approximately spherical and whose center of curvature is at M. Thus M is the real image of P. O is the center of curvature of SS'. (b) The diameter $B'OB$ intersects the chord A_1GC_1 normally.

Since u is a negative quantity, $(A_1G)^2$ is positive.

Now, $$BB_1 = v_1\tau \quad \text{and} \quad BB_2 = v_2\tau$$

Therefore, $$\frac{BB_1}{BB_2} = \frac{v_1}{v_2} = \frac{n_2}{n_1}$$

or, $$n_1 BB_1 = n_2 BB_2 \tag{13}$$

or, $$n_1(BG + GB_1) = n_2(BG - GB_2)$$

or, $$n_1\left[\frac{(A_1G)^2}{2R} - \frac{(A_1G)^2}{2u}\right] = n_2\left[\frac{(A_1G)^2}{2R} - \frac{(A_1G)^2}{2v}\right] \tag{14}$$

where we have used Eqs (10), (11) and (12). Thus,

$$\frac{n_2}{v} - \frac{n_1}{u} = \frac{n_2 - n_1}{R} \tag{15}$$

which may be rewritten in the form

$$\frac{n_2}{v} = \frac{n_1}{u} + \frac{n_2 - n_1}{R}$$

5.4 We assume a thin lens made of a material of refractive index n_2 to be placed in a medium of refractive index n_1 (see Fig. 5.6). Let the radii of curvatures of the first and the second surface be R_1 and R_2 respectively. Let v' be the distance of the image of the object P if the second surface were not present. Then,

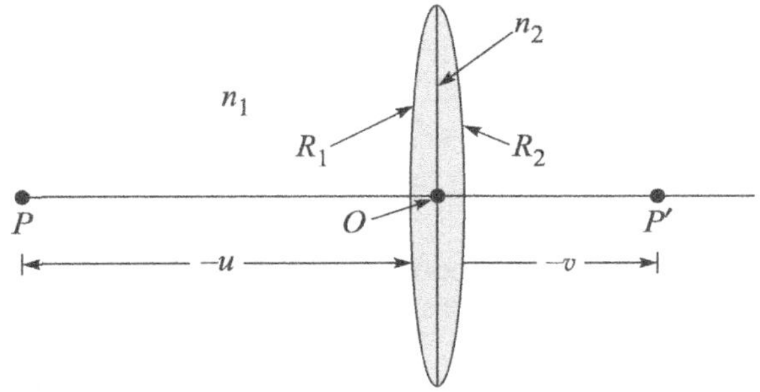

Fig. 5.6 A thin lens made of a medium of refractive index n_2 placed in a medium of refractive index n_1. The radii of curvatures of the two surfaces are R_1 and R_2. P is the image (at a distance v from the point O) of the point object P (at a distance $-u$ from the point O).

$$\frac{n_2}{v'} - \frac{n_1}{u} = \frac{(n_2 - n_1)}{R_1} \tag{16}$$

(Since the lens is assumed to be thin, all the distances are measured from the point O). This image now acts as an object to the spherical surface R_2 on the left of which is the medium of refractive index n_2 and on the right of which is the medium of refractive index n_1. Thus, if v is the distance of the final image point from O, then

$$\frac{n_1}{v} - \frac{n_2}{v'} = \frac{(n_1 - n_2)}{R_2} \tag{17}$$

Adding Eqs (12) and (13), we obtain

$$\frac{n_1}{v} - \frac{n_1}{u} = (n_2 - n_1)\left(\frac{1}{R_1} - \frac{1}{R_2}\right) \tag{18}$$

or,

$$\frac{1}{v} - \frac{1}{u} = \frac{1}{f} \tag{19}$$

where,

$$\frac{1}{f} = \frac{(n_2 - n_1)}{n_1}\left(\frac{1}{R_1} - \frac{1}{R_2}\right) \tag{20}$$

5.5 Let at $t = 0$, the source be at the point P_0 moving with a speed V in the x-direction (see Fig. 5.7). We wish to find out the wavefront at a later time τ. The disturbance emanating from the point P_0 traverses a distance $v\tau$ in time τ. Thus, from the point P_0 we draw a sphere of radius $v\tau$. We next consider the waves emanating from the source at a time τ_1 ($<\tau$). At time τ_1 let the source be at the position P_1; consequently,

$$P_0P_1 = V\tau_1$$

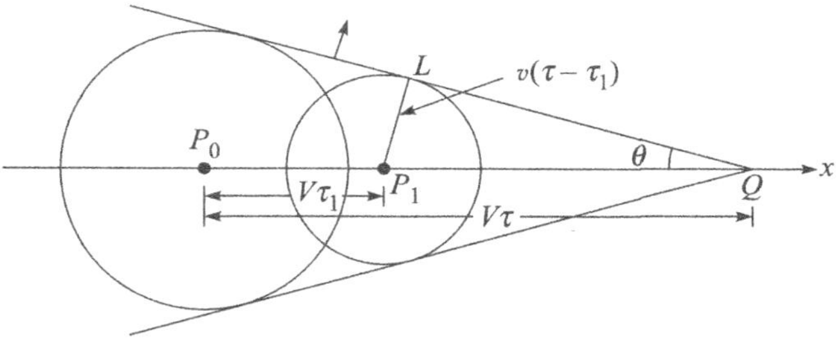

Fig. 5.7 Generation of a shock wavefront by a vibrating particle P_0 moving with a speed V, in a medium in which the velocity of propagation of the wave is v ($< V$).

In order to determine the shape of the wavefront at τ, we draw a sphere of radius $v(\tau - \tau_1)$ centered at P_1. Let the source be at the position Q at the instant τ. Then,

$$P_0Q = V\tau$$

We draw a tangent plane from the point Q, on the sphere whose origin is the point P_1. Since

$$P_1L = v(\tau - \tau_1) \quad \text{and} \quad P_1Q = V(\tau - \tau_1)$$

$$\sin\theta = \frac{P_1L}{P_1Q} = \frac{v}{V} \text{ (independent of } \tau_1) \tag{21}$$

Since θ is independent of τ_1, all the spheres drawn from any point on the line P_0Q will have a common tangent plane. This plane is known as the shock wavefront and propagates with a speed v.

It is interesting to point out that even when the source is not vibrating, if its speed is greater than the speed of sound waves, a shock wavefront is always set up. A similar phenomenon also occurs when a charged particle (like an electron) moves in a medium with a speed greater than the speed of light in that medium. The emitted light is known as Cerenkov radiation. If you ever see a swimming pool type reactor, you will find a blue glow coming out from it; this is because of the Cerenkov radiation emitted by the fast moving electrons.

5.6 Figure 5.8 shows the plane wavefront incident along the z-axis on the inhomogeneous medium. Since the refractive index decreases as x and y increase, the speed of the secondary wavelets emanating from portions of the incident wavefront will increase as we move away from the axis. Let us try to determine the shape of the wavefront at a time Δt; given that the wavefront at $t = 0$ is a plane wavefront A_1B_1 (see Fig. 5.8). We will have to draw spheres of radius $v(x, y)\,\Delta t$, centered at (x, y), where $v(x, y)$ is the velocity of the wave at the point (x, y), which increases as x and y increase. Thus the radii of the spheres increase as we move away from the axis and if we draw a common tangent to all these spheres then the resulting wavefront is shown in Fig. 5.8 as A_2B_2. It is at once evident that the wavefront which was initially plane has now become curved. If we again use the same procedure, then the shape of the wavefront at time $2\Delta t$ (say) is shown as A_3B_3. Thus, it is evident that in the present case the wavefront is getting focused. It should be borne in mind that since we are considering an inhomogeneous medium, the refractive index varies continuously with position. For the above construction to be valid, Δt should be small so that during this short interval the secondary wavelets may be assumed to be spherical.

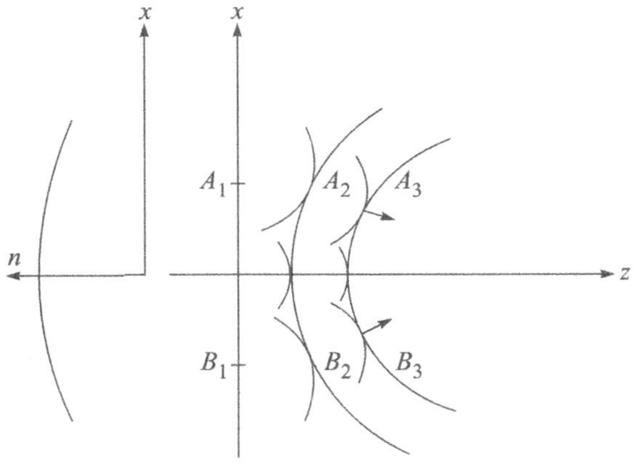

Fig. 5.8 The focusing of an incident plane wavefront in an inhomogeneous medium characterised by a refractive index variation given by Eq. (2).

5.7 Figure 5.9 shows the spherical mirror GOG' of radius of curvature R and a point source placed at a distance u from the mirror. Spherical waves from the object point P are incident on the mirror. For the object position drawn in the figure, by the time the spherical wavefront reaches the point O the remaining portion of the wavefront is still to reach the mirror. The portion at O gets reflected and when the incident wavefront reaches point G on the mirror, the reflected wave has already travelled to the point marked M. Now in the absence of the mirror the wavefront would have proceeded to the point S. Thus, $OS = OM$.

Now, using the same approximation as in Problem 5.5, we have

$$(GN)^2 = 2R \times (NO)$$
$$(GN)^2 = 2u \times (NS)$$
$$(GN)^2 = 2v \times (NM)$$

where we have assumed $PO = u$, $QO = v$ and $RO = R$.

Also, $$NS = NO + OS$$

and $$NM = NO - OM = NO - OS$$

From the above equations we obtain,

$$\frac{1}{u} + \frac{1}{v} = \frac{2}{R}$$

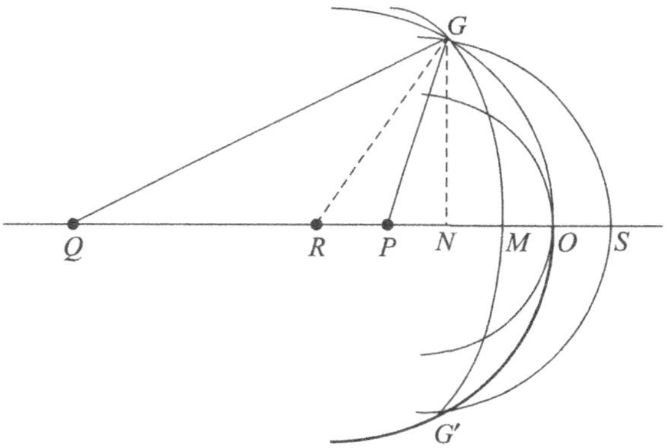

Fig. 5.9 Huygens' construction for reflection at a spherical mirror.

Interference—Division of Wavefront

6

A Quick Review

When two or more light waves superpose at any point in space then the total electric field is a superposition of the electric fields of the two waves at that point and depending on their phase difference, they may interfere constructively or destructively. This phenomenon of interference leads to many interesting applications.

Consider two coherent point sources S_1 and S_2 vibrating in phase. Let y_1 and y_2 be the corresponding displacements produced at the point P (see Fig. 6.1):

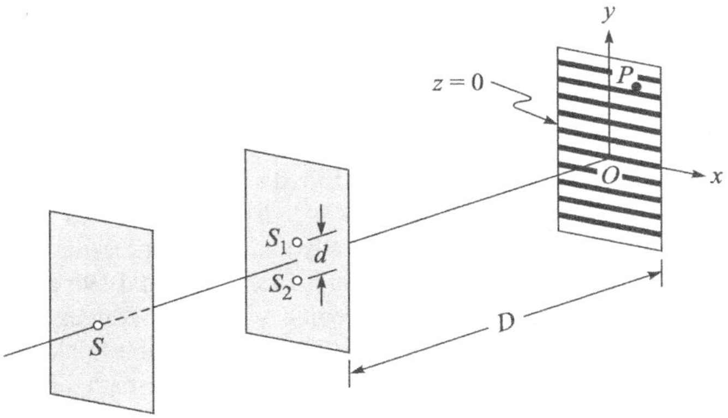

Fig. 6.1 Young's arrangement to produce interference pattern.

$$y_1 = a \cos \omega t = \text{Re}\,[e^{i\omega t}] \quad (1)$$

and

$$y_2 = a \cos (\omega t - \phi) = a\, \text{Re}\,[e^{i(\omega t - \phi)}] \quad (2)$$

where,

$$\phi = \frac{2\pi}{\lambda}(S_2P - S_1P) \quad (3)$$

is the phase difference between the two displacements and we have assumed the amplitude of the two displacements to be same; this will be very nearly true if the point P is far away from S_1 and S_2. Simple algebra will show that the resultant displacement will be given by

$$y = y_1 + y_2 = 2a\, \text{Re}\left[e^{i\left(\omega t - \frac{\phi}{2}\right)} \cos \frac{\phi}{2} \right] \quad (4)$$

Thus, the intensity distribution will be

$$I = 4I_0 \cos^2\left(\frac{\phi}{2}\right) \tag{5}$$

and therefore the resultant intensity will be maximum ($= 4I_0$) when

$$S_2P - S_1P = n\lambda; \quad n = 0, 1, 2, 3\ldots \quad \text{(Maxima)} \tag{6}$$

The resultant intensity will be zero when

$$S_2P - S_1P = \left(n + \frac{1}{2}\right)\lambda; \quad n = 0, 1, 2, 3\ldots \quad \text{(Minima)} \tag{7}$$

and we will obtain an interference pattern similar to that shown in Fig. 6.1. When the point P is far away so that we can put $S_2P + S_1P \approx 2D$, the interference fringes will be straight lines with fringe width given by

$$\beta = \frac{\lambda D}{d} \tag{8}$$

where the distances D and d are defined in Fig. 6.1.

PROBLEMS

6.1 In the Young's double-hole experiment (see Fig. 6.1), the distance between the two holes is 0.5 mm, $\lambda = 5 \times 10^{-5}$ cm and $D = 50$ cm. What will be the fringe width?

6.2 Figure 6.2 represents the layout of Lloyd's mirror experiment. S is a point source emitting waves of frequency 6×10^{14} sec^{-1}. A and B represent the two ends of a mirror placed horizontally and LOM represents the screen. The distances SP, PA, AB and BO are 1 mm, 5 cm, 5 cm and 190 cm respectively. (a) Determine the position of the region where the fringes will be visible and calculate the number of fringes. (b) Calculate the thickness of a mica sheet ($n = 1.5$) which should be introduced in the path of the direct ray so that the lowest fringe becomes the central fringe. The velocity of light is 3×10^{10} cm/sec. [**Ans.** (a) 2 cm, 40 fringes, (b) 38 μm]

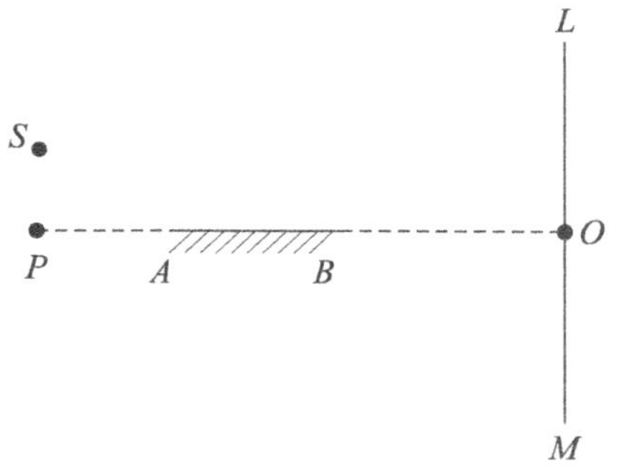

Fig. 6.2

6.3 (a) In the Fresnel's biprism arrangement, show that $d = 2(n-1)\,a\alpha$ where a represents the distance from the source to the base of the prism (see Fig. 6.3), α is the angle of the biprism and n the refractive index of the material of the biprism.

(b) In a typical biprism arrangement $b/a = 20$ and for sodium light ($\lambda = 5893$ Å) one obtains a fringe width of 0.1 cm; here b is the distance between the biprism and the screen. Assuming $n = 1.5$, calculate the angle α.

[**Ans.** $\approx 0.71°$]

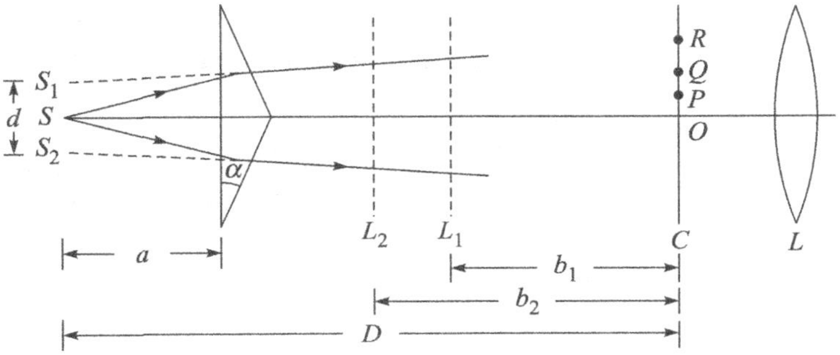

Fig. 6.3 Fresnel's biprism arrangement. *C* and *L* represent the positions of the crosswires and the eyepiece respectively. In order to determine *d* one introduce a lens between the biprism and the crossswires; L_1 and L_2 represent the two positions of the lens where the slits are clearly seen.

6.4 In the Young's double hole experiment a thin mica sheet ($n = 1.5$) is introduced in the path of one of the beams (see Fig. 6.4). If the central fringe gets shifted by 0.2 cm, calculate, the thickness of the mica sheet. Assume $d = 0.1$ cm, and $D = 50$ cm.

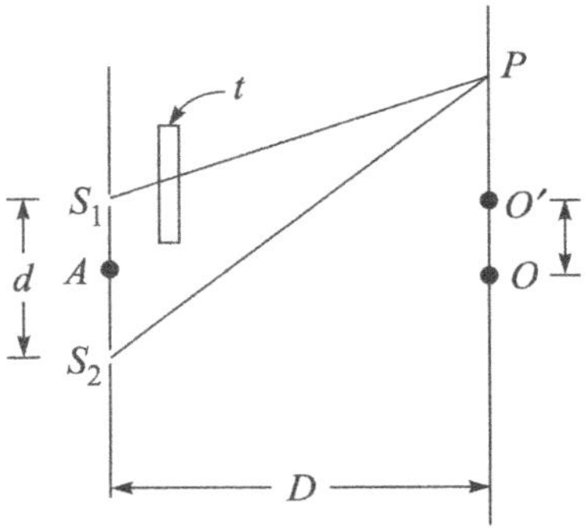

Fig. 6.4 If a thin transparent sheet (of thickness *t*) is introduced in one of the beams, the fringe pattern gets shifted.

6.5 In order to determine the distance between the slits in the Fresnel biprism experiment, one puts a convex lens in between the biprism and the eye piece.

Show that if $D > 4f$ one will obtain two positions of the lens where the image of the slits will be formed at the eye piece; here f is the focal length of the convex lens and D is the distance between the slit and the eye piece. If d_1 and d_2 are the distances between the images (of the slits) as measured by the eye piece, then show that $d = \sqrt{d_1 d_2}$. What would happen if $D < 4f$?

6.6 In the Young's double hole experiment, interference fringes are formed using sodium light which predominantly comprises of two wavelengths (5890 Å and 5896 Å). Obtain the regions on the screen where the fringe pattern will disappear. You may assume $d = 0.5$ mm and $D = 100$ cm.

6.7 If one carries out the Young's double hole interference experiment using microwaves of wavelength 3 cm, discuss the nature of the fringe pattern if $d = 0.1$ cm, 1 cm and 4 cm. You may assume $D = 100$ cm. Can you use Eq. (8) for the fringe width?

6.8 In the Fresnel's two mirror arrangement (see Fig. 6.5) show that the points S, S_1 and S_2 lie on a circle and $S_1S_2 = 2b\theta$ where $b = MS$ and θ is the angle between the mirrors.

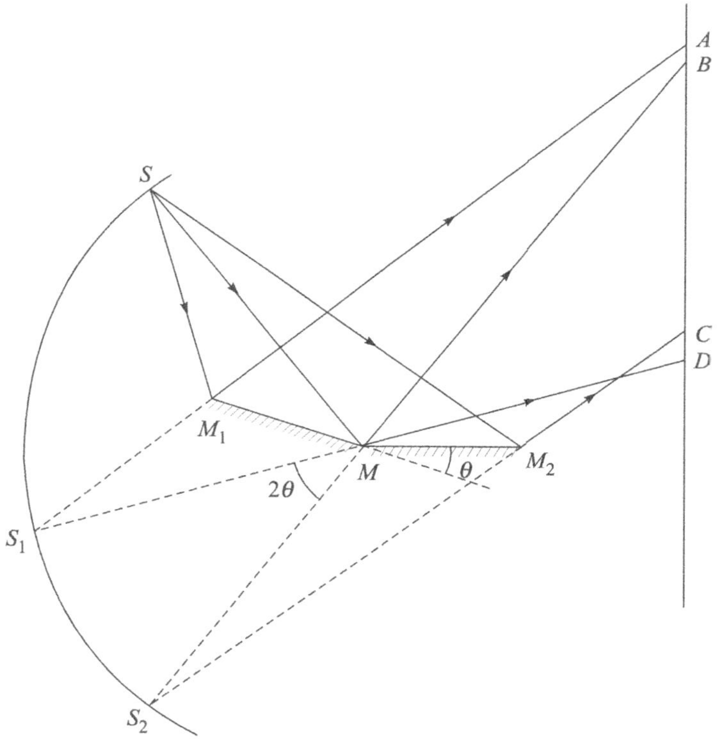

Fig. 6.5 Fresnel's two mirror arrangement.

6.9 In the double hole experiment using white light, consider two points on the screen, one corresponding to a path difference of 5000 Å and the other corresponding to a path difference of 40000 Å. Find the wavelengths (in the visible region) which correspond to constructive and destructive interference. What will be the colour of these points?

6.10 (a) Consider a plane which is normal to the line joining two point coherent sources S_1 and S_2 as shown in Fig. 6.6. If $S_1P - S_2P = \Delta$, then show that

$$y = \frac{1}{2\Delta}(d^2 - \Delta^2)^{1/2}\left[4D^2 + 4Dd + (d^2 - \Delta^2)\right]^{1/2}$$

$$\approx \frac{D}{\Delta}\sqrt{(d-\Delta)(d+\Delta)}$$

where the last expression is valid for $D >> d$.

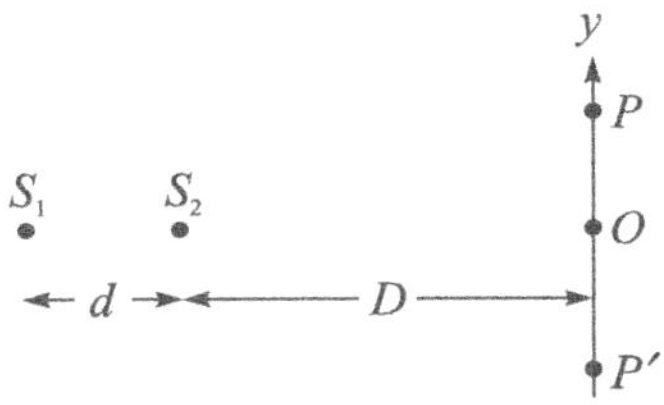

Fig. 6.6 S_1 and S_2 represent two coherent sources.

(b) For $\lambda = 0.5$ μm, $d = 0.4$ mm and $D = 20$ cm; $S_1O - S_2O = 800\,\lambda$. Calculate the value of $S_1P - S_2P$ for the point P to be first dark ring and first bright ring.

[**Ans.** 0.39975 mm, 0.3995 mm]

6.11 In continuation of the above problem calculate the radii of the first two dark rings for (a) $D = 20$ cm and (b) $D = 10$ cm.

[**Ans.** (a) ≈0.71 cm and 1.22 cm]

6.12 In continuation of the previous problem assume that $d = 0.5$ mm, $\lambda = 5 \times 10^{-5}$ cm and $D = 100$ cm. Thus the central (bright) spot will correspond to $n = 1000$. Calculate the radii of the first, second and third bright rings which will correspond to $n = 999$, 998 and $n = 997$ respectively.

6.13 Using Maxwells equations one can show that when a plane wave is incident from a medium of refractive index n_1 on a medium of refractive index n_2, the amplitude reflection and transmission coefficients are given by

$$r = \frac{\sin\theta_1\cos\theta_1 - \sin\theta_2\cos\theta_2}{\sin\theta_1\cos\theta_1 + \sin\theta_2\cos\theta_2}$$

and
$$t = \frac{2n_1\cos\theta_1}{n_2\cos\theta_1 + n_1\cos\theta_2} = \frac{2\cos\theta_1\sin\theta_2}{\sin\theta_1\cos\theta_1 + \sin\theta_2\cos\theta_2}$$

where θ_1 and θ_2 are the angles of incidence and of refraction respectively; the above expressions are valid when the electric field lies in the plane of incidence. Show that they satisfy Stokes' relations.

6.14 Assume a plane wave incident normally on a plane containing two holes separated by a distance d. If we place a convex lens behind the slits, show that the fringe width, as observed on the focal plane of the lens, will be $f\lambda/d$ where f is the focal length of the lens.

6.15 In the previous problem, show that if the plane (containing the holes) lies in the front focal plane of the lens, then the interference pattern will consist of exactly parallel straight lines. However, if the plane does not lie on the front focal plane, the fringe pattern will be hyperbolae.

6.16 In the Young's double hole experiment calculate I/I_{max} where I represents the intensity at a point where the path difference is $\lambda/5$.

6.17 Consider two plane waves incident on a screen as shown in Fig. 6.7. Calculate the fringe pattern on the screen.

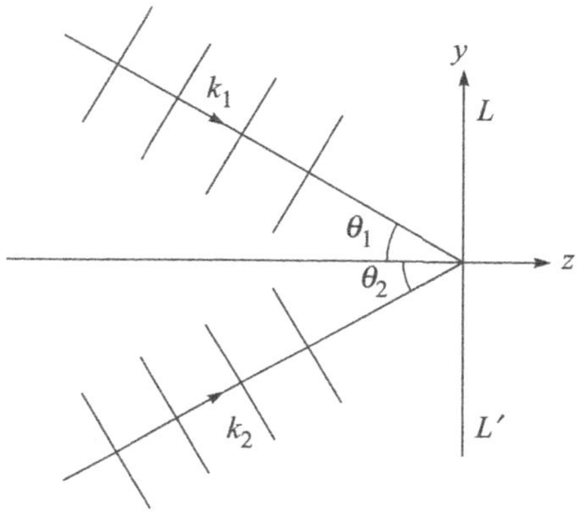

Fig. 6.7 Superposition of two plane waves on LL'.

6.18 Consider the interference pattern produced on PP' by the superposition of a plane wave incident normally and a spherical wave emanating from the point O (see Fig. 6.8). Show that the interference pattern will consist of circular fringes.

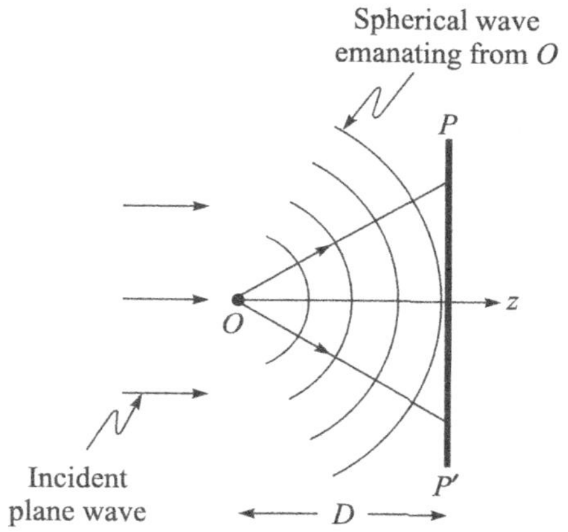

Fig. 6.8 Superposition of a plane wave and a spherical wave emanating from the point O.

6.19 Consider light from two distant stars incident on two slits S_1 and S_2 as shown in Fig. 6.9. Calculate the resultant intensity distribution on the screen.

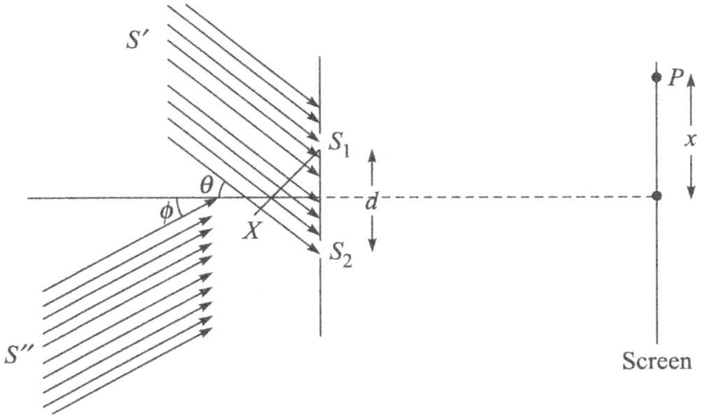

Fig. 6.9

6.20 A pair of point sources, S, S' emitting a wavelength of 500 nm with a spectral width $\Delta\lambda$, and separated by a distance of 1 mm are placed as shown in the figure. What is the condition on $\Delta\lambda$ so that one can observe an interference pattern around the point P, given that the screen is placed at a distance of 1 m from the midpoint of the sources?

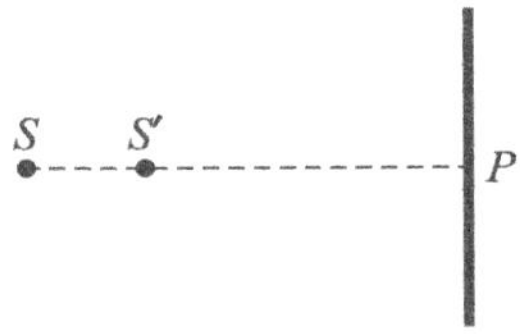

6.21 In the Young's double-hole experiment, interference fringes are formed using sodium light, which predominantly comprises two lines at 5890 Å and 5896 Å. Obtain the region on the screen closest to the axis where the fringe pattern will disappear. You may assume $d = 5$ mm and $D = 20$ cm.

6.22 In a Young's double hole experiment, in front of each slit we place a polariser. What will be the contrast of the interference pattern if (a) if the polarizer pass axes are perpendicular to each other and (b) if the pass axes are oriented at 45° with respect to each other?

6.23 In a Young's double hole experiment, the source S (see Fig. 6.1) is placed off axis at a distance a from the axis. If the distance of S from the screen containing the two slits is L ($L >> a, d$), what is the position of the zero order fringe on the screen?

SOLUTIONS

6.1 $\beta = \dfrac{\lambda D}{d} = \dfrac{5 \times 10^{-5} \times 50}{0.05} = 0.05$ cm.

6.2 (a) $\angle LAO = \angle SAP$ {Angle of incidence = angle of reflection}

$\angle NBO = \angle SBP$ (see Fig. 6.10)

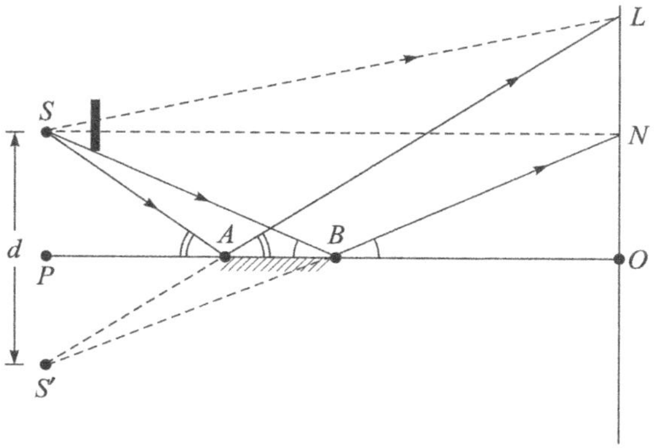

Fig. 6.10

$$\Rightarrow \quad LO = OA \tan \angle LAO = OA \tan \angle SAP = 195\left(\frac{0.1}{5}\right) = 3.9 \text{ cm}$$

$$NO = OB \tan \angle NBO = OB \tan \angle SBP = 190\left(\frac{0.1}{10}\right) = 1.9 \text{ cm}$$

Thus, $\quad LN = 2$ cm. Now $\lambda = \dfrac{c}{v} = \dfrac{3 \times 10^{10}}{6 \times 10^{14}} = 5 \times 10^{-5}$ cm.

Also, $\quad d = SS' = 2$ mm; $\quad D = 200$ cm

Therefore, $\quad$ Fringe width $= \beta = \dfrac{\lambda D}{d} = \dfrac{5 \times 10^{-5} \times 200}{0.2} = 0.05$ cm

$$\text{No. of fringe} = \frac{LN}{\beta} = \frac{2}{0.05} = 40.$$

(b) For central fringe at N, we should have

$$SN + (n-1)t = S'N$$

or
$$S'N - SN = (n-1)t = 0.5t$$

But
$$S'N - SN \approx ON \cdot \frac{d}{D}$$

Thus,
$$0.5t = 1.9 \times \frac{0.2}{200} \Rightarrow t = 38 \text{ µm}$$

6.3 (a) DF and EF are normals to the surface (see Fig. 6.11). Thus, $\angle DAE + \angle DFE = \pi \Rightarrow \angle DFE = \pi - \alpha$

Also, $\dfrac{\sin i}{\sin \theta} = n \Rightarrow \theta \approx \dfrac{i}{n}$ where all angles are assumed to be small so that $\sin \theta \approx \theta$ and all angles are measured in radians. Further

$$\frac{\sin \phi}{\sin t} = \frac{1}{n} \Rightarrow t \approx n\phi = n(\alpha - \theta) = n\left(\alpha - \frac{i}{n}\right) = n\alpha - i$$

Now, $i + \pi - \delta + t + \pi - \alpha = 2\pi$

Thus, $\delta = i + t - \alpha \approx i + n\alpha - i - \alpha$

or $\delta \approx (n - 1)\alpha.$

where α has been assumed to be very small.

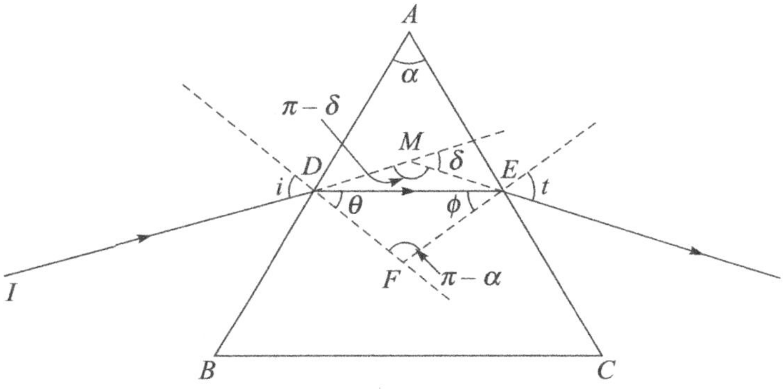

Fig. 6.11

If all rays are assumed to suffer the same deviation $\delta[= (n - 1)\alpha]$ then for small angles, the virtual image of S will be at S_1 as shown in the figure above (Fig. 6.12). Thus,

$$SS_1 \approx a\delta \approx a(n-1)\alpha \quad \text{and} \quad S_1S_2 = d \approx 2a(n-1)\alpha$$

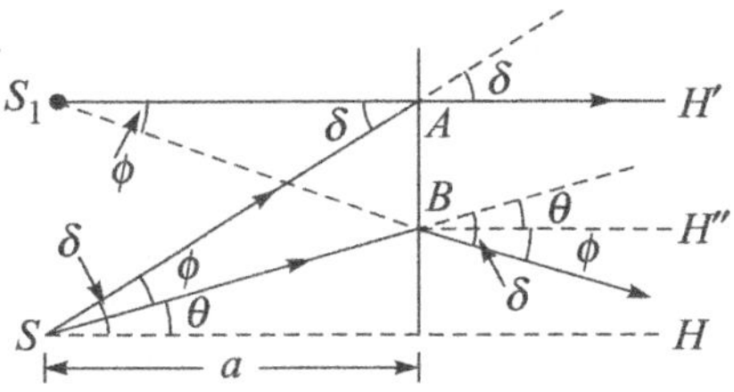

Fig. 6.12

In the figure, AH', BH'' and SH are all parallel lines.

(b) $$\beta = \frac{\lambda(b+a)}{d} = \frac{\lambda(b+a)}{2(n-1)a\alpha} = \frac{\lambda\left(\frac{b}{a}+1\right)}{2(n-1)\alpha}$$

Therefore, $$\alpha = \frac{\lambda\left(\frac{b}{a}+1\right)}{2(n-1)\beta} = \frac{5.893 \times 10^{-5} \times 21}{2 \times 0.5 \times 0.1} \approx 0.0124 \text{ radians} \approx 0.71°.$$

6.4 Shift, $\Delta = \frac{D}{d}(n-1)t$

$$\Rightarrow \quad 0.2 = \frac{50}{0.1}(0.5)t \Rightarrow t = 8\times10^{-4}\text{ cm} = 8\ \mu\text{m}.$$

6.5

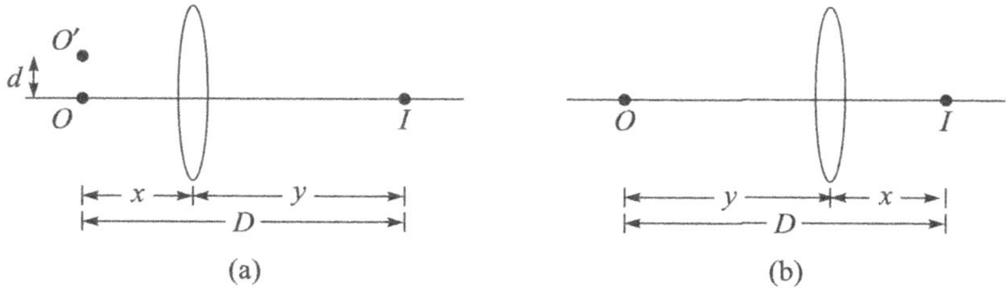

Fig. 6.13

In the first case $u = -x$. Thus,

$$\frac{1}{f} = \frac{1}{v} - \frac{1}{u} = \frac{1}{y} + \frac{1}{x} = \frac{1}{D-x} + \frac{1}{x}$$

$$\Rightarrow \quad Dx - x^2 = Df \Rightarrow x^2 - Dx + Df = 0$$

The above equation will have real roots if $D^2 > 4Df \Rightarrow D > 4f$

In case (a) the distance between two image points will be

$$d_1 = d \times \frac{y}{x}$$

Similarly, in case (b) the distance between two image points will be

$$d_2 = d \times \frac{x}{y}$$

Thus, $$d_1 d_2 = d^2 \Rightarrow d = \sqrt{d_1 d_2}$$

6.6 Let $\lambda_1 = 5.896 \times 10^{-5}$ cm and $\lambda_2 = 5.890 \times 10^{-5}$ cm. Thus, $\lambda_1 > \lambda_2$. For the point P (see Fig. 6.14) to correspond to a maximum intensity for λ_1, and minimum intensity for λ_2, we must have

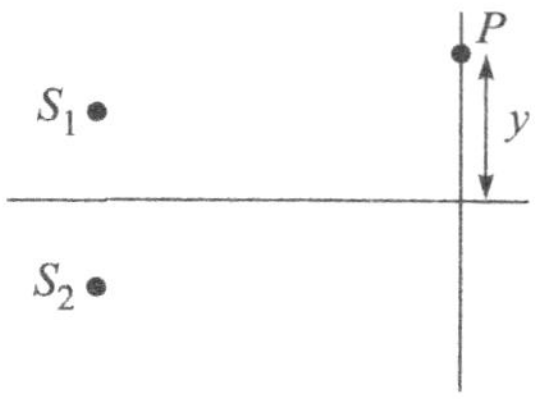

Fig. 6.14

$$\Delta = S_2P - S_1P = n\lambda_1 = \left(n + \frac{1}{2}\right)\lambda_2$$

Thus, $$\frac{\Delta}{\lambda_2} - \frac{\Delta}{\lambda_1} = \frac{1}{2}$$

$$\Rightarrow \quad S_2P - S_1P = \Delta = \frac{\lambda_1\lambda_2}{2(\lambda_1 - \lambda_2)} = \frac{5.890 \times 5.896 \times 10^{-10}}{2 \times 6 \times 10^{-8}} \approx 0.029\text{ cm}.$$

Since Δ is comparable to $d(= 0.05\text{ cm})$, we must use the accurate expression to calculate y.

$$y = \pm\left(\frac{\Delta^2}{d^2 - \Delta^2}\right)^{1/2}\left[D^2 + \frac{1}{4}(d^2 - \Delta^2)\right]^{1/2} \approx \pm 72 \text{ cm}$$

6.7 $S_2P - S_1P$ can never exceed d. Thus for $d = 0.1$ cm and $d = 1$ cm, the maximum values of $S_2P - S_1P$ are $\frac{\lambda}{30}$ and $\frac{\lambda}{3}$; thus even one dark fringe will not be observed. For $d = 4$ cm, the maximum value of $S_2P - S_1P$ will be $1.33\,\lambda$ so only one dark (hyperbolic) fringe will be observed on either side of the central maximum. Obviously, the central bright fringe will always occur at $y = 0$ corresponding to $\Delta = 0$.

6.8 Obviously since S_1 is the virtual image of S (formed by the mirror M_1M), $MS = MS_1$. Similarly, $MS = MS_2$. Thus if we draw a circle of radius MS (with M as the center) then S_1 and S_2 will lie on this circle. Further, if the angle between the mirrors is θ_1 then angle between the reflected rays (MB and MD) will be 2θ. Thus, the arc length $S_1S_2 = 2\theta b$.

6.9 The visible region corresponds to 4000 Å < λ < 7000 Å

(a) Path difference Δ = 5000 Å. Now, for constructive interference, $\Delta = n\lambda$ i.e.; constructive interference will occur for

$$\lambda = \frac{\Delta}{n}; \quad n = 1, 2, 3, \ldots$$
$$= 5000 \text{ Å}, 2500 \text{ Å}, 1667 \text{ Å}, \ldots$$

Only λ = 5000 Å lies in the visible region. Similarly, destructive interference will occur for

$$\lambda = \frac{\Delta}{n + \frac{1}{2}}; \quad n = 0, 1, 2, \ldots$$
$$= 10000 \text{ Å}, 3333 \text{ Å}, 2000 \text{ Å}, \ldots$$

Thus, no wavelength (in the visible region) corresponds to destructive interference.

(b) Path difference Δ = 40000 Å. Constructive interference will occur for

$$\lambda = \frac{\Delta}{n}; \quad n = 1, 2, \ldots$$
$$= 40000 \text{ Å}, 20000 \text{ Å}, 13333 \text{ Å}, 10000 \text{ Å}, 8000 \text{ Å}, 6667 \text{ Å}, 5714 \text{ Å}, 5000 \text{ Å}, 4444 \text{ Å}, 4000 \text{ Å}, 3636 \text{ Å}, \ldots$$

Thus, λ = 6667 Å, 5714 Å, 5000 Å, 4444 Å and 4000 Å (which lie in the visible region) would have constructive interference.

Destructive interference will occur for

$$\lambda = \frac{\Delta}{n + \frac{1}{2}}; \quad n = 0, 1, 2, \ldots$$
$$= 6154 \text{ Å}, 5333 \text{ Å}, 4706 \text{ Å}, 4211 \text{ Å}, \ldots$$

for n = 6, 7, 8 and 9 respectively.

6.10 (a) $S_1P = [(D + d)^2 + y^2]^{1/2}$; $S_2P = [D^2 + y^2]^{1/2}$ (see Fig. 6.15)

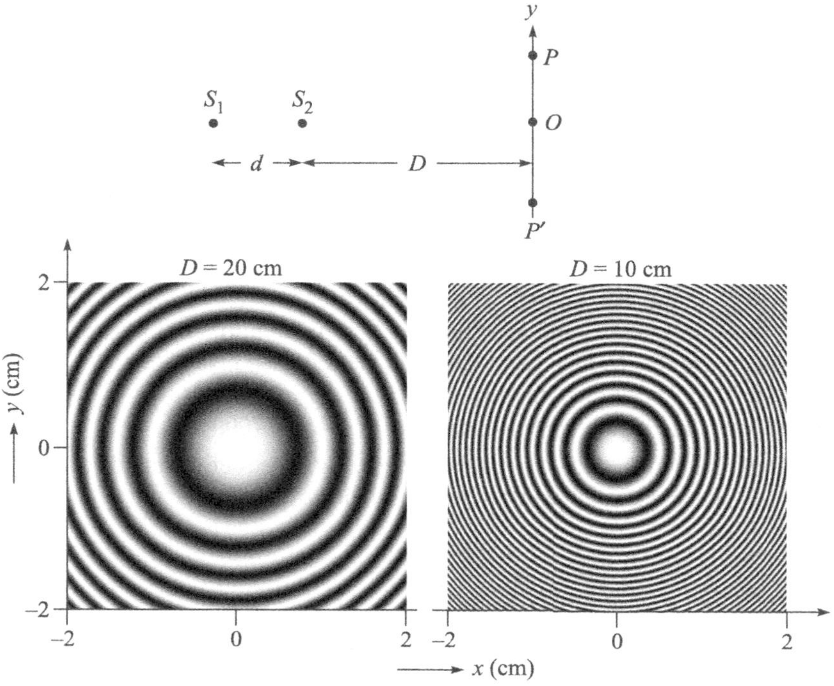

Fig. 6.15 (a) S_1 and S_2 represent two coherent sources, (b) and (c) show typical interference fringes observed on the screen PP' when $D = 20$ cm and $D = 10$ cm respectively.

Therefore, $S_1P - S_2P = \Delta$ implies $[(D + d)^2 + y^2]^{1/2} = \Delta + [D^2 + y^2]^{1/2}$

$$\Rightarrow \quad D^2 + 2Dd + d^2 + y^2 = \Delta^2 + D^2 + y^2 + 2\Delta[D^2 + y^2]^{1/2}$$

$$\Rightarrow \quad [2Dd + (d^2 - \Delta^2)]^2 = (2\Delta)^2\,[D^2 + y^2]$$

$$\Rightarrow \quad 4D^2d^2 + 4Dd(d^2 - \Delta^2) + (d^2 - \Delta^2)^2 = 4\Delta^2D^2 + 4\Delta^2y^2$$

$$\Rightarrow \quad 4\Delta^2y^2 = 4D^2(d^2 - \Delta^2) + 4Dd(d^2 - \Delta^2) + (d^2 - \Delta^2)^2$$

$$\Rightarrow \quad y = \frac{1}{2\Delta}(d^2 - \Delta^2)^{1/2}\,[4D^2 + 4Dd + (d^2 - \Delta^2)]^{1/2}$$

$$\approx \frac{D}{\Delta}\sqrt{(d - \Delta)(d + \Delta)} \quad [\text{when } D >> d]$$

(c) The central point O is such that $S_1O - S_2O = 0.4$ mm $= 800\,\lambda$; thus O will be a bright point. For the first dark ring

$$S_1P - S_2P = 799.5\,\lambda = 0.39975 \text{ mm}$$

and for the first bright ring

$$S_1P - S_2P = 799\,\lambda = 0.39950 \text{ mm}.$$

6.11 For the first dark ring, $\Delta = 0.39975$ mm

Thus, $$d - \Delta = \frac{\lambda}{2} = 2.5 \times 10^{-4} \text{ mm}; \quad d + \Delta \approx 2d = 0.8 \text{ mm}$$

$$\Rightarrow \qquad y = \frac{200}{0.39975}\sqrt{2.5\times10^{-4}\times 0.8} \approx 7.07 \text{ mm}$$

for the second dark ring, $\Delta = 798.5\,\lambda = 0.39925$ mm and

$$d - \Delta = \frac{3\lambda}{2} = 7.5 \times 10^{-4} \text{ mm}$$

$$\Rightarrow \qquad y = \frac{200}{0.4}\sqrt{7.5\times10^{-4}\times 0.8} \approx 12.2 \text{ mm}$$

(b) For $D = 10$ cm; $y \approx 3.54$ mm and 6.1 mm respectively.

Typical interference fringes are shown in Fig. 6.15.

6.12 $d = 0.5$ mm, $\lambda = 5 \times 10^{-4}$ mm; $D = 1000$ mm. For the first, second and third bright rings

$$S_1P - S_2P = 999\,\lambda,\ 998\,\lambda \quad \text{and} \quad 997\,\lambda$$

$$d = 0.5 \text{ mm} = 1000\,\lambda$$

Thus $$d - \Delta = \lambda,\ 2\,\lambda \quad \text{and} \quad 3\,\lambda \text{ respectively.}$$

$$y \approx \frac{D}{\Delta}\sqrt{(d-\Delta)(d+\Delta)} \approx \frac{D}{\Delta}\sqrt{2d\,(d-\Delta)}$$

$$\approx 44.7 \text{ mm, } 63.2 \text{ mm, } 77.4 \text{ mm}$$

6.13 Since $$r = \frac{\sin\theta_1\cos\theta_1 - \sin\theta_2\cos\theta_2}{\sin\theta_1\cos\theta_1 + \sin\theta_2\cos\theta_2}$$

For a plane wave incident from a medium of refractive index n_2 on a medium of refractive index n_1, the amplitude reflection and transmission coefficients are given by (we just have to interchange θ_1 and θ_2 because θ_2 is now the angle of incidence and θ_1 the angle of refraction

$$r' = \frac{\sin\theta_2\cos\theta_2 - \sin\theta_1\cos\theta_1}{\sin\theta_2\cos\theta_2 + \sin\theta_1\cos\theta_1} = -r$$

This is one of the Stokes relations. Further since

$$t = \frac{2\cos\theta_1\sin\theta_2}{\sin\theta_1\cos\theta_1 + \sin\theta_2\cos\theta_2}$$

we will have

$$t' = \frac{2\cos\theta_2\sin\theta_1}{\sin\theta_2\cos\theta_2 + \sin\theta_1\cos\theta_1}$$

Simple algebra will show that

$$1 + rr' = tt'$$

which is the other Stokes relation.

6.14 For the nth bright fringe (see Fig. 6.16)

Fig. 6.16 Interference pattern on the screen LL' for $\theta_1 = \theta_2 = \pi/6$ and $\lambda = 5000$ Å. The fringes are parallel to the x-axis.

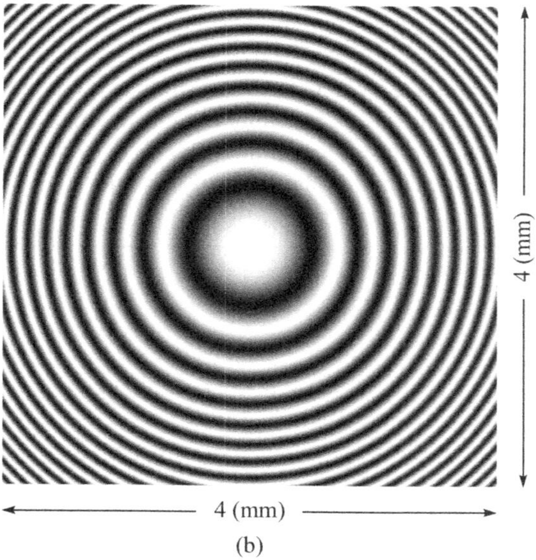

Fig. 6.17 Typical interference fringes observed on the screen PP'.

$$d \sin \theta = n\lambda$$

or

$$y_n = f \tan \theta \approx \frac{n\lambda f}{d}$$

Thus the fringe width is $\approx \dfrac{\lambda f}{d}$

Also, see Problem 10.13. Show that when the radiuses of the two holes become extremely small, we will obtain the result obtained here.

6.15 If the holes lie on the front focal plane then we will have plane waves coming out of the lens and when two plane waves (propagating in different directions) interfere we will have straight line interference pattern. However, if the plane (containing the holes) does not coincide with the front focal plane then the two real (or virtual) images can be assumed to be two point sources and the fringe pattern will be hyperbolae.

6.16 $I = 4I_0 \cos^2 \delta/2$ where $\delta = \dfrac{2\pi}{\lambda}(S_2P - S_1P) = \dfrac{2\pi}{\lambda}\Delta$, with Δ representing the path difference. Thus, $I_{max} = 4I_0$. Further, when $\Delta = \dfrac{\lambda}{5}$, $\delta = \dfrac{2\pi}{5}$, giving

$$\frac{I}{I_{max}} = \cos^2 \frac{2\pi}{10} \approx 0.65$$

6.17 The wave vectors of the two waves are given by

$$\mathbf{k_1} = -\hat{\mathbf{y}}k \sin \theta_1 + \hat{\mathbf{x}}k \cos \theta_1$$

and

$$\mathbf{k_2} = +\hat{\mathbf{y}}k \sin \theta_2 + \hat{\mathbf{x}}k \cos \theta_2$$

where $k = 2\pi/\lambda$ and θ_1 and θ_2 are the angles defined in Fig. 6.7. Thus, the electric fields of the two waves are described by the equations

$$\begin{aligned} E_1 &= E_{01} \cos (\mathbf{k_1} \cdot \mathbf{r} - \omega t) \\ &= E_{01} \cos (-ky \sin \theta_1 + kz \cos \theta_1 - \omega t) \end{aligned}$$

and

$$\begin{aligned} E_1 &= E_{01} \cos (\mathbf{k_2} \cdot \mathbf{r} - \omega t) \\ &= E_{01} \cos (ky \sin \theta_2 + kz \cos \theta_2 - \omega t) \end{aligned}$$

where we have assumed both electric fields are along the same direction (say along the x-axis). If we further assume that $E_{01} = E_{02} = E_0$ and $\theta_1 = \theta_2 = \theta$, then the resultant field is given by

$$E = E_0 \cos(ky \sin \theta) \cos (kz \cos \theta - \omega t)$$

Thus, the intensity distribution on the photographic plate LL' is given by

$$I = 4I_0 \cos^2 (ky \sin \theta)$$

and the fringe pattern will be strictly lines (parallel to the x-axis) with fringe width given by

$$\beta = \frac{\lambda}{2 \sin \theta}$$

Figure 6.16 shows the interference pattern as will be observed on the screen LL' for $\theta = \pi/6$ and $\lambda = 5000$ Å. Thus, $\beta = \lambda = 0.005$ mm.

6.18 The plane wave will be given by

$$E_1 = E_0 \cos(kz - \omega t + \phi)$$

and the spherical wave will be given by

$$E_2 = \frac{A_0}{r} \cos(kr - \omega t)$$

where r is the distance measured from the point O which is assumed to be the origin. Now, on the plane PP'' ($z = D$)

$$r = (x^2 + y^2 + D^2)^{1/2} \approx D\left[1 + \frac{x^2 + y^2}{2D^2}\right]$$

$$\approx D + \frac{x^2 + y^2}{2D}$$

where we have assumed $x, y << D$. On the plane $z = D$, the resultant field will be given by

$$E = E_1 + E_2$$

$$\approx E_0 \cos(kD - \omega t + \phi) + \frac{A_0}{D}\cos\left[kD + \frac{k}{2D}(x^2 + y^2) - \omega t\right]$$

Thus, $$\langle E^2 \rangle = \frac{1}{2}E_0^2 + \frac{1}{2}\left(\frac{A_0}{D}\right)^2 + E_0\frac{A_0}{D}\cos\left[\frac{k}{2D}(x^2 + y^2) - \phi\right]$$

If we assume that

$$\frac{A_0}{D} \approx E_0$$

i.e., the amplitude of the spherical wave (on the plane PP') is the same as the amplitude of the plane wave then

$$\langle E^2 \rangle \approx 2E_0^2 \cos^2\left(\frac{k}{4D}(x^2 + y^2) - \frac{1}{2}\phi\right)$$

and we would obtain circular interference fringes as shown in Fig. 6.17.

6.19 Consider a parallel beam of light (from a distant source S' like a star) incident (at an angle θ) on two slits S_1 and S_2 as shown in Fig. 6.9. Obviously the path difference between the waves emanating from the slits S_1 and S_2 will be given by

$$XS_2 = d \sin \theta$$

Therefore the intensity distribution on the screen due to S' will be given by

$$I = I_0 \cos^2 \frac{\delta}{2}$$

where, $$\delta = \frac{2\pi}{\lambda}[XS_2 + S_2P - S_1P]$$

$$= \frac{2\pi}{\lambda}[(S_2P - S_1P) + d \sin \theta]$$

$$= \frac{2\pi}{\lambda}\left[\frac{xd}{D} + d \sin \theta\right]$$

Thus, the intensity distribution (due to light coming from the distant source S') will be given by

$$I' = I_0 \cos^2\left(\frac{\pi}{\lambda}\left[\frac{xd}{D} + d\sin\theta\right]\right)$$

Similarly, if there is light incident from another distant source S'' (at an angle ϕ) then the corresponding intensity distribution on the screen will be given by

$$I'' = I_0 \cos^2\left(\frac{\pi}{\lambda}\left[\frac{xd}{D} - d\sin\phi\right]\right)$$

The resultant intensity distribution will be given by

$$I = I' + I''$$

6.20 When the path difference between the two interfering waves becomes equal to the coherence length, then the fringe contrast would be very poor. For a pair of wavelengths separated by a wavelength spacing $\Delta\lambda$, the fringe contrast would become poor when the path length is equal to $\lambda^2/2\Delta\lambda$. For the given arrangement, the path difference for fringes appearing around P is d. Hence, for good contast fringes, we must have

$$\Delta\lambda << \frac{\lambda^2}{2d} = 0.125 \text{ nm}$$

6.21 The closest point to the axis is when the maximas and minimas of the fringes produced by 589 nm will coincide with the minimas and maximas of the wavelength 589.6 nm. This will happen when

$$x_0 = \frac{\lambda^2}{2\Delta\lambda}\frac{D}{d} = 1.16 \text{ cm}$$

6.22 (a) If the polariser pass axes are perpendicular to each other, then we will not observe any interference pattern on the screen.

(b) If the pass axis of one of the polariser is oriented at 45° to the other polarizer, then only the polarisation component parallel to the polarisation state of light emerging from the other slit will form interference pattern. This will lead to poorer contrast in the interference pattern.

6.23 When the source S is placed away from the axis, the fringe pattern will get displaced due to additional phase difference generated between waves reaching S_1 and S_2 from S. Assuming $L >> a$, the path difference between the waves reaching S_1 and S_2 from S would be approximately given by $d(a/L)$. Hence the zero order fringe would appear at a distance x_0 from the axis such that

$$\frac{x_0 d}{D} = \frac{ad}{L}$$

or,

$$x_0 = \frac{a}{L}D$$

Interference by Division of Amplitude

7

A Quick Review

1. Consider a thin film of refractive index n_2 and thickness d as shown in Fig. 7.1. The optical path difference between the waves reflected from the upper surface of the film and from the lower surface of the film is given by

$$\Delta = 2n_2 d \cos \theta' \tag{1}$$

where θ' is the angle of refraction as shown in Fig. 7.1.

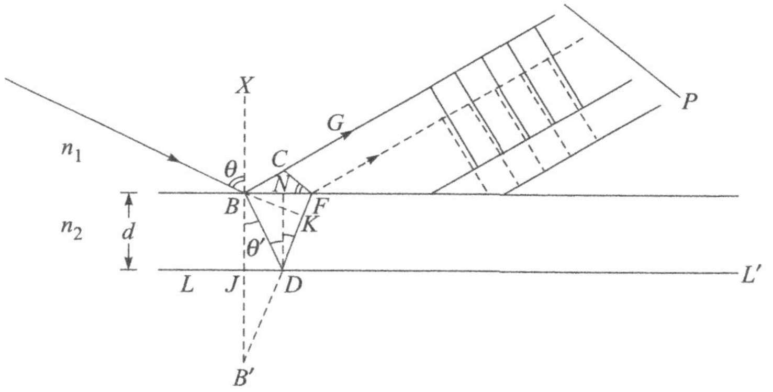

Fig. 7.1 Waves reflected from the lower surface of the film traverse an extra path than the waves reflected from the upper surface of the film.

2. If we consider a thin film of refractive index n_2 and thickness d coated on a medium of refractive index n_s placed in air, then light waves at a wavelength λ_0 incident normally on the film will undergo reflection at both the upper and lower interfaces. If the reflectivities at the interfaces are not large, then we can neglect multiple reflections of the light waves multiple. The path difference between the two reflected waves (one reflected from the upper surface and one from the lower surface) would be

$$\Delta = 2n_2 d \tag{2}$$

If we assume that $1 < n_2 < n_s$, then when

$$\Delta = m\lambda_0;\ m = 1, 2, 3, \ldots \tag{3}$$

we would have constructive interference and for

$$\Delta = \left(m + \frac{1}{2}\right)\lambda_0;\ m = 1, 2, 3, \ldots \tag{4}$$

we would have destructive interference. If the refractive index of the film is the geometric mean of the refractive indices of the two media surrounding the film, then the amplitudes of the two interfering beams are almost equal and we would have complete destructive interference. This is the principle behind anti-reflection coatings. The required minimum film thickness is

$$d = \frac{\lambda_0}{4n_2} \tag{5}$$

3. Figure 7.2 shows an arrangement for observing Newton's rings. Light from an extended source S is allowed to fall on a thin film of air formed between the plano-convex lens AOB and the plane glass plate POQ. M represents a traveling microscope. The optical path difference between the waves reflected from the lower surface of the lens and the upper surface of the glass plate is $2nt$ where n is the refractive index of the air film and t is the thickness of the film. For near normal incidence (and considering points very close to the point of contact), whenever the thickness of the film is satisfies the condition

$$2nt = \left(m + \frac{1}{2}\right)\lambda_0;\ m = 0, 1, 2, 3, \ldots \tag{6}$$

we will have maxima. Similarly the condition

$$2nt = m\lambda_0; \quad m = 0, 1, 2, 3, \ldots \tag{7}$$

will correspond to maxima. These conditions lead to the formation of Newton's rings (as viewed from the microscope).

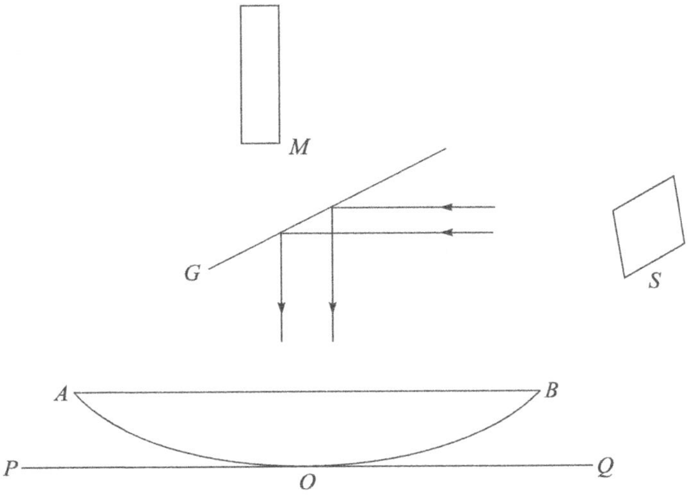

Fig. 7.2 An arrangement for observing Newton's rings. Light from an extended source S is allowed to fall on a thin film formed between the plano-convex lens AOB and the plane glass plate POQ. M represents a traveling microscope.

PROBLEMS

7.1 A glass plate of refractive index 1.6 is in contact with another glass plate of refractive index 1.8 along a line such that a wedge of 0.5° is formed. Light of wavelength 5000 Å is incident vertically on the wedge and the film is viewed from the top. Calculate the fringe spacing. The whole apparatus is immersed in an oil of refractive index 1.7. What will be the qualitative difference in the fringe pattern and what will be the new fringe width?

7.2 Two plane glass plates are placed on top of one another and on one side a cardboard is introduced to form a thin wedge of air. Assuming that a beam of wavelength 6000 Å is incident normally, and that there are 100 interference fringes per centimeter, calculate the wedge angle.

7.3 Consider a nonreflecting film of refractive index 1.38. Assume that its thickness is 9×10^{-6} cm. Calculate the wavelengths (in the visible region) for which the film will be nonreflecting. Repeat the calculations for the thickness of the film to be 45×10^{-6} cm. Show that both the films will be nonreflecting for a particular wavelength but only the former one will be suitable. Why?

7.4 In the Newton's rings arrangement, the radius of curvature of the curved side of the plano-convex lens is 100 cm. For $\lambda = 6 \times 10^{-5}$ cm what will be the radii of the 9th and 10th bright rings?

7.5 In the Newton's rings arrangement, the radius of curvature of the curved surface is 50 cm. The radii of the 9th and 16th dark rings are 0.18 cm and 0.2235 cm. Calculate the wavelength. [**Ans.** 5015 Å]

7.6 In the Newton's rings arrangement, if the incident light consists of two wavelengths 4000 Å and 4002 Å calculate the distance (from the point of contact) at which the rings will disappear. Assume that the radius of curvature of the curved surface is 400 cm. [**Ans.** 4 cm]

7.7 In the above problem if the lens is slowly moved upward, calculate the height of the lens at which the fringe system (around the center) will disappear. [**Ans.** 0.2 mm]

7.8 An equiconvex lens is placed on another equiconvex lens. The radii of curvature of the two surfaces of the upper lens are 50 cm and those of the lower lens are 100 cm. The waves reflected from the upper and lower surface of the air film (formed between the two lenses) interfere to produce Newton's rings. Calculate the radii of the dark rings. Assume $\lambda = 6000$ Å. [**Ans.** $0.0447\sqrt{m}$ cm]

7.9 In the Michelson interferometer arrangement, if one of the mirrors is moved by a distance 0.08 mm, 250 fringes cross the field of view. Calculate the wavelength. [**Ans.** 6400 Å]

7.10 In the Michelson interferometer experiment, calculate the various values of θ' (corresponding to bright rings) for $d = 5 \times 10^{-3}$ cm. Show that if d is decreased to 4.997×10^{-3} cm, the fringe corresponding to $m = 200$ disappears. What will be the corresponding values of θ'? Assume $\lambda = 5 \times 10^{-5}$ cm.

7.11 A soap film surrounded by air has an index of refraction of 1.34. Under normal illumination, if a region of the film reflects strongly a wavelength of 804 nm, what is the minimum thickness of the film?

7.12 A microscope lens of refractive index 1.55 is to be coated with a MgF_2 film (n = 1.38) to increase transmission of normally incident yellow light (λ = 5500 Å). What should be the minimum thickness of the film deposited on the lens?

7.13 The Michelson interferometer experiment is performed with a source that consists of two wavelengths: 4882 Å and 4886 Å. Through what distance does the mirror have to be moved between two positions of disappearance of the fringes?

7.14 White light is reflected normally from a soap film of uniform thickness in air. An interference maximum is observed at 0.6 μm and a minimum at 0.45 μm with no minimum or maximum in between. Assuming n = 1.33 for the film, calculate the thickness of the film.

7.15 Newtons rings are observed under white light illumination. What will be the colour of the innermost ring?

7.16 As a soap bubble evaporates, one can see a change of colour. What colour will the soap bubble appear when it is about to burst?

SOLUTIONS

7.1 Fringe width $\beta \approx \dfrac{\lambda}{2n_f\phi}$; $\phi = 0.5° = \dfrac{0.5\times\pi}{180}$ radians $\approx 8.73\times10^{-3}$ radians.

When $n_f = 1$

$$\beta = \frac{5\times10^{-5}}{2\times8.73\times10^{-3}} \approx 0.0286 \text{ mm}$$

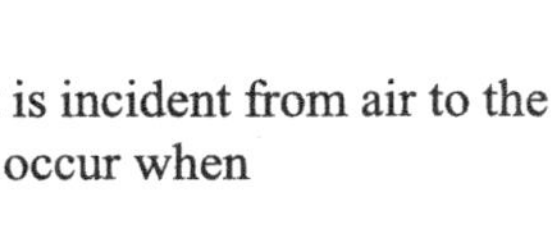

Fig. 7.3

When $n_f = 1.7$

$$\beta = \frac{5\times10^{-5}}{2\times1.7\times10^{-3}} \approx 1.68\times10^{-3} \text{ cm} = 0.0168 \text{ mm}$$

In the first case, phase change will occur when light is incident from air to the glass of refractive index 1.6; thus bright fringe will occur when

$$2n_f t_f = \left(m + \frac{1}{2}\right)\lambda$$

where t_f is the thickness of the (air) film. On the other hand, when $n_f = 1.7$, no phase change will occur at either of the reflections and bright fringe will occur when

$$2n_f t_f = m\lambda.$$

7.2 $\beta = \dfrac{1 \text{ cm}}{100} = 0.01$ cm

Now $\beta = \dfrac{\lambda}{2n\phi} \Rightarrow \phi = \dfrac{\lambda}{2n\beta} = \dfrac{6\times10^{-5}}{2\times1\times0.01} = 0.003$ radians $\approx 0.17°$.

7.3 The film will be non-reflecting when [see Eq. (4)]

$$2n_f d = \Delta = \left(m + \frac{1}{2}\right)\lambda_0; \quad m = 1, 2, \ldots$$

$$\Rightarrow \qquad \lambda_0 = \frac{4n_f d}{(2m+1)}; \quad m = 0, 1, 2, \ldots$$

For $n_f = 1.38$ and $d = 9 \times 10^{-6}$ cm, the film is non-reflecting when

$$\lambda_0 = \frac{4.968 \times 10^{-5} \text{ cm}}{2m+1} = \underbrace{4968\,\text{Å}}_{\text{visible}},\ \underbrace{1660\,\text{Å}}_{\text{UV}}$$

for $m = 0$ and 1. Only the first wavelength (4968 Å) lies in the visible region.

For $n_f = 1.38$ and $d = 45 \times 10^{-6}$ cm, the film is nonreflecting when

$$\lambda = \frac{24.84 \times 10^{-5} \text{ cm}}{2m+1} = \underbrace{24840\,\text{Å}, 8280\,\text{Å}}_{\text{IR}},\ \underbrace{4968\,\text{Å}, 3549\,\text{Å}}_{\text{visible}},\ \underbrace{2760\,\text{Å}}_{\text{UV}}$$

Thus only 2 wavelengths are in the visible region; actually 3549 Å is almost in the ultraviolet region. Thus, both films will be nonreflecting for $\lambda_0 =$ 4968 Å; but the first film will be preferable because for the second case, the film will have high reflection for a wavelengths between 3549 Å and 4968 Å.

7.4 $R = 100$ cm and $\lambda = 6 \times 10^{-5}$ cm. If r_m represents the radius of the mth bright right ring then:

$$r_m^2 = \left(m - \frac{1}{2}\right)\lambda R; \quad m = 1, 2, 3, \ldots \quad \text{or} \quad r_m \approx 0.07746\sqrt{m - \frac{1}{2}} \text{ cm}$$

Thus, the radii of the 9th and 10th bright rings will be given by

$$r_9 \approx 0.226 \text{ cm} \quad \text{and} \quad r_{10} \approx 0.239 \text{ cm}.$$

7.5 $$\lambda = \frac{D_{16}^2 - D_9^2}{4(16-9)R} = \frac{(0.447)^2 - (0.36)^2}{4 \times 7 \times 50} \text{ cm} \approx 5015 \text{ Å}.$$

7.6 $\lambda_1 = 4000$ Å, $\lambda_2 = 4002$ Å. The height t at which the fringes disappear will be given by

$$\frac{2t}{\lambda_1} - \frac{2t}{\lambda_2} = \frac{1}{2} \Rightarrow t = \frac{\lambda_1 \lambda_2}{4(\lambda_2 - \lambda_1)} = \frac{4 \times 10^{-5} \times 4.002 \times 10^{-5}}{4 \times 2 \times 10^{-8}} \text{ cm} \approx 0.02 \text{ cm}.$$

The distance from the point of contact will be given by

$$r^2 \approx 2Rt \quad \Rightarrow \quad r = \sqrt{2 \times 400 \times 0.02} \approx 4 \text{ cm}.$$

7.7 The height t will be given by

$$\frac{2t}{\lambda_1} - \frac{2t}{\lambda_2} = \frac{1}{2} \quad \Rightarrow \quad t \approx 0.20 \text{ mm}$$

7.8 $2t_1R_1 = r_m^2$

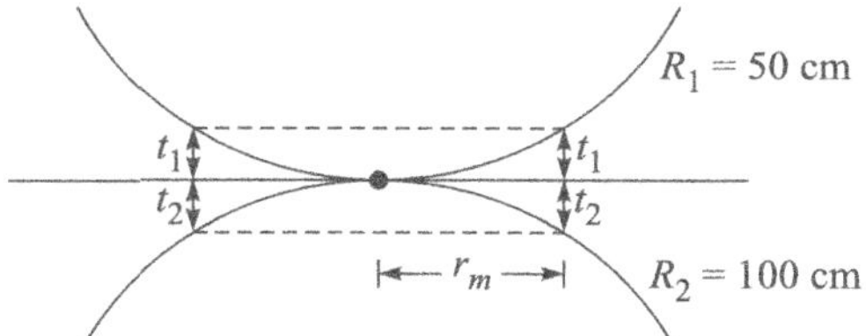

For dark ring

Fig. 7.4

$$2t_2R_2 = r_m^2$$

$$\Rightarrow \quad 2(t_1 + t_2) = r_m^2\left[\frac{1}{R_1} + \frac{1}{R_2}\right] = m\lambda$$

$$\Rightarrow \quad r_m^2 = m\lambda\frac{R_1R_2}{R_1 + R_2} = m \times 6 \times 10^{-5} \times \frac{50 \times 100}{150} = (2 \times 10^{-3} m)\ \text{cm}^2$$

$$\Rightarrow \quad r_m = 0.0447\sqrt{m}\ \text{cm}$$

7.9 $2(d_1 - d_2) = 250\,\lambda$

$$\lambda = \frac{2 \times 0.008}{250}\ \text{cm} = 6.4 \times 10^{-5}\ \text{cm} = 6400\ \text{Å}$$

7.10 The condition for a bright ring is given by

$$2d\cos\theta = \left(m + \frac{1}{2}\right)\lambda \quad \Rightarrow \quad \theta = \cos^{-1}\left[\frac{\left(m + \frac{1}{2}\right)\lambda}{2d}\right] = \cos^{-1}\left[\frac{\left(m + \frac{1}{2}\right)}{200}\right]$$

The central fringe corresponds to $2d = 10 \times 10^{-3}$ cm $= 200\,\lambda$ which will be dark. Thus the first, second and third bright rings will be given by

$$\theta = \cos^{-1}\left[\frac{199.5}{200}\right],\ \cos^{-1}\left[\frac{198.5}{200}\right] \quad \text{and} \quad \cos^{-1}\left[\frac{197.5}{200}\right]$$

$$\approx 4.05°,\ 7.02° \quad \text{and} \quad 9.07°.$$

corresponding to m = 199, 198 and 197 respectively. If we decrease d to 4.997×10^{-3} cm, then the central fringe will correspond to

$$2d = 9.994 \times 10^{-3}\ \text{cm} = 199.88\,\lambda$$

and will be neither bright nor dark. Obviously, the $m = 200$ fringe disappears!

7.11 For maximum reflection the waves reflected from the two surfaces of the soap film must be in phase. Since there is a phase difference of p between the waves due to the phase change on reflection from a denser medium, the wavelength corresponding to maximum reflectivity would be

$$\frac{2\pi}{\lambda_0}2nt = \pi$$

where n is the refractive index of the film and t its thickness. Substituting the values given in the problem we obtain the thickness of the soap film to be t = 150 nm.

7.12 The required thickness would be $\lambda_0/4n$ which on substitution of the values gives us a thickness of 99.6 nm.

7.13 The maximas for a wavelength λ_i correspond to

$$\frac{2\pi}{\lambda_i}2(\Delta d) = 2m\pi;\ m = 0, 1, 2, \ldots$$

where Δd is the difference in length of the two arms of the interferometer. Thus starting from $\Delta d = 0$, for the values of Δd given below the maximas of wavelengths λ_1 and λ_2 will again coincide when

$$\frac{2\pi}{\lambda_1}2(\Delta d) = 2m\pi;$$

$$\frac{2\pi}{\lambda_2}2(\Delta d) = 2(m+1)\pi;$$

Subtracting one equation from the other, we obtain

$$\Delta d = \frac{\lambda_1\lambda_2}{2(\lambda_1 - \lambda_2)}$$

which gives us $\Delta d = 0.298$ mm.

7.14 The phase difference between the two reflected waves at $\lambda_1 = 0.6$ μm should be an odd multiple of π. Hence,

$$\frac{2\pi}{\lambda_1}2nt = (2m+1)\pi;$$

Since at $\lambda_2 = 0.45$ μm there is a minimum and there are no other maximas or minimas in between we should have

$$\frac{2\pi}{\lambda_2}2nt = (2m+2)\pi;$$

From the above two equations we obtain for the thickness of the soap film $t \sim 338$ nm.

7.15 Since the radii of the rings are proportional to the wavelength, starting from the central dark fringe, the first bright fringe of blue colour will have a smaller radius than that of red colour. Hence the innermost fringe will be bluish in colour.

7.16 When a soap bubble is about to burst, it will have the smallest thickness. Since there is a phase change of π on reflection at the air-bubble interface, the bubble will appear black as it is about to burst.

Multiple Beam Interferometry

8

A Quick Review

8.1 MULTIPLE REFLECTIONS FROM A PLANE PARALLEL FILM

Consider the incidence of a plane wave on a film of thickness h (and of refractive index n_f) placed in air. The wave will undergo multiple reflections at the two interfaces. We assume that the reflectivities of the two surfaces are equal and given by R; then the intensity transmittance of the film would be given by

$$T = \frac{1}{1 + F\sin^2\left(\frac{\delta}{2}\right)} \tag{1}$$

where,

$$\delta = \frac{2\pi}{\lambda_0}\Delta = \frac{4\pi n_f h \cos\theta_f}{\lambda_0} \tag{2}$$

represents the phase difference accumulated during one back and forth propagation of the wave through the film (see Fig. 8.1) with θ_f representing the angle made by the waves inside the film. Further,

$$\Delta = 2n_f h \cos\theta_f \tag{3}$$

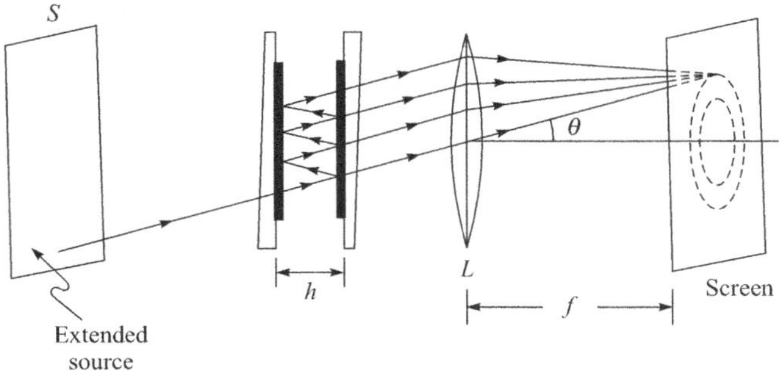

Fig. 8.1 The Fabry-Perot etalon.

is the corresponding path difference and

$$F = \frac{4R}{(1-R)^2} \tag{4}$$

is called the coefficient of finesse. Note that when

$$\delta = 2m\pi; \quad m = 1, 2, 3, \ldots \tag{5}$$

then $T = 1$ and all the incident light gets transmitted. If R is close to unity, then the transmittance drops very quickly as δ changes. The changes in δ could be brought about by changes in the thickness of the medium between the two highly reflecting surfaces (see Problem 8.2) or by changes in the wavelength of the incident radiation or the angle of illumination (see Problem 8.4). Thus in transmission, this produces very sharp interference fringes. This interference phenomenon is referred to as multiple beam interference. The Fabry Perot interferometer and the Fabry Perot etalon are based on this principle; the Fabry-Perot interferometer consists of two plane glass (or quartz) plates which are coated on one side with a partially reflecting metallic film[1] (of aluminum or silver) of about 80% reflectivity. These two plates are kept in such a way that they enclose a plane parallel slab of air between their coated surfaces. If the reflecting glass plates are held parallel to each other at a fixed separation, we have what is known as a Fabry-Perot etalon. Equations (2) and (5) may be combined to give

$$2n_f h \cos\theta_f = m\lambda_0; \quad m = 1, 2, 3, \ldots \tag{6}$$

If $\Delta\delta$ represents the FWHM (full width at half maximum) then

$$\Delta\delta \approx \frac{4}{\sqrt{F}} = \frac{2(1-R)}{\sqrt{R}} \tag{7}$$

where we have assumed $\Delta\delta \ll 1$, which is true in almost all cases. We assume normal incidence ($\theta_f = 0$) and write $h = h_0 + x$. We also assume $\lambda_0 = 5 \times 10^{-5}$ cm and $h_0 = 10$ cm; thus

$$\delta = 80000\pi\left(1 + \frac{x}{h_0}\right)$$

(see Problem 8.2). Figure 8.2 shows the variation of the transmission coefficient T with x for different values of F.

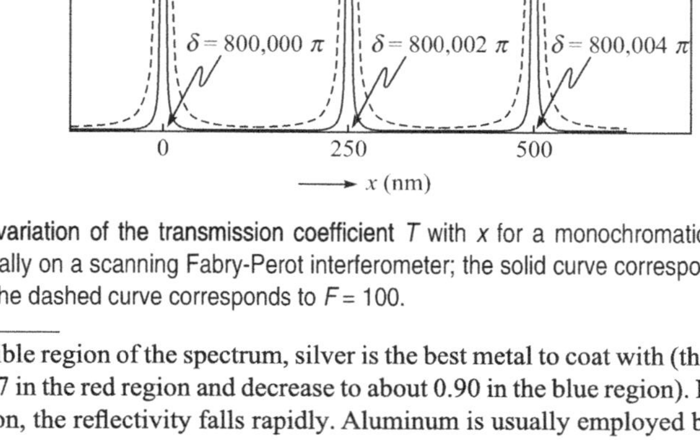

Fig. 8.2 The variation of the transmission coefficient T with x for a monochromatic beam incident normally on a scanning Fabry-Perot interferometer; the solid curve corresponds to $F = 1000$ and the dashed curve corresponds to $F = 100$.

[1] In the visible region of the spectrum, silver is the best metal to coat with (the reflectivity is about 0.97 in the red region and decrease to about 0.90 in the blue region). But beyond the blue region, the reflectivity falls rapidly. Aluminum is usually employed below 4000 Å.

8.2 MODES OF THE FABRY-PEROT CAVITY

We consider a polychromatic beam incident normally ($\theta_f = 0$) on a Fabry-Perot etalon with air between the reflecting plates–see Fig. 8.3. In terms of the frequency

$$\nu = \frac{c}{\lambda_0}, \tag{8}$$

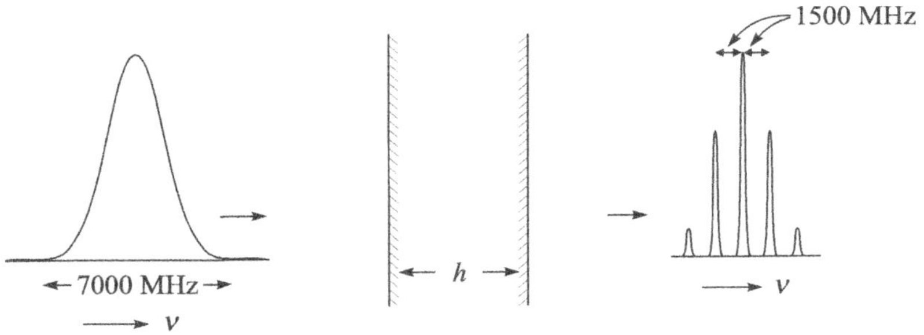

Fig. 8.3 A beam having a spectral width of about 7000 MHz (around $\nu_0 = 6 \times 10^{14}$ Hz) is incident normally on a Fabry-Perot etalon with $h = 10$ cm and $n_f = 1$. The output has 5 narrow spectral lines.

Equation (6) tells us that transmission resonance will occur when

$$\nu = \nu_m = m \frac{c}{2h} \tag{9}$$

where m is an integer. The above equation represents the different (longitudinal) modes of the Fabry-Perot cavity.

8.3 RESOLVING POWER

The resolving power of a Fabry-Perot interferometer is given by

$$\text{Resolving Power} = \left|\frac{\nu}{\Delta \nu}\right| = \frac{\pi h \nu \sqrt{F}}{c} \tag{10}$$

or, in terms of the wavelength

$$\text{Resolving Power} = \left|\frac{\lambda_0}{\Delta \lambda_0}\right| = \frac{\pi h \sqrt{F}}{\lambda_0} \tag{11}$$

PROBLEMS

8.1 Consider a beam having a central frequency of

$$\nu = \nu_0 = 6 \times 10^{14} \text{ Hz}$$

and a spectral width[2] of 7000 MHz incident on a Fabry-Perot cavity (see Fig. 8.3). Assume $h = 10$ cm, calculate the frequencies of the output beam and the corresponding mode numbers.

8.2 In a scanning Fabry-Perot interferometer, we vary the separation h between the mirrors and measure the intensity variation on the focal plane of the lens L as shown in Fig. 8.4. We write

$$h = h_0 + x \tag{12}$$

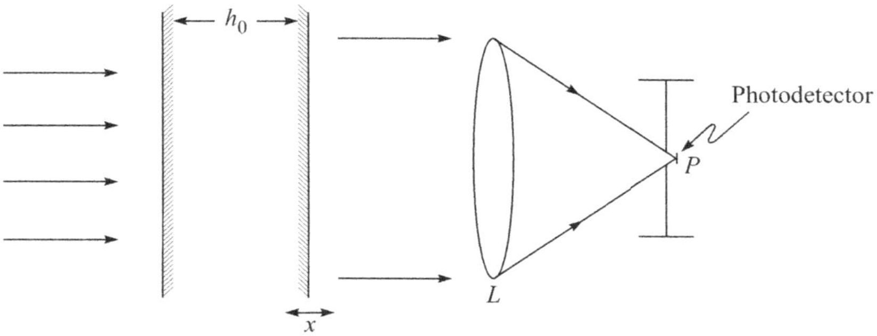

Fig. 8.4 A scanning Fabry-Perot interferometer. The intensity variation is recorded (by a photodetector) on the focal plane of the lens *L*.

Consider a monochromatic beam ($\nu = \nu_0 = 6 \times 10^{14}$ Hz) incident normally on the interferometer with $h_0 = 10$ cm and $n_f = 1$. Calculate and plot the intensity variation at the point P as a function of x.

8.3 Consider a Fabry-Perot etalon with $n_f = 1$, $h = 1$ cm and $F = 400$. Calculate the reflectivity of each mirror.

8.4 In continuation of the previous problem, plot the intensity variation with θ for $\lambda_0 = 5000$ Å and 4999.98 Å.

8.5 Calculate the resolving power of a Fabry-Perot interferometer made of reflecting surfaces of reflectivity 0.85 and separated by a distance 1 mm at $\lambda_0 = 4880$ Å.

8.6 Calculate the minimum spacing between the plates of a Fabry-Perot interferometer which would resolve two lines with $\Delta\lambda_0 = 0.1$ Å at $\lambda_0 = 6000$ Å. Assume the reflectivity to be 0.8.

8.7 Consider a monochromatic beam of wavelength 6000 Å incident (from am extended source) on a Fabry-Perot etalon with $n_f = 1$, $h = 1$ cm and $F = 200$. Concentric rings are observed on the focal plane of a lens of focal length 20 cm.

(a) Calculate the reflectivity of each mirror.

---

2. For $\nu_0 = 6 \times 10^{14}$ Hz, $\lambda_0 = 5000$ Å and a spectral width of 7000 MHz would imply $\left|\frac{\Delta\lambda_0}{\lambda_0}\right| = \frac{\Delta\nu}{\nu_0} = \frac{7 \times 10^9}{6 \times 10^{14}} \approx 1.2 \times 10^{-5}$ giving $\Delta\lambda_0 \approx 0.06$ Å. Thus a frequency spectral width of 7000 MHz (around $\nu_0 = 6 \times 10^{14}$ Hz) implies a wavelength spread of only 0.06 Å.

(b) Calculate the radii of the first four bright rings. What will be the corresponding value of m?

(c) Calculate the angular width of each ring where the intensity falls by half and the corresponding FWHM (in mm) of each ring.

8.8 Consider now two wavelengths 6000 Å and 5999.9 Å incident on a Fabry-Perot etalon with the same parameters as given in the previous problem. Calculate the radii of the first three bright rings corresponding to each wavelength. What will be the corresponding values of m? Will the lines be resolved?

8.9 Consider a monochromatic beam of wavelength 6000 Å incident normally on a scanning Fabry-Perot interferometer with $n_f = 1$ and $F = 400$. The distance between the two mirrors is written as $h = h_0 + x$. With $h_0 = 10$ cm, calculate

(a) The first three values of x for which we will have unit transmittivity and the corresponding value of m.

(b) Also calculate the FWHM Δh for which the transmittivity will be half.

(c) What would be the value of Δh if F was 200?

[**Ans.** (a) $x \approx 200$ nm ($m = 333334$), 500 nm ($m = 333335$); (b) $\Delta h \approx 8$ nm]

8.10 In continuation of the previous problem, consider now two wavelengths $\lambda_0 (= 6000$ Å$)$ and $\lambda_0 + \Delta\lambda$ incident normally on the Fabry-Perot interferometer with $n_f = 1$, $F = 400$ and $h_0 = 10$ cm. What will be the value of $\Delta\lambda$ so that $T = 1/2$ occurs at the same value of h for both the wavelengths.

8.11 Consider a laser beam incident normally on the Fabry-Perot interferometer as shown in Fig. 8.4.

(a) Assume $h_0 = 0.1$ m, $c = 3 \times 10^8$ m/s, $\nu = \nu_0 = 5 \times 10^{14}\ s^{-1}$. Plot T as a function of x (-100 nm $< x < 400$ nm) for $F = 200$ and $F = 1000$.

(b) Show that if $\nu = (\nu_0 \pm p$ 1500 MHz; $p = 1, 2, \ldots)$ we will have the same T vs. x curve; 1500 MHz is known as the free spectral range (FSR). What will be the corresponding values of δ?

8.12 When a parallel beam of light is normally incident on a FP interferometer (with air between the highly reflecting surfaces) the intensity distribution at the output of the interferometer is as shown below.

(a) Determine the separation between the reflecting surfaces of the interferometer.

(b) If the minimum transmission is 0.0025, what is the reflectivity of the mirrors?

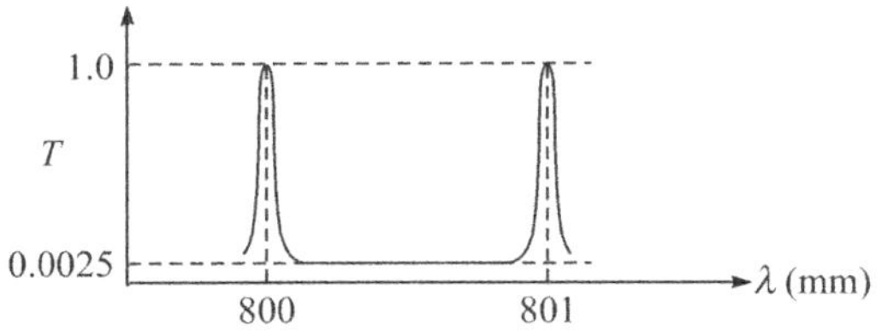

Fig. 8.5

8.13 Show that in the system of fringes formed in transmitted light by multiple reflection (i.e., in the system of fringes formed in a Fabry-Perot interferometer), the ratio of the intensity of the maxima to that mid way between two maxima is $(1 + R)^2/(1 - R)^2$.

SOLUTIONS

8.1 The frequency spacing of two adjacent modes would be given by

$$\delta\nu = \frac{c}{2h} = 1500 \text{ MHz}$$

For an incident beam having a central frequency of

$$\nu = \nu_0 = 6 \times 10^{14} \text{ Hz}$$

and a spectral width[3] of 7000 MHz the output beam will have frequencies

$$\nu_0,\ \nu_0 \pm \delta\nu \quad \text{and} \quad \nu_0 \pm 2\,\delta\nu$$

as shown in Fig. 8.3. One can readily calculate from Eq. (8) that the five lines correspond to

$$m = 399998, 399999, 400000, 400001 \quad \text{and} \quad 400002.$$

8.2 Since $\nu = \nu_0 = 6 \times 10^{14}$ Hz, $h_0 = 10$ cm, $n_f = 1$ and $\cos\theta_f = 1$, we get

$$\delta = \frac{4\pi\nu_0(h_0 + x)}{c} = 800000\,\pi\left(1 + \frac{x}{h_0}\right)$$

Thus, transmittivity resonances will occur for

$$\delta = 800000\,\pi,\ 800002\,\pi,\ 800004\,\pi, \ldots$$

which will occur when

$$x = 0{,}250 \text{ nm},\ 500 \text{ nm}, \ldots$$

respectively. The two curves in Fig. 8.2 correspond to $F = 100$ and $F = 1000$. Notice that the transmission resonances become sharper as we increase the value of F.

8.3 From the equation

$$F = \frac{4R}{(1 - R)^2}$$

we readily obtain

$$R^2 - 2\left(1 + \frac{2}{F}\right)R + 1 = 0 \quad \Rightarrow \quad R = 1 + \frac{2}{F} - \frac{2}{F}\sqrt{F + 1}$$

$F = 400$ implies $R \approx 0.905$; i.e., each mirror of the etalon has about 90% reflectivity.

---

3. For $\nu_0 = 6 \times 10^{14}$ Hz, $\lambda_0 = 5000$ Å and a spectral width of 7000 MHz would imply $\left|\frac{\Delta\lambda_0}{\lambda_0}\right| = \frac{\Delta\nu}{\nu_0} = \frac{7 \times 10^9}{6 \times 10^{14}} \approx 1.2 \times 10^{-5}$ giving $\Delta\lambda_0 \approx 0.06$ Å. Thus a frequency spectral width of 7000 MHz (around $\nu_0 = 6 \times 10^{14}$ Hz) implies a wavelength spread of only 0.06 Å.

8.4 In Fig. 8.6 we have plotted the intensity variation with θ for $\lambda_0 = 5000$ Å and 4999.98 Å. The actual fringe pattern (as obtained on the focal plane of a lens of focal length 25 cm) is shown in Fig. 8.7. Now, for

$$\lambda_0 = \lambda_1 = 5000 \text{ Å}$$

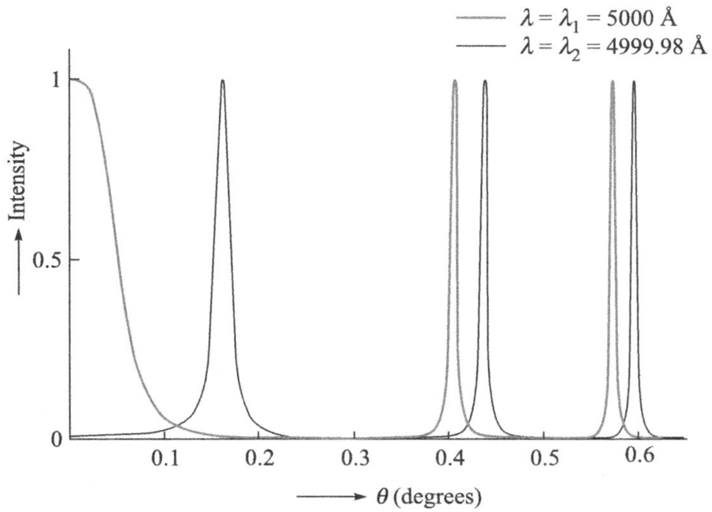

Fig. 8.6 The variation of intensity with θ for a Fabry-Perot interferometer with $n_f = 1$, $h = 1.0$ cm and $F = 400$, corresponding to $\lambda_0 = 5000$ Å ($= \lambda_1$) and $\lambda_0 = 4999.98$ Å ($= \lambda_2$).

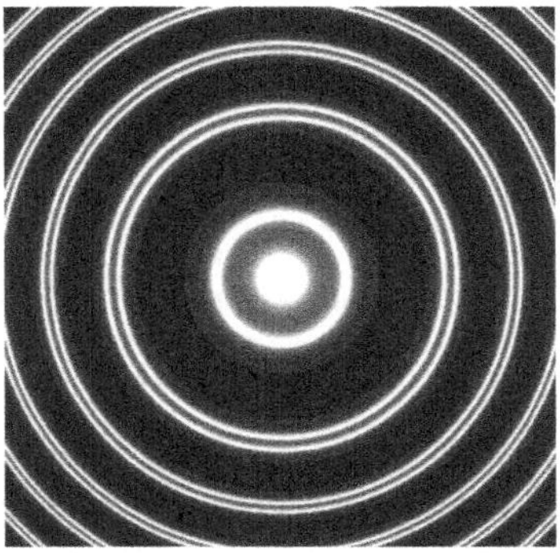

Fig. 8.7 The (computer generated) ring pattern as obtained (on the focal plane of a lens) in a Fabry-Perot etalon with $n_f = 1$, $h = 1.0$ cm and $F = 400$, corresponding to $\lambda_0 = 5000$ Å ($= \lambda_1$) and $\lambda_0 = 4999.98$ Å ($= \lambda_2$).

Equations (2) and (5) give us

$$\theta_f = \cos^{-1}\left(\frac{m}{40000}\right)$$

Thus, bright rings will form at

$$\theta_f = 0°, 0.41°, 0.57°, 0.70°, \ldots$$

corresponding to m = 40000, 39999, 39998, 39997,... respectively. This is shown as the thick curve in Fig. 8.6. On the other hand, for

$$\lambda_0 = \lambda_2 = 4999.98 \text{ Å}$$

we get
$$\theta_f = \cos^{-1}\left(\frac{m}{40000.16}\right)$$

Thus bright rings will form at

$$\theta_f = 0.162°, 0.436°, 0.595°, \ldots$$

corresponding to m = 40000, 39999 and 39998 respectively. This is shown as the thin curve in Fig. 8.6. The corresponding ring patterns as obtained on the focal plane of the lens is shown in Fig. 8.7. From the figure, we can see that the two spectral lines having a small wavelength difference of 0.02 Å are quite well resolved by the etalon. In the figure, the central bright spot and the first ring corresponds respectively to λ_0 = 5000 Å and λ_0 = 4999.98 Å; both corresponding to m = 40000. The next two closely spaced rings correspond to m = 39999 for the two wavelengths.

8.5 $h = 1$ mm $= 0.1$ cm; $\lambda = 4.88 \times 10^{-5}$ cm;

$$R = 0.85 \Rightarrow F = \frac{4R}{(1-R)^2} \approx 151$$

$$\text{Resolving Power} = \frac{\lambda}{\Delta\lambda} = \frac{\pi h\sqrt{F}}{\lambda} \approx 0.8 \times 10^5.$$

8.6 $R = 0.8 \Rightarrow F = 80$; $\Delta\lambda = 0.1$ Å. Thus,

$$\frac{\lambda}{\Delta\lambda} = \frac{6\times10^{-5}}{0.1\times10^{-8}} = \frac{\pi\times h\times\sqrt{80}}{6\times10^{-5}} \Rightarrow h_{\min} \approx 0.13 \text{ cm.}$$

8.7 $\lambda = 6 \times 10^{-5}$ cm, $n_2 = 1$, $h = 1$ cm and $F = 200$

(a) $(1-R)^2 \cdot F = 4R \Rightarrow 1 - 2R + R^2 = 0.02R$

$$\Rightarrow \qquad R^2 - 2.02R + 1 = 0 \Rightarrow R \approx 0.868$$

(b) Bright rings will occur when

$$\theta = \cos^{-1}\left[\frac{m\lambda}{2n_2h}\right] = \cos^{-1}\left[\frac{m}{33333.333}\right]$$

Thus, the first four bright rings will correspond to m = 33333, 33332, 33331 and 33330; the corresponding values of θ_2 are approximately 0.256°, 0.512°, 0.678° and 0.810° respectively. The corresponding radii will be approximately

$$f\theta \approx 0.89 \text{ mm}, 1.79 \text{ mm}, 2.36 \text{ mm} \quad \text{and} \quad 2.83 \text{ mm}$$

respectively; in the above equation θ_2 has to be in radians.

(c) We write $\mu = \cos\theta$. Thus,

$$\Delta\mu \approx \frac{c}{\pi\nu h\sqrt{F}} \Rightarrow \sin\theta\,\Delta\theta \approx \frac{\lambda}{\pi h\sqrt{F}} \approx 1.35\times10^{-6}$$

$$\Rightarrow \qquad \Delta\theta \approx \frac{1.35\times10^{-6}}{\sin\theta}\text{ radians} \approx \frac{7.735\times10^{-5}}{\sin\theta}\text{degrees}$$

$$\approx 0.017°, 0.0087°, 0.0066°, 0.0055°.$$

8.8 For $\lambda_0 = 6000$ Å, the results have been obtained in the previous problem. For $\lambda_0 = 5999.9$ Å bright rings will occur when

$$\theta = \cos^{-1}\left[\frac{m\lambda}{2n_f h}\right] = \cos^{-1}\left[\frac{m}{33333.889}\right]$$

$$\approx 0.418°, 0.610°, 0.754° \quad \text{and} \quad 0.875°.$$

for m = 33333, 33332, 33331 and 33330, respectively. If we compare the above values and the values of θ obtained for $\lambda = 6\times10^{-5}$ cm with the values of $\Delta\theta$, we can easily see that the lines are well resolved.

8.9 $T = \dfrac{1}{1+F\sin^2\left(\dfrac{\delta}{2}\right)}$; $\quad \delta = \dfrac{4\pi}{\lambda_0} n_f h\cos\theta_f$

$\lambda_0 = 6\times10^{-5}$ cm, $n_2 = 1$, $h = h_0 + x$ with $h_0 = 10$ cm and for normal incidence $\theta_f = 0$; thus for $T = 1$, we must have

$$\delta = \frac{4\pi}{6\times10^{-5}}\times10\times\left(1+\frac{x}{h_0}\right) = 2m\pi$$

$$\Rightarrow \qquad m = \frac{10^6}{3}\left(1+\frac{x}{h_0}\right) = 333333.333\left(1+\frac{x}{h_0}\right)$$

Thus for $x = 0$, m is not an integer and $T \neq 1$. The first maximum will occur when

$$m = 333334 = \frac{10^6}{3}\left(1+\frac{x}{h_0}\right)$$

or,
$$2 = 10^6\frac{x}{h} \Rightarrow x \approx 2\times10\times10^{-6}\text{ cm} = 200\text{ nm}$$

The second and third maxima will occur when

$$m = 333335 = \frac{10^6}{3}\left(1+\frac{x}{h_0}\right) \Rightarrow x \approx 500\text{ nm}$$

and
$$m = 333336 = \frac{10^6}{3}\left(1+\frac{x}{h_0}\right) \Rightarrow x \approx 800\text{ nm}$$

respectively.

(b) The FWHM is given by

$$\Delta\delta = 4\sin^{-1}\left[\frac{1}{\sqrt{F}}\right]$$

For normal incidence, $\delta = \dfrac{4\pi}{\lambda_0} h \; (n_2 = 1)$. Thus,

$$\Delta\delta = \frac{4\pi}{\lambda_0} h = 4 \sin^{-1}\left[\frac{1}{\sqrt{F}}\right]$$

$$\Rightarrow \qquad \Delta h = \frac{\lambda_0}{\pi} \sin^{-1}\left[\frac{1}{\sqrt{F}}\right] \approx 9.5 \text{ nm} \quad \text{for} \quad F = 400$$

(c) For $F = 200$, $\Delta h \approx 13.5$ nm.

8.10 If Δh represents the FWHM (see Fig. 8.8), then for $\lambda_0 = \lambda_1$

$$\Delta\delta = \frac{4\pi}{\lambda_1} \Delta h = 4 \sin^{-1}\left[\frac{1}{\sqrt{F}}\right] \approx \frac{4}{\sqrt{F}}$$

$$\Rightarrow \quad \Delta h \approx \frac{\lambda_1}{\pi\sqrt{F}}$$

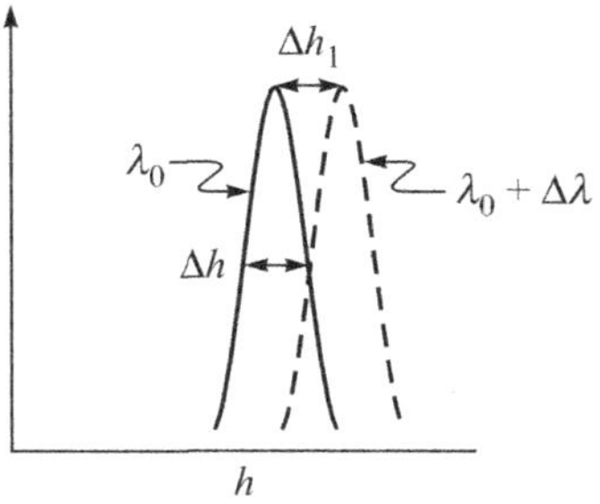

Fig. 8.8

For the $T = \dfrac{1}{2}$ point to occur at the same value of h, we must have

$$\Delta h = \Delta h_1$$

where,
$$\Delta h_1 = -\frac{h_1}{\nu_1} \Delta\nu_1 = \frac{h_1}{\lambda_1} \Delta\lambda_1$$

Thus,
$$\frac{h_1}{\lambda_1} \Delta\lambda_1 \approx \frac{\lambda_1}{\pi\sqrt{F}} \Rightarrow \Delta\lambda_1 \approx \frac{\lambda_1^2}{\pi h\sqrt{F}}$$

or,
$$\Delta\lambda_1 \approx \frac{(6 \times 10^{-5})^2}{\pi \times 10 \times \sqrt{400}} \approx 5.7 \times 10^{-12} \text{ cm} = 5.7 \times 10^{-4} \text{ Å}$$

8.11 $\delta = \dfrac{4\pi}{\lambda_0} n_f (h_0 + x) \cos\theta_f$

$T = 1$, when

$$2m\pi = \delta = \frac{4\pi\nu}{3 \times 10^8} \times 0.1 \times \left(1 + \frac{x}{h_0}\right) = 333333.33 \times 2\pi \times \left(1 + \frac{x}{h_0}\right)$$

or,
$$m = 333333.33\left(1 + \frac{x}{h_0}\right)$$

Thus $\quad m = 333332, 333333, 333334$ and 333335 for

$x \approx -400$ nm, -100 nm, $+200$ nm and $+500$ nm

respectively. Obviously, $T = 1$ when

$$\frac{4\pi\nu}{c} h_0 \times \frac{\Delta x}{h_0} = 2p\pi \quad \Rightarrow \quad \Delta x = \frac{pc}{2\nu} = 300p$$

where, $p = \pm 1, \pm 2, \pm 3, \ldots$. Similarly, if we change the frequency by $\Delta\nu$ such that

$$\frac{4\pi\Delta\nu}{c} \times h_0 = 2p\pi \quad \Rightarrow \quad \Delta\nu = \frac{cp}{2h_0} = p \times 1500 \text{ MHz}$$

then we will have the same T vs. x curve; the value of δ will change by $2p\pi$.

8.12 (a) If the adjacent wavelengths of maximum transmission are given by λ_1 and λ_2, then we have

$$2n_2 d = m\lambda_1 = (m+1)\lambda_2$$

Eliminating m from the two equations we obtain $d \sim 0.32$ mm.

(b) The minimum transmittivity is given by $1/(1 + F)$. Using this equation we can obtain the value of F and thus the reflectivity as 0.905.

8.13 The maximum transmission of a Fabry-Perot interferometer is unity while the minimum corresponds to $1/(1 + F)$. Using the expression for the coefficient of Finesse, we obtain the desired result.

Fraunhofer Diffraction: I

9

A Quick Review

9.1 FRAUNHOFER DIFFRACTION BY A SINGLE SLIT

A plane wave (of wavelength λ) is incident normally on a long narrow slit (of width b) and the Fraunhofer diffraction pattern is observed on the focal plane of the lens as shown in Fig. 9.1. The resultant field on the focal plane of the lens is given by

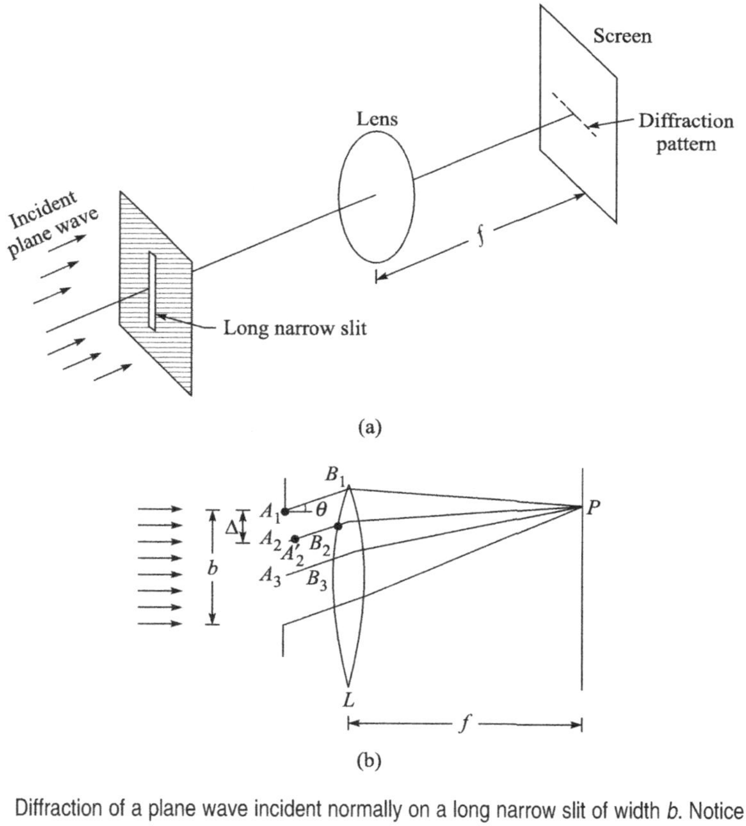

Fig. 9.1 Diffraction of a plane wave incident normally on a long narrow slit of width b. Notice that the spreading occurs along the width of the slit.

$$E = A\frac{\sin\beta}{\beta}\cos(\omega t - \beta) \tag{1}$$

where,
$$\beta = \frac{\pi b \sin\theta}{\lambda} \tag{2}$$

and θ is the angle of diffraction along the width of the slit. The corresponding intensity distribution on the focal plane of the lens is given by (see Fig. 9.2)

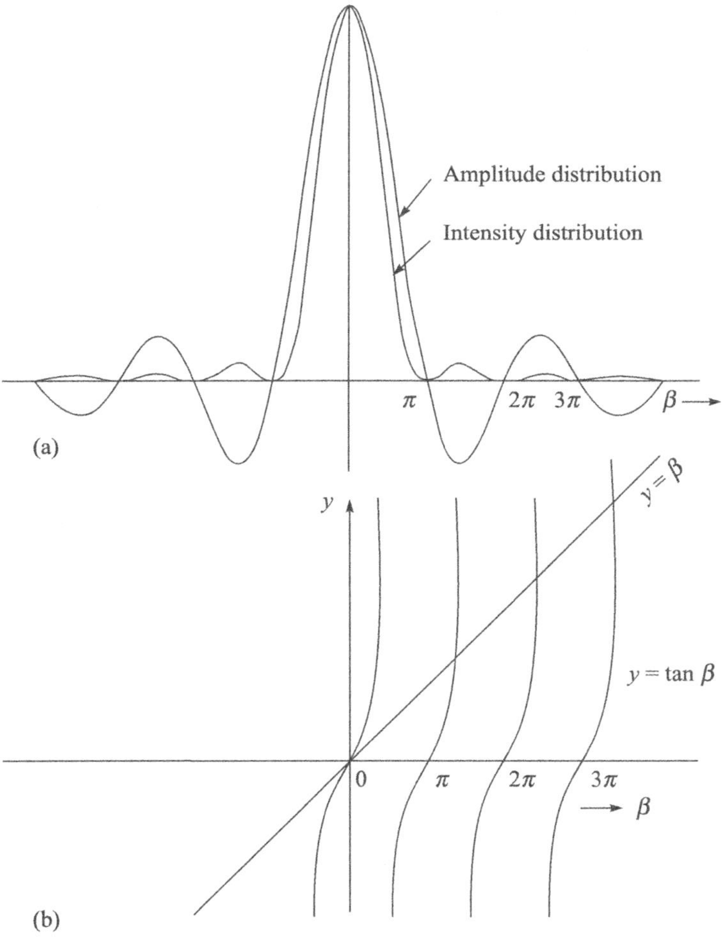

Fig. 9.2 (a) The intensity distribution corresponding to the single slit Fraunhofer diffraction pattern. (b) Graphical method for determining the roots of the equation tan $\beta = \beta$.

$$I = I_0\frac{\sin^2\beta}{\beta^2} \tag{3}$$

where, I_0 represents the intensity at $\theta = 0$. The intensity is zero when

$$\beta = m\pi \quad \Rightarrow \quad b\sin\theta = m\lambda;\ m = \pm 1, \pm 2, \pm 3, \ldots \tag{4}$$

The maxima correspond to the roots of the transcendental equation [see Fig. 9.2(b)]:

$$\tan \beta = \beta \tag{5}$$

which occur at

$$\beta \approx 1.43\pi,\ \beta \approx 2.46\pi, \ldots \tag{6}$$

9.2 FRAUNHOFER DIFFRACTION BY MULTIPLE SLITS

We next consider a plane wave (of wavelength λ) incident normally on N parallel slits, each of width b, and the distance between two consecutive slits is assumed to be d (see Fig. 9.3). The resultant field on the focal plane of the lens is given by

$$E = A\frac{\sin\beta}{\beta}[\cos(\omega t - \beta) + \cos(\omega t - \beta - \Phi_1) + \ldots + \cos\{\omega t - \beta - (N-1)\Phi_1\}]$$

$$= A\frac{\sin\beta}{\beta}\frac{\sin N\gamma}{\sin\gamma}\cos\left[\omega t - \beta - \frac{1}{2}(N-1)\Phi_1\right] \tag{7}$$

where, $\gamma = \dfrac{\Phi_1}{2} = \dfrac{\pi d \sin\theta}{\lambda}$ (8)

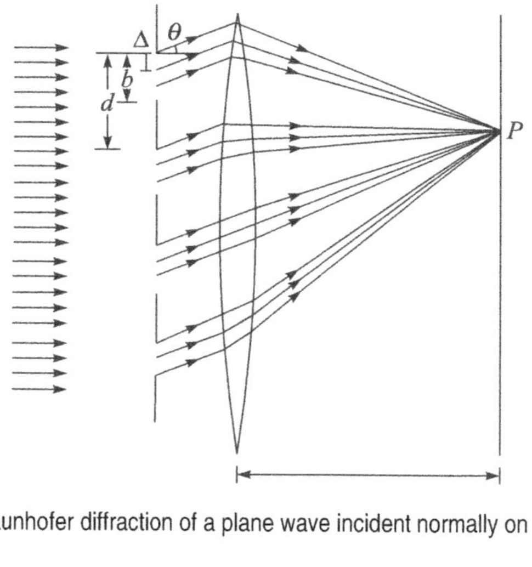

Fig. 9.3 Fraunhofer diffraction of a plane wave incident normally on a multiple slit.

The corresponding intensity distribution on the focal plane of the lens is given by

$$I = I_0\frac{\sin^2\beta}{\beta^2}\frac{\sin^2 N\gamma}{\sin^2\gamma} \tag{9}$$

Principal maxima occur when,

$$\gamma = m\pi \ \Rightarrow \ d\sin\theta = m\lambda;\ m = 0, \pm 1, \pm 2, \pm 3, \ldots \tag{10}$$

Between two principal maxima, the intensity vanishes when,

$$\gamma = \frac{p\pi}{N};\ p = \pm 1, \pm 2, \pm 3, \ldots \quad \text{but} \quad p \neq 0, \pm N, \pm 2N, \pm 3N, \ldots \tag{11}$$

These are referred to as secondary minima. In a diffraction grating, N is usually a very large number, as such there are many minima between two principal maxima.

In addition, there are few diffraction minima which Returning to Eq. (9), we see that for $N = 1$ we obtain the single slit diffraction pattern. For $N = 2$ we obtain

$$I = 4I_0 \frac{\sin^2\beta}{\beta^2} \cos^2\gamma \tag{12}$$

which is a product of the single slit diffraction pattern and the two point interference pattern.

9.3 RESOLVING POWER OF A GRATING

The resolving power of a grating is based on the Rayleigh criterion (see Fig. 9.4) and is given by

$$R = \frac{\lambda}{\Delta\lambda} = mN \tag{13}$$

where N represent the total number of lines in the grating and m represents the order of the spectrum.

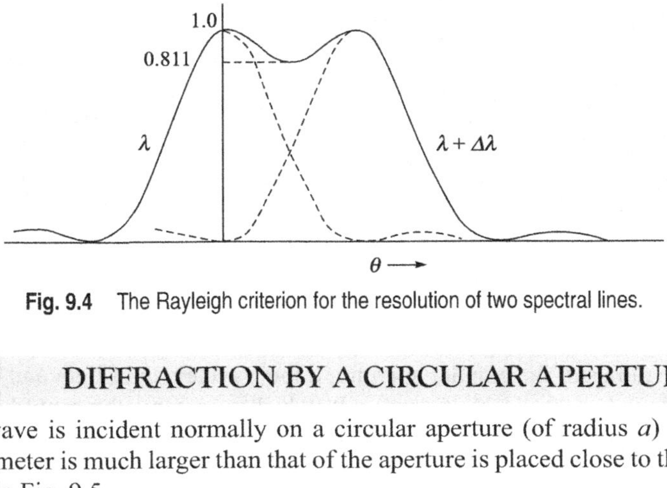

Fig. 9.4 The Rayleigh criterion for the resolution of two spectral lines.

9.4 DIFFRACTION BY A CIRCULAR APERTURE

A plane wave is incident normally on a circular aperture (of radius a) and a lens whose diameter is much larger than that of the aperture is placed close to the aperture as shown in Fig. 9.5.

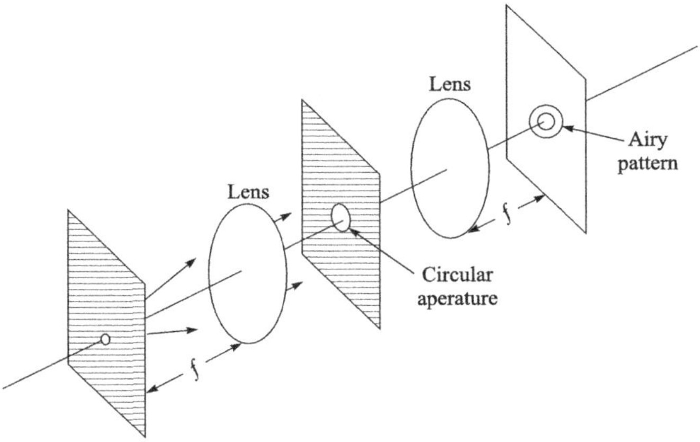

Fig. 9.5 Experimental arrangement for observing the Fraunhofer diffraction pattern by a circular aperture.

The Fraunhofer diffraction pattern is observed on the focal plane of the lens. Because of the rotational symmetry of the system, the diffraction pattern will consist of concentric dark and bright rings; this diffraction pattern (as observed on the back focal plane of the lens) is known as the Airy pattern (see Fig. 9.6) and the corresponding intensity distribution is given by (see Problem 9.9)

$$I = I_0\left[\frac{2J_1(v)}{v}\right]^2 \tag{14}$$

where,

$$v = \frac{2\pi}{\lambda}a\sin\theta \tag{15}$$

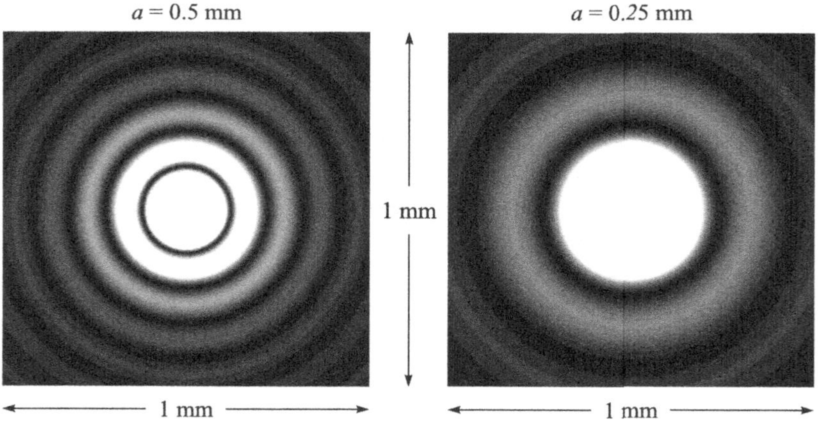

Fig. 9.6 Computer generated Airy patterns; (a) and (b) correspond to $a = 0.5$ mm and $a = 0.25$ mm respectively at the focal plane of a lens of focal length 20 cm ($\lambda = 0.5$ μm).

a being the radius of the circular aperture, λ the wavelength of light and θ the angle of diffraction; I_0 is the intensity at $\theta = 0$ (which represents the central maximum) and $J_1(v)$ is known as the Bessel function of the first order. On the focal plane of the convex lens

$$v \approx \frac{2\pi}{\lambda}a\frac{(x^2+y^2)^{1/2}}{f} \tag{16}$$

where f is the focal length of the lens. For those not familiar with Bessel functions, we may mention that the variation of $J_1(v)$ is somewhat like a damped sine curve [see Fig. 9.7] and although $J_1(0) = 0$, we have

$$\underset{v\to 0}{\text{Lt}}\ \frac{2J_1(v)}{v} = 1 \tag{17}$$

similar to the relation

$$\underset{x\to 0}{\text{Lt}}\ \frac{\sin x}{x} = 1 \tag{18}$$

Other zeros of $J_1(v)$ occur at

$$v = 3.832,\ 7.016,\ 10.174,\ldots \tag{19}$$

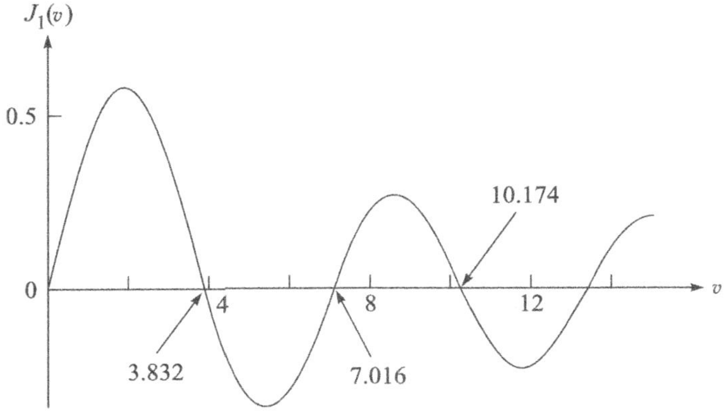

Fig. 9.7 The variation of $J_1(v)$ with v.

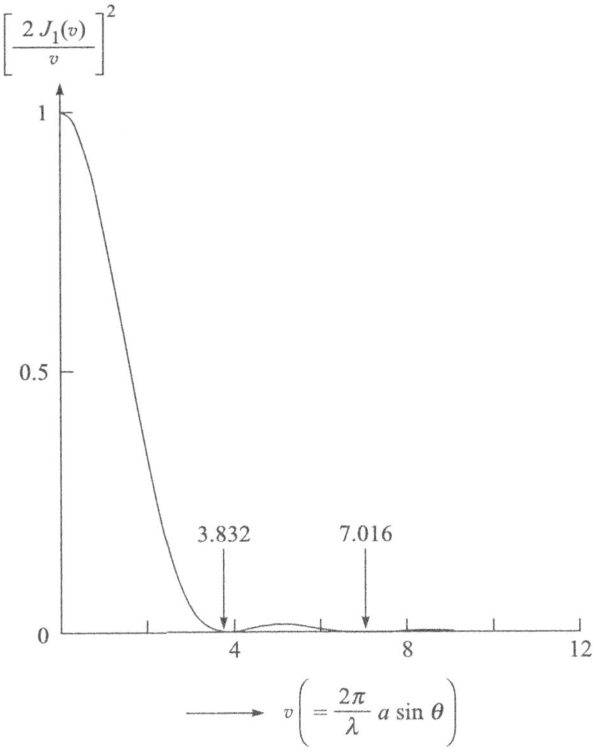

Fig. 9.8 The intensity variation associated with the Airy pattern.

In Fig. 9.8 we have plotted the function

$$\left[\frac{2J_1(v)}{v}\right]^2$$

which represents the intensity distribution corresponding to the Airy pattern. Thus, the successive dark rings in the Airy pattern [see Fig. 9.6] will correspond to

$$v = \frac{2\pi}{\lambda} a \sin\theta = 3.832,\ 7.016,\ 10.174, \ldots \tag{20}$$

or

$$\sin\theta = \frac{3.832\lambda}{2\pi a},\ \frac{7.016\lambda}{2\pi a}, \ldots \tag{21}$$

If f represents the focal length of the convex lens, then the

$$\text{Radii of the dark rings} = f\tan\theta \approx \frac{3.832\lambda f}{2\pi a},\ \frac{7.016\lambda f}{2\pi a}, \ldots \tag{22}$$

where we have assumed θ to be small so that $\tan\theta \approx \sin\theta$. In Figs. 9.6(a) and (b) we have shown the Airy patterns corresponding to the radius of the circular aperture being 0.5 mm and 0.25 mm respectively; both figures correspond to $\lambda = 5000$ Å and $f = 20$ cm. Thus,

$$\text{Radius of the first dark ring} \approx 0.12 \text{ mm and } 0.24 \text{ mm}$$

corresponding to $a = 0.5$ mm and 0.25 mm respectively.

9.5 LIMIT OF RESOLUTION

Consider two point sources, such as stars (so that we can consider plane waves entering the aperture) being focused by a telescope objective of diameter D. Each point source will produce its Airy pattern as schematically shown in Fig. 9.9. The diameters of the Airy rings will be determined by the diameter of the objective, its focal length and the wavelength of light. According to the Rayleigh criterion for the two objects to be resolved, the central spot of one pattern should fall on the first dark ring of the second, and this would happen when the angular separation of the two distant objects is given by

$$\Delta\theta = \frac{1.22\lambda}{D} \tag{23}$$

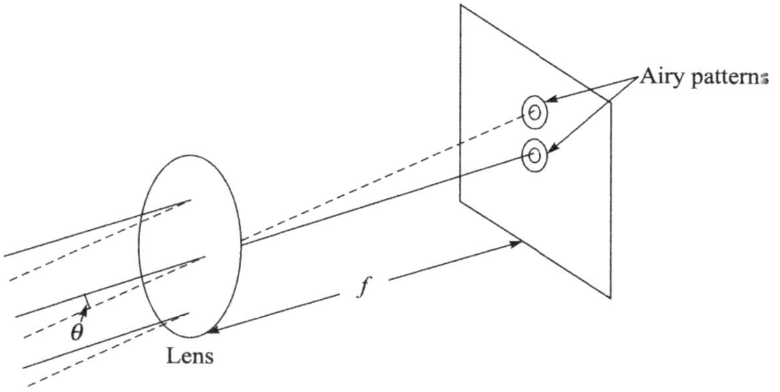

Fig. 9.9 The image of two distant objects on the focal plane of a convex lens. If the diffraction patterns are well separated, they are said to be resolved.

PROBLEMS

9.1 A plane wave (λ = 5000 Å) falls normally on a long narrow slit of width 0.5 mm. Calculate the angles of diffraction corresponding to the first three minima. Repeat the calculations corresponding to a slit width of 0.1 mm. Interpret physically the change in the diffraction pattern.

[**Ans.** 0.057°, 0.115°, 0.17°; 0.29°, 0.57°, 0.86°]

9.2 A convex lens of focal length 20 cm is placed after a slit of width 0.6 mm. If a plane wave of wavelength 6000 Å falls normally on the slit, calculate the separation between the second minima on either side of the central maximum.

[**Ans.** $\approx$0.08 cm]

9.3 In the above problem calculate the ratio of the intensity of the principal maximum to the first maximum on either side of the principal maximum.

[**Ans.** $\sim$21]

9.4 A circular aperture of radius 0.01 cm is placed in front of a convex lens of focal length of 25 cm and illuminated by a parallel beam of light of wavelength 5×10^{-5} cm. Calculate the radii of the first three dark rings.

[**Ans.** 0.76, 1.4, 2.02 mm]

9.5 Consider a plane wave incident on a convex lens of diameter 5 cm and of focal length 10 cm. If the wavelength of the incident light is 6000 Å, calculate the radius of the first dark ring on the focal plane of the lens. Repeat the calculations for a lens of same focal length but diameter 15 cm. Interpret the results physically. [**Ans.** 1.46×10^{-4} cm, 4.88×10^{-5} cm]

9.6 Consider a set of two slits each of width $b = 5 \times 10^{-2}$ cm and separated by a distance d = 0.1 cm, illuminated by a monochromatic light of wavelength 6.328×10^{-5} cm. If a convex lens of focal length 10 cm is placed beyond the double slit arrangement, calculate the positions of the minima inside the first diffraction minimum. [**Ans.** 0.0316 mm, 0.094 mm]

9.7 Show that when $b = d$, the resulting diffraction pattern corresponds to a slit of width $2b$.

9.8 Show that the first order and second order spectra will never overlap when the grating is used for studying a light beam containing wavelength components from 4000 Å to 7000 Å.

9.9 Consider a diffraction grating of width 5 cm with slits of width 0.0001 cm separated by a distance of 0.0002 cm. What is the corresponding grating element? How many orders would be observable at $\lambda = 5.5 \times 10^{-5}$ cm? Calculate the width of principal maximum. Would there be any missing orders?

9.10 For the diffraction grating of the above problem, calculate the dispersion in different orders. What will be the resolving power in each order?

9.11 A grating (with 15,000 lines per inch) is illuminated by white light. Assuming that white light consists of wavelengths lying between 4000 and 7000 Å, calculate the angular widths of first and the second order spectra.

9.12 A grating (with 15,000 lines per inch) is illuminated by sodium light. The grating spectrum is observed on the focal plane of a convex lens of focal length 10 cm. Calculate the separation between the D_1 and D_2 lines of sodium. (The wavelengths of D_1 and D_2 lines are 5890 and 5896 Å respectively.)

9.13 Calculate the resolving power in the second order spectrum of a 1 inch grating having 15,000 lines.

9.14 Consider a wire grating of width 1 cm having 1000 wires. Calculate the angular width of the second order principal maximum and compare the value with the one corresponding to a grating having 5000 lines in 1 cm. Assume $\lambda = 5.5 \times 10^{-5}$ cm.

9.15 In the minimum deviation position of a diffraction grating the first order spectrum corresponds to an angular deviation of 30°. If $\lambda = 6 \times 10^{-5}$ cm, calculate the grating element.

9.16 Calculate the diameter of a telescope lens if a resolution of 0.1 seconds of arc is required at $\lambda = 6 \times 10^{-5}$ cm.

9.17 Assuming that the resolving power of the eye is determined by diffraction effects only, calculate the maximum distance at which two objects separated by a distance of 2 m can be resolved by the eye. (Assume pupil diameter to be 2 mm and $\lambda = 6000$ Å.)

9.18 A pinhole camera is essentially a rectangular box with a tiny pinhole in front. An inverted image of the object is formed on the rear of the box. Consider a parallel beam of light incident normally on the pinhole. If we neglect diffraction effects then the diameter of the image will increase linearly with the diameter of the pinhole. On the other hand, if we assume Fraunhofer diffraction, then the diameter of the first dark ring will go on increasing as we reduce the diameter of the pinhole. Find the pinhole diameter for which the diameter of the geometrical image is approximately equal to the diameter of the first dark ring in the Airy pattern. Assume $\lambda = 6000$ Å and a separation of 15 cm between the pinhole and the rear of the box. [**Ans.** 0.47 mm]

9.19 Calculate the Fraunhofer diffraction pattern produced by a double slit arrangement with slits of widths b and $3b$, with their centers separated by a distance $6b$.

9.20 Plot the function

$$\frac{\sin^2 N\gamma}{\sin^2 \gamma}$$

for $N = 5$ and $N = 12$ and find the values of γ corresponding to the secondary minima.

9.21 Assuming ideal conditions, estimate the linear separation of two objects on the surface of the moon that can just be resolved by an observer on the earth using naked eye assuming a pupil diameter 5 mm. Assume a wavelength of 550 nm.

9.22 Show that for a diffraction grating with $d = 2b$, where b is the width of each slit and d is the spacing between the slits, all even order diffraction maxima will be absent in the diffraction pattern.

9.23 A grating with 200 lines (slits) per millimeter and of width 2 cm is fully illuminated by light consisting of wavelengths 600 nm and 600.1 nm.

(a) What is the lowest diffraction order where the two wavelengths will be resolved?

(b) If the slits are of width 3 μm in the grating, then calculate the ratio of intensities of the 600 nm wavelength in the second order to that in the first order.

9.24 For a diffraction grating illuminated normally by a plane wave of wavelength 535 nm, the highest order spectrum observed is equal to five. If there is a principal maximum along a diffraction angle of 35°, find the period of the grating?

9.25 Find the distance between the images of the two stars which are just resolved by a lens of focal length 3 m and diameter 10 cm. (Assume λ = 5500 Å). Assuming the diameter of the pupil of the human eye to be 5 mm, show whether the two images can be resolved by the eye or not when viewed from a distance if 25 cm from the focal plane of the lens.

9.26 Consider a telescope having an objective of diameter 5 cm and focal length 30 cm.

(a) What is the minimum angular resolution of the telescope? (Assume λ = 0.5 μm.b).

(b) Assuming an eye pupil diameter of 4 mm, calculate the focal length of the eye piece required to fully utilise the objective resolution.

(c) What is the corresponding angular magnification of the telescope?

9.27 A parallel laser beam with a diameter of 2 mm and a power of 10 W falls on a convex lens of diameter 25 mm and focal length 10 mm. If the wavelength of the laser beam is 500 nm, estimate the average intensity at the focused spot.

9.28 Parallel light from 2 incoherent light sources of equal intensity falls on a long narrow slit of width 'a' and the Fraunhofer diffraction pattern is observed on the back focal plane of a lens. If the two sources are just resolved, calculate the drop in intensity (with respect to the maximum) midway between the maxima.

SOLUTIONS

9.1 For minima $b \sin \theta_m = m\lambda \quad \Rightarrow \quad \theta_m = \sin^{-1}\left(\frac{m\lambda}{\theta}\right)$

Thus, for $b = 0.05$ cm

$$\theta_1 = \sin^{-1}\left(\frac{5 \times 10^{-5}}{0.05}\right) \approx 0.057°$$

$$\theta_2 = \sin^{-1}\left(\frac{10 \times 10^{-5}}{0.05}\right) \approx 0.115°$$

for $b = 0.01$ cm

$$\theta_1 = \sin^{-1}\left(\frac{5 \times 10^{-5}}{0.01}\right) \approx 0.29°$$

$$\theta_2 = \sin^{-1}\left(\frac{10 \times 10^{-5}}{0.01}\right) \approx 0.57°$$

$$\theta_3 = \sin^{-1}\left(\frac{15\times10^{-5}}{0.05}\right) \approx 0.17° \qquad \theta_3 = \sin^{-1}\left(\frac{15\times10^{-5}}{0.01}\right) \approx 0.86°.$$

9.2 Second minima occurs at $\sin\theta = \frac{2\lambda}{a} \Rightarrow \theta = 2\times10^{-3}$ rad.

Therefore, the angular separation between the two 'second' minima, lying on either side of central maxima = 4×10^{-3} rad. Thus, on the screen

The separation between the two minima = $4\times10^{-3}\times20 = 0.08$ cm.

9.3 $I = I_0\frac{\sin^2\beta}{\beta^2} \qquad \beta \equiv \frac{\pi}{\lambda}b\sin\theta$

First maximum occurs at $\beta = 1.43\,\pi$. If I_1 is the intensity at the first maximum then

$$\frac{I_0}{I_1} = \frac{(1.43\pi)^2}{\sin^2(1.43\pi)} \approx 21$$

9.4 Radii of the dark rings

$$r = f\tan\theta = \frac{3.832\,\lambda f}{2\pi a}, \frac{7.016\,\lambda f}{2\pi a}, \frac{10.174\,\lambda f}{2\pi a}$$

$$= 0.076 \text{ cm}, 0.14 \text{ cm}, 0.202 \text{ cm}$$

9.5 **Radius of the first dark ring** $= \frac{3.832\,\lambda f}{\pi D} = \frac{3.832\times6\times10^{-5}\times10}{5\pi}$

$= 1.46\times10^{-4}$ cm.

For $D = 15$ cm, **Radius of the first dark ring** $= \frac{3.832\times6\times10^{-5}\times10}{15\pi}$

$= 4.88\times10^{-5}$ cm.

In the second case the beam gets diffracted to a lesser extent because the lens offers a larger aperture for the same wavelength.

9.6 For $N = 2$, the interference term is given by

$$\frac{\sin^2 N\gamma}{\sin^2\gamma} = 4\cos^2\gamma$$

where, $\gamma = \frac{\pi d\sin\theta}{\lambda}$.

Interference minima will occur when

$$\sin\theta = \frac{\left(m+\frac{1}{2}\right)\lambda}{d} = \left(m+\frac{1}{2}\right)\frac{6.328\times10^{-5}}{0.1} = 6.328\times10^{-4}\left(m+\frac{1}{2}\right)$$

Thus $\sin\theta = 3.164\times10^{-4}$ and 9.492×10^{-4} for $m = 0$ and 1 respectively. Thus, $x = f\tan\theta \approx 0.0316$ mm and 0.094 mm

9.7 $I = 4I_0\frac{\sin^2\beta}{\beta^2}\cos^2\gamma = 4I_0\frac{\sin^2\beta}{\beta^2}\cos^2\beta = I_0\frac{\sin^2 2\beta}{\beta^2}\times\frac{4}{4} = 4I_0\frac{\sin^2 2\beta}{(2\beta)^2}$

9.8 The first and second order maxima are given by $d\sin\theta_{1,2} = \lambda$ and 2λ respectively. For the visible regions of λ

$$4\times 10^{-5} < d\sin\theta_1 < 7\times 10^{-5} \quad \text{and} \quad 8\times 10^{-5} < d\sin\theta_2 < 14\times 10^{-5}$$

Clearly θ_1 and θ_2 are disjoint for any value of d. The second and third order spectra will overlap.

9.9 The grating element is $d = 0.0002$ cm.

$$\text{Now} \qquad d\sin\theta_m = m\lambda \Rightarrow \frac{m\lambda}{d} = \sin\theta_m \le 1$$

$$\Rightarrow \qquad m \le \frac{d}{\lambda} = \frac{2\times 10^{-4}}{5.5\times 10^{-5}} = 3.6$$

Thus, we observe only three orders at $\lambda = 5.5\times 10^{-5}$ cm.

The number of lines in the grating $= \dfrac{5\text{ cm}}{2\times 10^{-4}\text{ cm}} = 25000$, thus, the width of principal maximum will be given by

$$\Delta\theta_m = \frac{\lambda}{Nd\cos\theta_m} = \frac{5.5\times 10^{-5}}{25000\times 2\times 10^{-4}} \qquad (\theta_m \approx 0)$$
$$= 1.1\times 10^{-5}\text{ rad}$$

Missing orders correspond to the following two equations being simultaneously satisfied $b\sin\theta_n = n\lambda$ and $d\sin\theta_m = m\lambda$ with $\theta_n = \theta_m$. Thus, $\dfrac{b}{d} = \dfrac{n}{m} \Rightarrow$ $m = 2n$, so every second principle maxima would be absent.

9.10 $d\sin\theta_m = m\lambda$; $d = 2\times 10^{-4}$ cm $\Rightarrow \sin\theta_m = \dfrac{\lambda}{d}, \dfrac{2\lambda}{d}, \dfrac{3\lambda}{d} = 0.275, 0.55, 0.825$

for $m = 1, 2$ and 3. Thus, $\theta_m = 15.96°, 33.37°, 55.59°$.

$$\text{Dispersion } \frac{\Delta\theta}{\Delta\lambda} = \frac{m}{d\cos\theta_m} \approx 5.2\times 10^{-5}, 1.2\times 10^{-4} \text{ and } 2.7\times 10^{-4} \text{ radians/Å}$$

for $m = 1, 2$ and 3 respectively.

$R = mN = 25000$, 50000 and 75000 for the first, second and third order respectively.

9.11 $d = \dfrac{2.54}{15000} = 1.69\times 10^{-4}$ cm $\Rightarrow \sin\theta_1 = \dfrac{\lambda}{d} = \dfrac{4\times 10^{-5}}{1.693\times 10^{-4}}$ to $\dfrac{7\times 10^{-5}}{1.693\times 10^{-4}}$

$= 0.236$ to 0.413

$$\Rightarrow \qquad \theta_1 = 13.7° \text{ to } 24.4° \Rightarrow \Delta\theta_1 = 10.7°$$

$$\sin\theta_2 = \frac{2\lambda}{d} = \frac{8\times 10^{-5}}{1.693\times 10^{-4}} \text{ to } \frac{14\times 10^{-5}}{1.693\times 10^{-4}} = 0.473 \text{ to } 0.827$$

$$\Rightarrow \qquad \theta_2 = 28.20° \text{ to } 55.79°$$

$$\Rightarrow \qquad \Delta\theta_2 = 27.6°$$

9.12 $d \sin \theta_{1,2} = \lambda_{1,2}$

$$d = \frac{2.54}{15000} \approx 1.693 \times 10^{-4} \text{ cm} \Rightarrow \sin \theta_1 = \frac{5.89 \times 10^{-5}}{1.693 \times 10^{-4}} \approx 0.3479$$

$$\Rightarrow \quad \theta_1 \approx 20.36°$$

Thus,

$$\Delta\theta \approx \frac{m}{d \cos \theta} \Delta\lambda \approx \frac{1}{1.693 \times 10^{-4} \times 0.937} \times 6 \times 10^{-8} \approx 3.78 \times 10^{-4} \text{ radians}$$

$$\Rightarrow \quad \text{separation} \approx f\Delta\theta \approx 3.78 \times 10^{-3} \text{ cm}$$

9.13 $R = 2N = 30000.$

9.14 $d = \dfrac{1 \text{ cm}}{1000} = 10^{-3} \text{ cm} \Rightarrow \sin \theta_2 = \dfrac{2\lambda}{d} = \dfrac{10^{-4}}{10^{-3}} = 0.1.$

Now, the width of principal maxima $\Delta\theta_2 = \dfrac{\lambda}{Nd \cos \theta_m}$

$$\Rightarrow \quad \Delta\theta_2 = \frac{\lambda}{Nd\sqrt{1 - \sin^2 \theta_m}} = \frac{5 \times 10^{-5}}{1000 \times 10^{-3}\sqrt{0.99}} = 5.02 \times 10^{-5} \text{ rad}$$

If $\quad N = 5000$ lines, $d = 2 \times 10^{-4}$ cm

$$\Rightarrow \quad \sin \theta_2 = \frac{2\lambda}{d} = \frac{10^{-4}}{2 \times 10^{-4}} = 0.5 \Rightarrow \cos \theta_2 \approx 0.866$$

$$\Rightarrow \quad \Delta\theta_2 = \frac{5 \times 10^{-5}}{0.866} = 5.77 \times 10^{-5} \text{ rad}$$

9.15 $2d \sin \dfrac{\delta}{2} = \lambda \Rightarrow d = \dfrac{6 \times 10^{-5}}{2 \times \sin 15°} \text{ cm} = 1.16 \times 10^{-4} \text{ cm}.$

9.16 $\dfrac{1.22\lambda}{D} = \dfrac{0.1}{3600} \times \dfrac{\pi}{180} = 4.85 \times 10^{-7} \Rightarrow D \approx 150 \text{ cm}.$

9.17 Let the distance be x.

$$\Rightarrow \quad \frac{1.22\lambda}{D} = \frac{2}{x}; D = 2 \text{ mm} = 2 \times 10^{-3} \text{ m}; \lambda = 6 \times 10^{-7} \text{ m}$$

$$\Rightarrow \quad x = \frac{2}{1.22 \times 6 \times 10^{-7}} \times 2 \times 10^{-3} \approx 5.5 \text{ km}$$

Thus, two objects (separated by 2 m) at a distance of about 5 km should be resolvable by the eye.

9.18 The first dark ring in the Airy pattern occurs at

$$\theta \approx \frac{1.22\lambda}{D}$$

The diameter of the first dark ring $\approx 2 \times 15 \times \theta \approx \dfrac{2 \times 15 \times 1.22 \times 6 \times 10^{-5}}{D}$

Thus the required condition is

$$D \approx \frac{2 \times 15 \times 1.22 \times 6 \times 10^{-5}}{D}$$

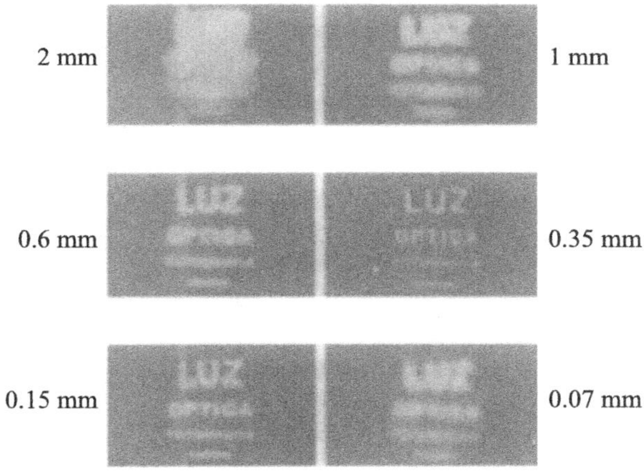

Fig. 9.10 The image formed in a pinhole camera for different diameters of the pinhole. [Ref: http://www.cs.berkeley.edu/~daf/book/chapter-4.pdf]

$$\Rightarrow \qquad D \approx 0.047 \text{ cm} = 0.47 \text{ mm} \quad \text{(see Fig. 9.10)}$$

9.19 $AC = 6b - \dfrac{3b}{2} + \dfrac{b}{2} = 5b$ (see Fig. 9.11)

The field at the angle θ is

$$E = a[\cos \omega t + \cos(\omega t - \phi) + \dots + \cos(\omega t - (n-1)\phi)] + a[\cos(\omega t - \Phi_1) + \cos(\omega t - \phi - \Phi_1) + \dots + \cos(\omega t - (3n-1)\phi - \Phi_1)]$$

$$= a\frac{\sin n\phi/2}{\sin \phi/2} \cos\left[\omega t - \frac{1}{2}(n-1)\phi\right] + 3a\frac{\sin 3n\phi/2}{3\sin \phi/2} \cos\left[\omega t - \frac{1}{2}(3n-1)\phi - \Phi_1\right]$$

$$= A\frac{\sin\beta}{\beta}\cos\left(\omega t - \beta + \frac{\phi}{2}\right) + 3A\frac{\sin 3\beta}{3\beta} \cos\left(\omega t - 3\beta - \Phi_1 + \frac{\phi}{2}\right)$$

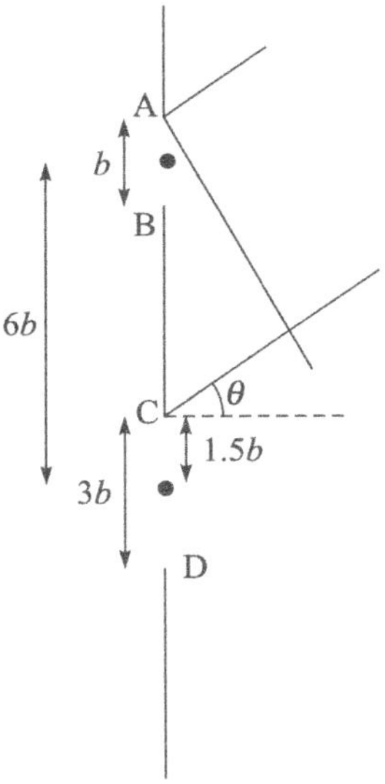

Fig. 9.11

where $\beta = \dfrac{1}{2}n\phi = \dfrac{\pi b \sin\theta}{\lambda}$; $\Phi_1 = \dfrac{2\pi}{\lambda} \cdot 5b\sin\theta$ $= 10\beta$

$$\text{Thus,} \quad E = \frac{A\sin\beta}{\beta}\left[\cos\left(\omega t - \beta + \frac{\phi}{2}\right) + \frac{\sin 3\beta}{\sin\beta}\cos\left(\omega t - 13\beta + \frac{\phi}{2}\right)\right]$$

$$= \frac{A\sin\beta}{\beta}\left[C_1\cos\left(\omega t + \frac{\phi}{2}\right) + C_2\sin\left(\omega t + \frac{\phi}{2}\right)\right]$$

where, $C_1 = \cos\beta + (3 - 4\sin^2\beta)\cos 13\beta$

and $C_2 = \sin\beta + (3 - 4\sin^2\beta)\sin 13\beta$

$$I = I_0 \frac{\sin^2\beta}{\beta^2}(C_1^2 + C_2^2)$$

9.20 See Fig. 9.12

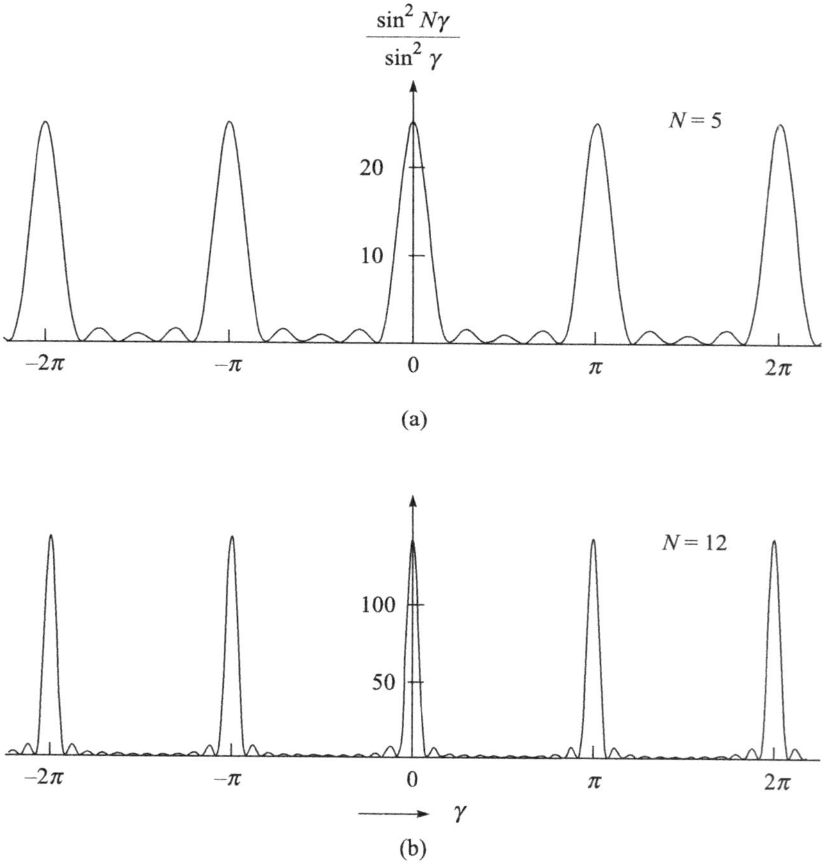

Fig. 9.12 The variation of the function $\sin^2(N\gamma)/\sin^2\gamma$ with γ for $N = 5$ and 12. As N becomes larger, the function would become more and more sharply peaked at $\gamma = 0, \pm\pi, \pm 2\pi, \pm 3\pi, \ldots$.

9.21 The angular separation that can be resolved is given by

$$\theta = \frac{1.22\lambda}{d} \approx \frac{0.67\times 10^{-6}}{5\times 10^{-3}} = 0.13\times 10^{-3}$$

Assuming the distance of moon from the earth to be given by 3.84×10^5 km, the linear separation that can be resolved comes out to be $3.84\times 10^5 \times 0.13\times 10^{-3} \approx 50$ km.

9.22 The intensity pattern in a diffraction grating is given by Eq. (9). In this case since $d = 2b$, we have $\beta = \gamma/2$. Even order grating spectra correspond to $\gamma = 2m\pi (m = 1, 2, 3\ldots)$. For these angles we would have $\beta = m\pi (m = 1, 2, 3\ldots)$. For these values of b, the diffraction term gives zero and hence all even orders will be absent. Note that along these directions the amplitudes of all the slits add constructively but there is no diffracted amplitude along these directions as they coincide with the diffraction minima of each slit.

9.23 (a) The number of illuminated slits equals $200 \times 20 = 4000$. Using the formula for the resolving power of the grating we find that $m > 1.5$. This means that the two lines will be resolved in all orders other than the first order.

(b) The ratio of the intensities in the second order to first order will be given by

$$\frac{I_2}{I_1} = \left(\frac{\sin\left\{\dfrac{\pi b \sin\theta_2}{\lambda}\right\}}{\left\{\dfrac{\pi b \sin\theta_2}{\lambda}\right\}} \right)^2 \times \left(\frac{\left\{\dfrac{\pi b \sin\theta_1}{\lambda}\right\}}{\sin\left\{\dfrac{\pi b \sin\theta_1}{\lambda}\right\}} \right)^2$$

where θ_1 and θ_2 are the angles at which the first order and the second order spectra appear. Note that the interference term got cancelled in taking the ratio. Now the angles at which the two orders appear satisfy the following equations:

$$d \sin\theta_1 = \lambda$$
$$d \sin\theta_2 = 2\lambda$$

Also from the values given in the problem, $d = 5$ μm and $b = 3$ μm. Using these values in the equation for the ratio we obtain

$$\frac{I_2}{I_1} \approx 0.096$$

9.24 It is given that the maximum order seen is 6. Hence, the value of d must lie between 5λ and 6λ which implies that 2.675 μm $< d <$ 3.21 μm. Now since there is an order appearing at 35°, we must have

$$d = m\frac{0.535}{\sin 35^\circ} \approx 0.93m$$

The value of m satisfying the condition on d is 3. Hence this must be the third order and in such a case the value of d will be 2.799 μm.

9.25 The separation between the two spots will be

$$l = \frac{1.22\lambda}{d} f \approx 20.13 \text{ μm}$$

The angle subtended by the two spots on the eye placed at a distance of 25 cm will be 0.805×10^{-4} rad. The resolving power of the eye for an opening of 5 mm is 1.34×10^{-4} rad. Hence the eye will not be able to resolve the two images although they are resolved by the telescope.

9.26 (a) The angular resolution of the telescope will be

$$\theta_{\text{tel}} = \frac{1.22\lambda}{d} = \frac{1.22 \times 0.5 \times 10^{-6}}{5 \times 10^{-2}} \approx 1.22 \times 10^{-5} \text{ rad}$$

(b) The angular resolution of the eye will be

$$\theta_{\text{eye}} = \frac{1.22\lambda}{d} = \frac{1.22 \times 0.5 \times 10^{-6}}{4 \times 10^{-3}} \approx 1.525 \times 10^{-4} \text{ rad}$$

(c) The required angular magnification is 1.525/0.122 ~ 12.5.

9.27 The radius of the focused spot would be $a \sim 3.05$ μm. Thus, the intensity of the focused spot would be $P/\pi a^2 \sim 3.4 \times 10^{11}$ W/m^2.

9.28 The diffraction pattern of a single slit with a single source is given by Eq. (3). The two sources would be just resolved when the maximum of one falls on the first zero of the other pattern. Since the first zero appears at $\sin\theta = \lambda/b$, the intensity midway would be

$$I_{\text{mid}} = 2\left(\frac{\sin\left[\frac{\pi b}{\lambda}\frac{\lambda}{2b}\right]}{\left[\frac{\pi b}{\lambda}\frac{\lambda}{2b}\right]}\right)^2 = \frac{8}{\pi^2} \approx 0.81$$

Fraunhofer Diffraction II: The Diffraction Integral

10

A Quick Review

For an electromagnetic wave propagating in the $+z$ direction, the transverse components of the electric field (E_x or E_y) satisfy the scalar wave equation

$$\nabla^2 \psi = \varepsilon \mu_0 \frac{\partial^2 \psi}{\partial t^2} \tag{1}$$

If we assume the time dependence of the form $e^{-i\omega t}$ and write

$$\psi = U(x, y, z)\, e^{-i\omega t} \tag{2}$$

we would obtain

$$\nabla^2 U + k^2 U = 0 \tag{3}$$

where,
$$k = \omega\sqrt{\varepsilon \mu_0} = \frac{\omega}{v} \tag{4}$$

and U represents one of the Cartesian components of the electric field. The solution of Eq. (3) can be written as

$$U(x, y, z) = \int_{-\infty}^{+\infty}\int_{-\infty}^{+\infty} F(k_x, k_y)\, e^{i(k_x x + k_y y + k_z z)}\, dk_x dk_y \tag{5}$$

where,
$$k_z = \pm\sqrt{k^2 - k_x^2 - k_y^2} \tag{6}$$

For waves making small angles with the z axis we may write

$$k_z = \sqrt{k^2 - k_x^2 - k_y^2} \approx k\left[1 - \frac{k_x^2 + k_y^2}{2k^2}\right] \tag{7}$$

If we use the above approximation, we obtain (see Problem 10.1)

$$U(x, y, z) = \frac{1}{i\lambda z} e^{ikz} \iint U(\xi, \eta, 0) \exp\left[\frac{ik}{2z}\left\{(x-\xi)^2 + (y-\eta)^2\right\}\right] d\xi d\eta \tag{8}$$

where the integral is over the area of the aperture on the plane $z = 0$ (see Fig. 10.1). The above equation can be rewritten in the form

$$U(x, y, z) \approx \frac{1}{i\lambda z} e^{ikz} \exp\left\{\frac{ik}{2z}(x^2 + y^2)\right\} \iint U(\xi, \eta, 0)$$
$$\exp\left\{\frac{ik}{2z}(\xi^2 + \eta^2)\right\} e^{-i(u\xi + v\eta)}\, d\xi d\eta \tag{9}$$

where,
$$u = \frac{2\pi x}{\lambda z} \quad \text{and} \quad v = \frac{2\pi y}{\lambda z} \tag{10}$$

are known as spatial frequencies. Both Eqs. (8) and (9) are usually referred to as the *Fresnel diffraction integral.* In the next chapter we will use the above integrals to calculate the Fresnel diffraction pattern. Now, if we assume z to be so large that the function

$$\exp\left\{\frac{ik}{2z}(\xi^2 + \eta^2)\right\}$$

[inside the integral in Eq. (9)] can be replaced by unity, then we would obtain

$$U(x, y, z) \approx \frac{1}{i\lambda z} e^{ikz} \exp\left\{\frac{ik}{2z}(x^2 + y^2)\right\} \iint U(\xi, \eta, 0)\, e^{-i(u\xi + v\eta)}\, d\xi d\eta \tag{11}$$

which represents the Fraunhofer diffraction pattern. The integral on the right hand side is the two dimensional Fourier transform of the function $U(\xi, \eta, 0)$. Thus Eq. (11) gives the very important result that the *Fraunhofer diffraction pattern is the Fourier transform of the aperture function.*

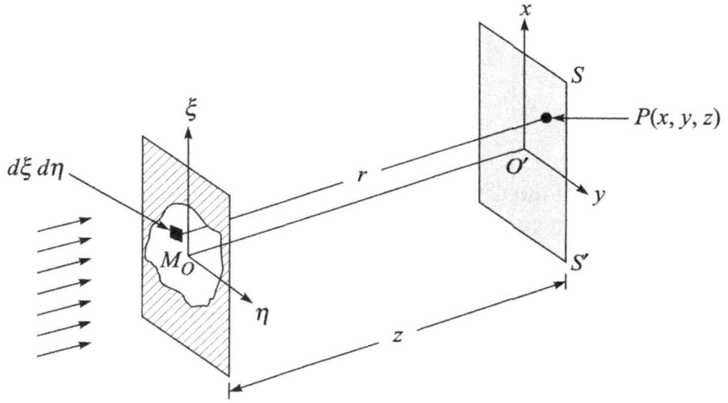

Fig. 10.1 A plane wave incident normally on an aperture. The diffraction pattern is observed on the screen SS′.

PROBLEMS

10.1 Start with Eq. (5), use Eq. (7) and derive Eq. (8).

10.2 Consider a plane wave incident normally on a long narrow slit of width b (along the ξ-axis) placed on the aperture plane (see Fig. 10.2). For such a case, we will have

$$\left.\begin{aligned} U(\xi, \eta, 0) &= A \quad |\xi| < \frac{b}{2} \\ &= 0 \quad |\xi| > \frac{b}{2} \end{aligned}\right\} \tag{12}$$

for all values of η. Calculate the corresponding Fraunhofer diffraction pattern.

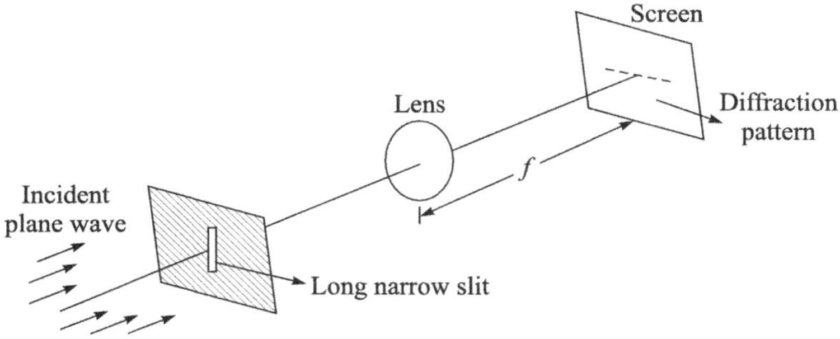

Fig. 10.2 Diffraction of a plane wave incident normally on a long narrow slit of width *b*. Notice that the spreading occurs along the width of the slit.

10.3 Consider a plane wave incident normally on two long narrow slits each of width b (along the ξ-axis) separated by distance d (see Fig. 10.3). For such a case, we will have

$$\left.\begin{aligned} U(\xi, \eta, 0) &= A \quad -\frac{d}{2} - \frac{b}{2} < \xi < -\frac{d}{2} + \frac{b}{2} \text{ and for } \frac{d}{2} - \frac{b}{2} < \xi < \frac{d}{2} + \frac{b}{2} \\ &= 0 \quad \text{elsewhere} \end{aligned}\right\} \tag{13}$$

for all values of η. Calculate the corresponding Fraunhofer diffraction pattern.

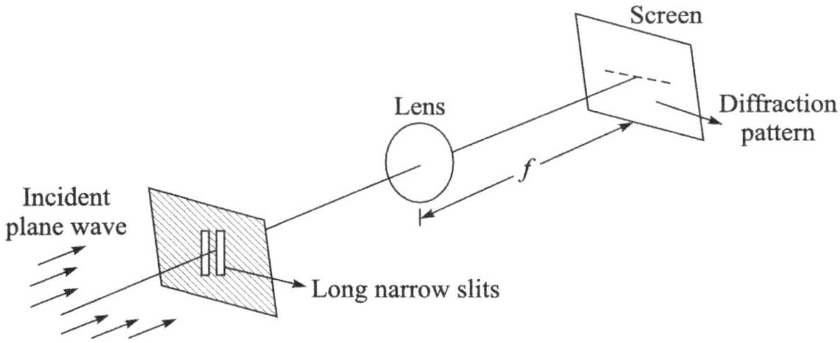

Fig. 10.3 Diffraction of a plane wave incident normally on two long narrow slits of width *b*. Notice that the spreading occurs along the width of the slit.

10.4 (a) In the previous problem, choose the origin at the bottom of the first slit so that

$$\left.\begin{aligned} U(\xi, \eta, 0) &= A \quad 0 < \xi < b \text{ and for } d < \xi < d + b \\ &= 0 \quad \text{elsewhere} \end{aligned}\right\} \tag{14}$$

for all values of η. Show that there will only be change of phase and the corresponding intensity distribution will remain the same.

(b) Assume $d = b$ and physically interpret the final result.

10.5 Extend the analysis of the previous problem to N equidistant slits each of width b so that

$$U(\xi, \eta, 0) = A \quad 0 < \xi < b; d < \xi < d + b; \ldots; (N-1)d < \xi < (N-1)d + b \qquad = 0 \quad \text{elsewhere} \tag{15}$$

for all values of η. Show that there will only be change of phase and the corresponding intensity distribution will remain the same.

10.6 Consider a plane wave incident normally on a rectangular aperture of width b (along the ξ-axis) and width a (along the η-axis) placed on the aperture plane. For such a case, we will have

$$\begin{aligned} U(\xi, \eta, 0) &= A \quad |\xi| < \frac{b}{2} \text{ and } |\eta| < \frac{a}{2} \\ &= 0 \quad \text{everywhere else} \end{aligned} \tag{16}$$

for all values of η. Calculate the corresponding Fraunhofer diffraction pattern.

10.7 Assume a plane wave with $\lambda = 5 \times 10^{-5}$ cm incident normally on a rectangular aperture of dimensions 0.2 mm × 0.3 mm. A convex lens (of focal length 20 cm) is placed immediately after the aperture. The screen is at the focal plane of the lens. Calculate the positions of the first three maxima and minima on the x-axis (implying $\phi = 0$) and also on the y-axis (implying $\theta = 0$).

10.8 In continuation of Problem 10.4, we consider two slits of unequal widths so that

$$\begin{aligned} U(\xi, \eta, 0) &= A \quad 0 < \xi < b_1 \text{ and for } d < \xi < d + b_2 \\ &= 0 \quad \text{elsewhere} \end{aligned} \tag{17}$$

for all values of η. Obtain an expression for the Fraunhofer diffraction pattern.

(b) Assume $d = b_1$ and show that the final result will correspond to a single slit of width $b_1 + b_2$.

10.9 Consider a plane wave incident normally on a circular aperture of radius a (see Fig. 10.4). Calculate the corresponding Fraunhofer diffraction.

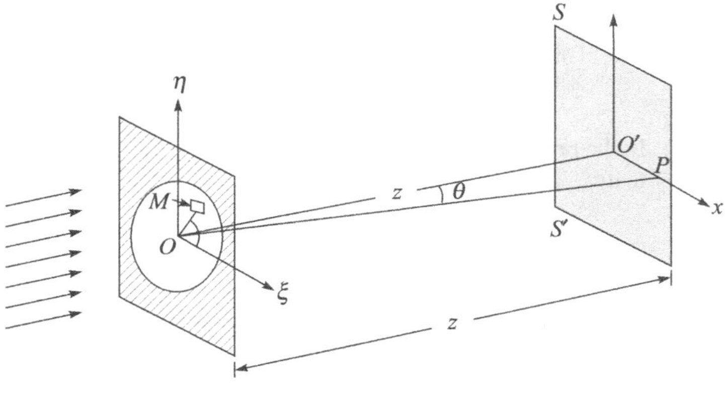

Fig. 10.4(a) Diffraction of a plane wave incident on a circular aperture of radius a.

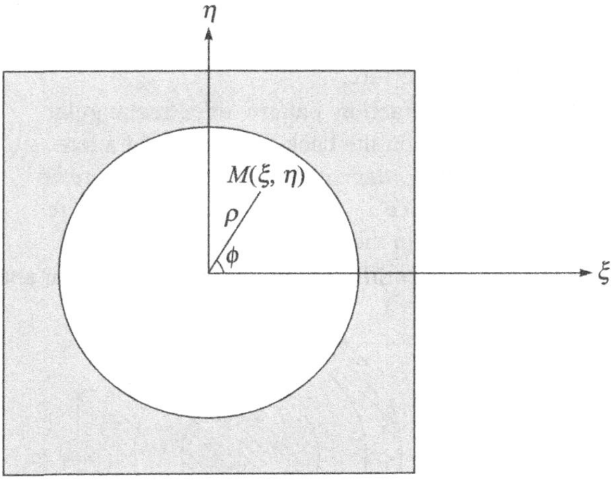

Fig. 10.4(b) Cylindrical coordinates (ρ, ϕ) on the plane of the circular aperture.

10.10 The Fraunhofer diffraction pattern of a circular aperture (of radius 0.5 mm) is observed on the focal plane of a convex lens of focal length 20 cm. Calculate the radii of the first and the second dark rings. Assume $\lambda = 5.5 \times 10^{-5}$ cm.

10.11 A long narrow single slit of width $a = 50$ μm is covered by a thin slide of glass of refractive index 1.5 and thickness 0.5 μm so that it covers half the slit as shown in the figure.

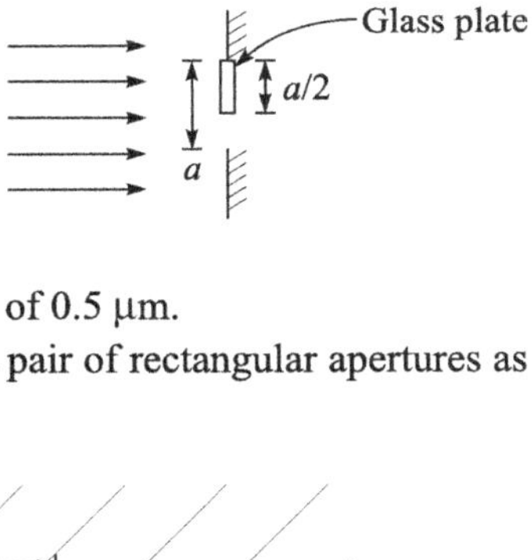

Calculate the corresponding Fraunhofer diffraction pattern for an incident wavelength of 0.5 μm.

10.12 Obtain the Fraunhofer diffraction pattern of a pair of rectangular apertures as shown below.

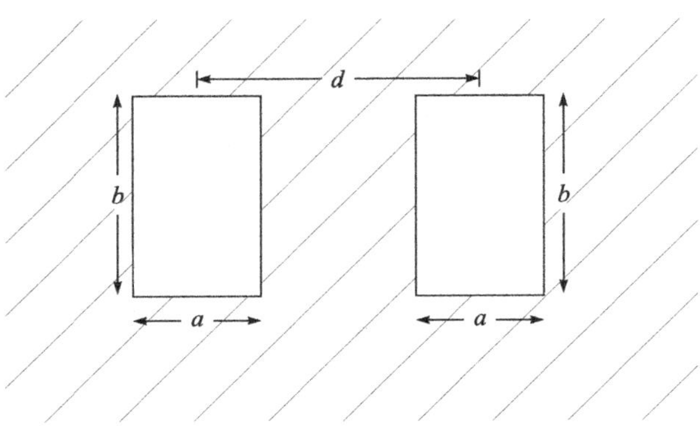

Fig. 10.5

10.13 Consider a pair of circular apertures of radii a as shown in the figure. Obtain the Fraunhofer diffraction pattern of such an aperture.

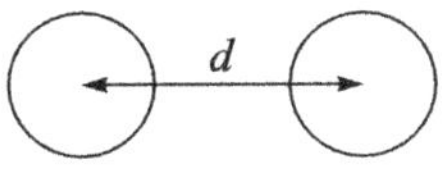

Fig. 10.6

10.14 Consider four identical circular apertures of radii a as shown in the figure. Obtain the Fraunhofer diffraction pattern of such an aperture.

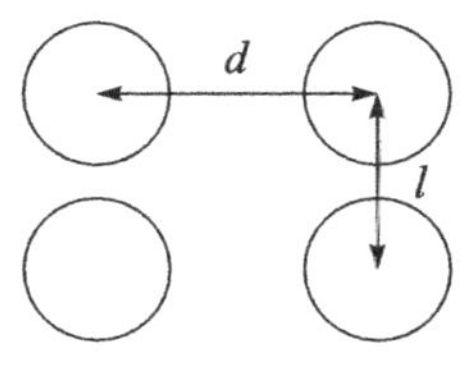

Fig. 10.7

10.15 The Fraunhofer diffraction pattern of a rectangular aperture is observed on the back focal plane of a lens. Show, *without actual integration*, that if the aperture is displaced by a distance x_0 along the x-direction, there would be no change in the observed intensity pattern.

10.16 Obtain the Fraunhofer diffraction pattern of an annular aperture shown below.

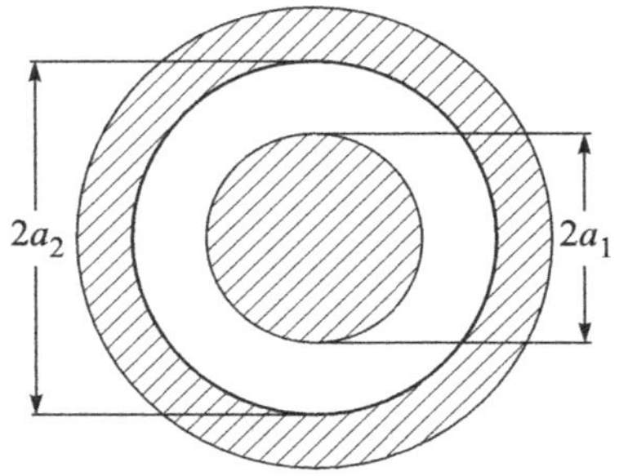

Fig. 10.8

10.17 A long narrow slit of width b is illuminated obliquely by a plane parallel beam at an angle θ_0 with the axis as shown in the figure. Obtain the corresponding Fraunhofer diffraction pattern and the positions of maxima and minimum.

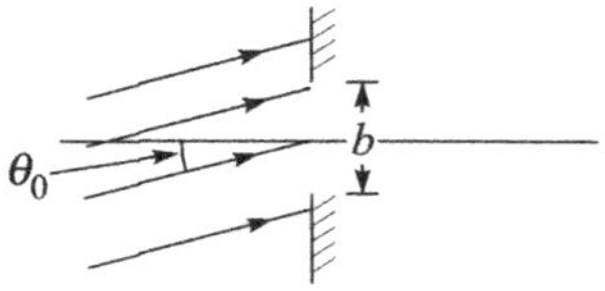

Fig. 10.9

SOLUTIONS

10.1 For waves making small angles with the z axis we may write

$$k_z = \sqrt{k^2 - k_x^2 - k_y^2} \approx k\left[1 - \frac{k_x^2 + k_y^2}{2k^2}\right] \tag{18}$$

Substituting in Eq. (5) we obtain

$$U(x, y, z) = e^{ikz} \iint F(k_x, k_y) \exp\left[i\left(k_x x + k_y y - \frac{k_x^2 + k_y^2}{2k} z\right)\right] dk_x dk_y \tag{19}$$

Thus, the field distribution on the plane $z = 0$ will be given by

$$U(x, y, z = 0) = \iint F(k_x, k_y)\, e^{i(k_x x + k_y y)}\, dk_x dk_y \tag{20}$$

Thus, $U(x, y, z = 0)$ is the Fourier transform of $F(k_x, k_y)$. The inverse transform will give us

$$F(k_x, k_y) = \frac{1}{(2\pi)^2} \iint U(\xi, \eta, 0)\, e^{-i(k_x \xi + k_y \eta)}\, d\xi d\eta \tag{21}$$

Substituting the above expression for $F(k_x, k_y)$ in Eq. (19), we get

$$U(x, y, z) = \frac{e^{ikz}}{4\pi^2} \iint U(\xi, \eta, 0)\, I_1 I_2\, d\xi d\eta \tag{22}$$

where,

$$I_1 = \int_{-\infty}^{+\infty} \exp[ik_x (x - \xi)] \exp\left[-\frac{ik_x^2}{2k} z\right] dk_x$$

$$= \sqrt{\frac{4\pi^2}{i\lambda z}} \exp\left[\frac{ik(x-\xi)^2}{2z}\right] \tag{23}$$

and we have used the following integral

$$\int_{-\infty}^{+\infty} e^{-\alpha x^2 + \beta x} dx = \sqrt{\frac{\pi}{\alpha}} \exp\left[\frac{\beta^2}{4\alpha}\right] \tag{24}$$

Similarly,

$$I_2 = \int_{-\infty}^{+\infty} \exp[ik_y (y - \eta)] \exp\left[-\frac{ik_y^2}{2k} z\right] dk_y$$

$$= \sqrt{\frac{4\pi^2}{i\lambda z}} \exp\left[\frac{ik(y-\eta)^2}{2z}\right] \tag{25}$$

Thus,

$$U(x, y, z) = \frac{1}{i\lambda z} e^{ikz} \iint U(\xi, \eta, 0)$$

$$\exp\left[\frac{ik}{2z}\{(x-\xi)^2 + (y-\eta)^2\}\right] d\xi d\eta \tag{26}$$

10.2 If we use Eq. (12) in Eq. (11), we obtain

$$U(x, y, z) = \frac{A}{i\lambda z} e^{ikz} \exp\left\{\frac{ik}{2z}(x^2 + y^2)\right\} \int_{-b/2}^{+b/2} e^{-iu\xi} d\xi \int_{-\infty}^{+\infty} e^{-iv\eta} d\eta \tag{27}$$

Now,

$$\delta(v) = \frac{1}{2\pi} \int_{-\infty}^{+\infty} e^{-iv\eta} d\eta \tag{28}$$

and

$$\int_{-b/2}^{+b/2} e^{-iu\xi} d\xi = \frac{1}{-iu} e^{-iu\xi}\Big|_{-b/2}^{+b/2} = \frac{2}{u} \frac{e^{iub/2} - e^{-iub/2}}{2i} = b\frac{\sin\beta}{\beta} \tag{29}$$

where,

$$\beta = \frac{ub}{2} = \frac{\pi b x}{\lambda z} \approx \frac{\pi b \sin\theta}{\lambda} \tag{30}$$

and $\sin\theta \approx \frac{x}{z}$; θ representing the angle of diffraction along the x-direction. Thus

$$U(x, y, z) = \frac{Ab}{i\lambda z} e^{ikz} \exp\left\{\frac{ik}{2z}(x^2 + y^2)\right\}\left(\frac{\sin\beta}{\beta}\right) 2\pi\delta(v) \tag{31}$$

Because of the δ function, the intensity is zero except on the x-axis (see Fig. 10.2); thus the intensity distribution along the x-axis will be

$$I = I_0 \frac{\sin^2\beta}{\beta^2} \tag{32}$$

We thus obtain the single slit diffraction pattern.

10.3 If we use Eq. (13) in Eq. (11) we obtain

$$U(x, y, z) = \frac{A}{i\lambda z} e^{ikz} \exp\left\{\frac{ik}{2z}(x^2 + y^2)\right\} G \int_{-\infty}^{+\infty} e^{-iv\eta}\, d\eta \tag{33}$$

where,

$$G = \int_{-\frac{d}{2}-\frac{b}{2}}^{-\frac{d}{2}+\frac{b}{2}} e^{-iu\xi}\, d\xi + \int_{\frac{d}{2}-\frac{b}{2}}^{\frac{d}{2}+\frac{b}{2}} e^{-iu\xi}\, d\xi \tag{34}$$

Carrying out the straight forward integrations, we get

$$G = b\frac{\sin\beta}{\beta}(e^{i\gamma} + e^{-i\gamma}) = \left(b\frac{\sin\beta}{\beta}\right)(2\cos\gamma) \tag{35}$$

where,

$$\gamma = \frac{ud}{2} = \frac{\pi dx}{\lambda z} \approx \frac{\pi d\sin\theta}{\lambda} \tag{36}$$

Once again, the intensity is zero except on the x-axis and the intensity distribution along the x-axis will be

$$I = I_0 \frac{\sin^2\beta}{\beta^2}(4\cos^2\gamma) \tag{37}$$

The term $\left(\frac{\sin^2\beta}{\beta^2}\right)$ represents the diffraction pattern produced by a single slit and the term $\cos^2\gamma$ represents the interference pattern produced by 2 point sources separated by distance d.

10.4 (a) Instead of Eq. (34), we will have

$$G = \int_0^b e^{-iu\xi}\, d\xi + \int_d^{d+b} e^{-iu\xi}\, d\xi \tag{38}$$

Carrying out the integrations we get

$$G = \left(e^{-i\beta}\, b\frac{\sin\beta}{\beta}\right)(1 + e^{-iud/2})$$

or,

$$G = \left(e^{-i\beta}\, b\frac{\sin\beta}{\beta}\right)(e^{-i\gamma} 2\cos\gamma) \tag{39}$$

where, β and γ have defined through Eqs. (30) and (36). Thus, the expression for G is the same as given by Eq. (35) except for (unimportant) phase factors. The intensity distribution will therefore be the same as given by Eq. (37).

(b) When $d = b$, we will have $\gamma = \beta = \frac{\pi b\sin\theta}{\lambda}$ and

$$G = \left(e^{-i\beta}\, b\frac{\sin\beta}{\beta}\right)(e^{-i\gamma} 2\cos\gamma) = e^{-i2\beta}\left(2b\frac{\sin 2\beta}{2\beta}\right)$$

The corresponding intensity distribution will be given by

$$I = I_0 \frac{\sin^2 2\beta}{(2\beta)^2}$$

And we will have single slit diffraction pattern corresponding to a single slit of width $2b$.

10.5 Instead of Eq. (38), we will now have

$$G = \int_0^b e^{-iu\xi}\, d\xi + \int_d^{d+b} e^{-iu\xi}\, d\xi + \ldots + \int_{(N-1)d}^{(N-1)d+b} e^{-iu\xi}\, d\xi \tag{40}$$

Carrying out the integrations, we get

$$G = \left(e^{-i\beta}\, b \frac{\sin\beta}{\beta} \right) (1 + e^{-iud/2} + \ldots + e^{-i(N-1)ud/2})$$

$$= \left(e^{-i\beta}\, b \frac{\sin\beta}{\beta} \right) \frac{e^{-iNud/2}}{e^{-iud/2}} \frac{\sin\left(\frac{Nud}{2}\right)}{\sin\left(\frac{ud}{2}\right)}$$

$$= \left(e^{-i\beta}\, b \frac{\sin\beta}{\beta} \right) \left(e^{-i(N-1)\gamma} \frac{\sin N\gamma}{\sin\gamma} \right)$$

Thus, the corresponding intensity distribution will be

$$I = I_0 \frac{\sin^2\beta}{\beta^2} \frac{\sin^2 N\gamma}{\sin^2\gamma} \tag{41}$$

where the first term $\left(\frac{\sin^2\beta}{\beta^2}\right)$ represents the diffraction pattern produced by a single slit and the second term $\left(\frac{\sin^2 N\gamma}{\sin^2\gamma}\right)$ represents the interference pattern produced by N equally spaced point sources.

(b) When $d = b$, we will have $\gamma = \beta = \frac{\pi b \sin\theta}{\lambda}$ and

$$I = I_0 \frac{\sin^2\beta}{\beta^2} \frac{\sin^2 N\beta}{\sin^2\beta} = I_0 N^2 \frac{\sin^2 N\beta}{(N\beta)^2}$$

and we will have single slit diffraction pattern corresponding to a single slit of width Nb.

10.6 The Fraunhofer diffraction of a plane wave incident normally on a rectangular aperture will be given by

$$U(x, y, z) = \frac{A}{i\lambda z} e^{ikz} \exp\left\{\frac{ik}{2z}(x^2 + y^2)\right\} \int_{-b/2}^{+b/2} e^{-iu\xi}\, d\xi \int_{-a/2}^{+a/2} e^{-iv\eta}\, d\eta \tag{42}$$

where we have chosen the origin to be at the center of the rectangular aperture. Carrying out the integration as in the previous section we obtain

$$U(x, y, z) = \frac{Aba}{i\lambda z} e^{ikz} \exp\left\{\frac{ik}{2z}(x^2 + y^2)\right\}\left(\frac{\sin\beta}{\beta}\right)\left(\frac{\sin\gamma}{\gamma}\right) \tag{43}$$

where β is given by Eq. (30),

$$\gamma = \frac{va}{2} = \frac{\pi a y}{\lambda z} \approx \frac{\pi a \sin\phi}{\lambda} \tag{44}$$

and $\sin\phi \approx \frac{y}{z}$; ϕ representing the angle of diffraction along the y-direction. Thus, we may write for the intensity distribution

$$I(P) = I_0 \frac{\sin^2\gamma}{\gamma^2} \frac{\sin^2\beta}{\beta^2} \tag{45}$$

The above equation represents the Fraunhofer diffraction pattern by a rectangular aperture. The intensity distribution due to a square aperture ($a = b$) is shown in Fig. 10.10; the figure corresponds to $a = b = 0.01$ cm, $z = 100$ cm, and we have assumed $\lambda = 5 \times 10^{-5}$ cm.

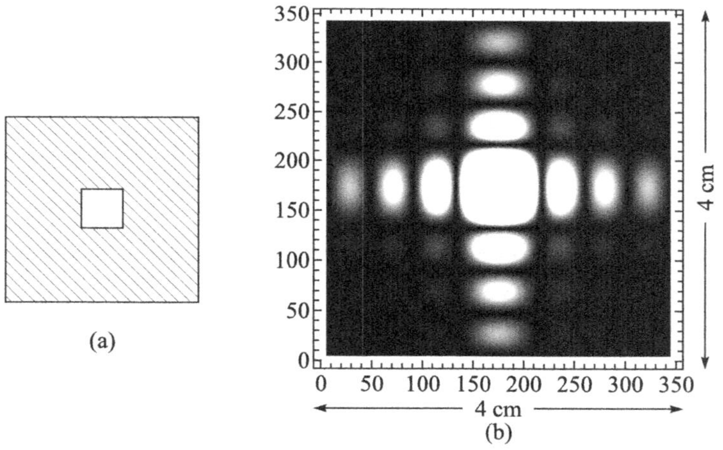

Fig. 10.10 (a) A square aperture of side 0.01 cm. (b) The corresponding (computer generated) Fraunhofer diffraction pattern on a screen at a distance of 100 cm from the aperture; $\lambda = 5 \times 10^{-5}$ cm.

10.7 Along the x-axis, minima will occur at $b \sin\theta = m\lambda$ or at

$$x = f\tan\theta \approx f\sin\theta = \frac{m\lambda f}{b} = \frac{m}{20}\text{ (cm)}$$
$$\approx 0.05, 0.10, 0.15, \ldots \text{ cm}$$

Along the y-axis, minima will occur at $b \sin\theta = m\lambda$ or at

$$y = f\tan\theta \approx f\sin\theta = \frac{m\lambda f}{a} = \frac{m}{30}\text{ (cm)}$$
$$\approx 0.033, 0.067, 0.1, \ldots \text{ cm}$$

10.8 (a) Instead of Eq. (38), we will have

$$G = \int_0^{b_1} e^{-iu\xi}\, d\xi + \int_d^{d+b_2} e^{-iu\xi}\, d\xi$$

$$= \left(e^{-i\beta_1}\, b_1 \frac{\sin \beta_1}{\beta_1} \right) + \left(e^{-iud}\, e^{-i\beta_2}\, b_2 \frac{\sin \beta_2}{\beta_2} \right)$$

where, $\beta_1 = \frac{1}{2} u b_1$ and $\beta_2 = \frac{1}{2} u b_2$. The corresponding intensity distribution (on the x-axis) will be proportional to

$$\left| b_1 \frac{\sin \beta_1}{\beta_1} + e^{-i\phi}\, b_2 \frac{\sin \beta_2}{\beta_2} \right|^2$$

where $\phi = \beta_1 - \beta_2 - ud$

10.9 Obviously on the plane of the circular aperture, it will be convenient to choose cylindrical coordinates [see Fig. 10.4(b)]

$$\xi = \rho \cos \phi \quad \text{and} \quad \eta = \rho \sin \phi \tag{46}$$

Further, because of the circular symmetry of the system the diffraction pattern will be of the form of concentric circular rings with their centers at the point O'. Consequently, we may calculate the intensity distribution only along the x-axis (i.e., at points for which $y = 0$) and in the final result replace x by $\sqrt{x^2 + y^2}$. Now, when $y = 0$

$$v = 0 \quad \text{and} \quad \sin \theta \approx \frac{x}{z} \tag{47}$$

where θ is the angle that OP makes with the z-axis. Thus,

$$u = \frac{2\pi x}{\lambda z} = k \sin \theta \tag{48}$$

and Eq. (11) becomes

$$U(P) = \frac{A}{i\lambda z} e^{ikz} \exp\left\{ \frac{ikr^2}{2z} \right\} \int_0^a \int_0^{2\pi} e^{-ik\rho \sin\theta \cos\phi}\, \rho\, d\rho\, d\phi \tag{49}$$

Thus,

$$U(P) = \frac{A}{i\lambda z} e^{ikz} \exp\left\{ \frac{ikr^2}{2z} \right\} \frac{1}{(k \sin \theta)^2} \int_0^{ka\sin\theta} \zeta\, d\zeta \int e^{-i\zeta\cos\phi}\, d\phi$$

$$= \frac{A}{i\lambda z} e^{ikz} \exp\left\{ \frac{ikr^2}{2z} \right\} \frac{2\pi}{(k \sin \theta)^2} \int_0^{ka\sin\theta} \zeta J_0(\zeta)\, d\zeta \tag{50}$$

where $\zeta = k\rho \sin \theta$ and use has made of the following well known relation[1]

$$J_0(\zeta) = \frac{1}{2\pi} \int_0^{2\pi} e^{\pm i\zeta\cos\phi}\, d\phi \tag{51}$$

1. The identities associated with Bessel functions can be found in most books on mathematical physics; see, e.g., Ref. Ar1, Ir1, Gh4; Ref. Ab1 gives detailed tables of Bessel functions.

If we further use the relation

$$\frac{d}{d\zeta}[\zeta J_1(\zeta)] = \zeta J_0(\zeta) \tag{52}$$

then Eq. (36) becomes

$$\begin{aligned} U(P) &= \frac{A}{i\lambda z} e^{ikz} \exp\left\{\frac{ikr^2}{2z}\right\} \frac{2\pi}{(k\sin\theta)^2}\Big[\zeta J_1(\zeta)\Big]\Big|_0^{ka\sin\theta} \\ &= \frac{A}{i\lambda z} e^{ikz} \exp\left\{\frac{ikr^2}{2z}\right\} \pi a^2 \left[\frac{2J_1(v)}{v}\right] \end{aligned} \tag{53}$$

where $v = ka\sin\theta$. Thus, the intensity distribution would be given by

$$I(P) = I_0\left[\frac{2J_1(v)}{v}\right]^2 \tag{54}$$

where I_0 is the intensity at the point O' [see Fig. 10.4(a)]. This is the famous Airy pattern.

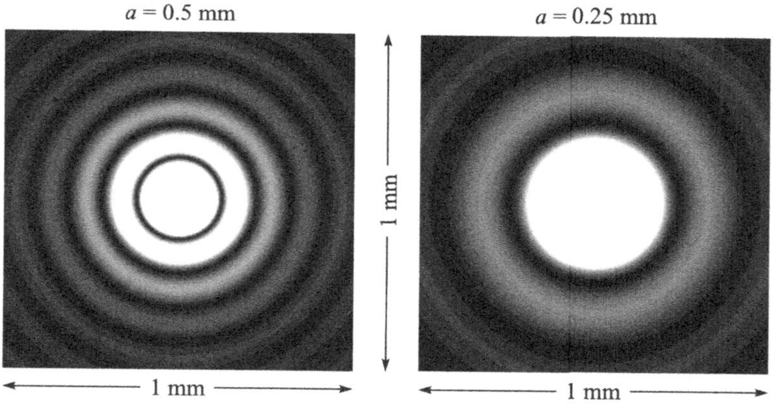

Fig. 10.11 Computer generated Airy patterns; (a) and (b) correspond to a = 0.5 mm and a = 0.25 mm respectively at the focal plane of a lens of focal length 20 cm (λ = 0.5 μm).

10.10 The radii of the dark rings are given by

$$\begin{aligned} r_n &= f\tan\theta \approx \frac{3.832\lambda f}{2\pi a}, \frac{7.016\lambda f}{2\pi a}, \frac{10.174\lambda f}{2\pi a}, \ldots \\ &\approx 0.134 \text{ mm}, 0.246 \text{ mm}, 0.356 \text{ mm} \end{aligned}$$

10.11 The glass slide introduces an additional phase difference of $\frac{2\pi}{\lambda}(n-1)t = \pi$ between the portion that traverses the glass slide and the portion of the wavefront that passes through the slit that is not covered by the slide. Thus, the Fraunhofer diffraction pattern of the given aperture would be proportional to

$$\int_0^{a/2} e^{-iu\xi}\, d\xi - \int_{a/2}^{a} e^{-iu\xi}\, d\xi$$

Integrating and simplifying we obtain the intensity variation in the Fraunhofer diffraction pattern to be given by $\sin^2\left(\frac{ua}{4}\right)$. Note that in this case the intensity on the axis (corresponding to $u = 0$) is zero.

10.12 Let the center of the first aperture (A_1) be chosen as the origin and let the line connecting the two apertures be along the x-axis. Then the coordinates of the origin of the second aperture (A_2) would be (d, 0, 0). Thus, the Fraunhoffer diffraction due to the pair of apertures would be proportional to

$$F(u, v) = \iint f(\xi, \eta)\, e^{-i(u\xi + v\eta)}\, d\xi\, d\eta \tag{55}$$

where $u = \dfrac{2\pi x}{\lambda z}$ and $v = \dfrac{2\pi y}{\lambda z}$

And the integral is performed over the two apertures. Hence, we have

$$F(u, v) = \iint_{A_1} e^{-i(u\xi + v\eta)}\, d\xi\, d\eta + \iint_{A_2} e^{-i(u\xi + v\eta)}\, d\xi d\eta$$

In the second integral we replace ξ by $\zeta = \xi - d$ and since the two apertures are identical, the limits of integration of (ξ, η) in the first integral and the limits of integration of (ξ, η) in the second integral would be the same and both integrals would represent the Fraunhofer diffraction integral of a rectangular aperture (see Eq. 45). Thus, we obtain

$$F(u, v) = A\,[1 + e^{-iud}]\, \frac{\sin\beta}{\beta} \frac{\sin\gamma}{\gamma} \tag{56}$$

where A is the amplitude at the center of the diffraction pattern produced by a single rectangular aperture and $\beta = ua/2$, $\beta = vb/2$ with angles of diffraction θ and ϕ along the x- and y-directions respectively. Hence the intensity pattern would be proportional to

$$F(u, v) = 4I_0 \cos^2 \frac{\pi x d}{\lambda z} \left(\frac{\sin\beta}{\beta}\right)^2 \left(\frac{\sin\gamma}{\gamma}\right)^2 \tag{57}$$

10.13 Let the center of the first aperture (A_1) be chosen as the origin and let the line connecting the two apertures be along the x-axis. Then the coordinates of the origin of the second aperture (A_2) would be (d, 0, 0). Thus, the Fraunhofer diffraction due to the pair of apertures would be given by Eq. (55). For the two apertures, the integral is performed over the two apertures. Hence, we have

$$F(u, v) = \iint_{A_1} e^{-i(u\xi + v\eta)}\, d\xi d\eta + \iint_{A_2} e^{-i(u\xi + v\eta)}\, d\xi\, d\eta$$

In the second integral we replace ξ by $\zeta = \xi - d$ and since the two apertures are identical, the limits of integration of (ξ, η) in the first integral and the limits of integration of (ξ, η) in the second integral would be the same and both integrals would represent the Fraunhofer diffraction integral of a circular aperture of radius a (see Solution 10.9). Thus, we obtain

$$F(u, v) = A[1 + e^{-iud}]\left[\frac{2J_1(v)}{v}\right] \quad (58)$$

where A is the amplitude at the center of the diffraction pattern produced by a single circular aperture and where $v = ka\sin\theta$. Thus the intensity pattern observed would be proportional to

$$F(u, v) = 4I_0 \cos^2 \frac{ud}{2}\left(\frac{2J_1(v)}{v}\right)^2 \quad (59)$$

10.14 Proceeding in an identical fashion as in Problem 10.13 we will obtain for the intensity pattern

$$F(u, v) = 16I_0 \cos^2 \frac{ud}{2} \cos^2 \frac{vd}{2}\left(\frac{2J_1(v)}{v}\right)^2 \quad (60)$$

10.15 The Fraunhofer diffraction pattern is given by the Fourier transform of the aperture function (see Eq. (55)). Now if the aperture is displaced by distances x_0 and y_0 along the x- and y-directions respectively then the corresponding Fraunhofer diffraction pattern will be given by

$$\begin{aligned}\tilde{F}(u, v) &= \iint f(\xi - x_0, \eta - y_0)\, e^{-i(u\xi + v\eta)}\, d\xi\, d\eta \\ &= e^{-i(ux_0 + vy_0)} \iint f(\zeta, \sigma)\, e^{-i(u\zeta + v\sigma)}\, d\zeta\, d\sigma \end{aligned} \quad (61)$$

Thus, the Fraunhofer diffraction pattern differs from the pattern of the undisplaced aperture only by a phase factor. Hence, the intensity patterns of the undisplaced and the displaced apertures will be identical.

10.16 For an aperture shown in the figure, the Fraunhofer diffraction pattern is given by

$$U = C \int_{ka_1 \sin\theta}^{ka_2 \sin\theta} J_0(\zeta)\,\zeta\, d\zeta \quad (62)$$

where $\zeta = ka\sin\theta$. Carrying out the integration as in the case of a circular aperture, we obtain for the intensity pattern,

$$I(\theta) = \frac{I_0}{(1-\kappa^2)^2}\left[\frac{2J_1(ka_2\sin\theta)}{ka_2\sin\theta} - \kappa^2 \frac{2J_1(k\kappa a_2\sin\theta)}{k\kappa a_2\sin\theta}\right]^2 \quad (63)$$

where I_0 represents the intensity along the axis and $\kappa = a_1/a_2$. Note that $\kappa = 0$ corresponds to a circular aperture.

10.17 Assuming the slit width to be along the x-axis, we can write for the incident plane wave will have a phase distribution given by $e^{-ikx\sin\theta_0}$ on the plane of the slit. Hence, the Fraunhofer diffraction pattern will be given by

$$F(u) = \int_{-b/2}^{b/2} e^{+ik\xi\sin\theta_0} e^{-iu\xi}\, d\xi$$

On integrating and simplifying we obtain the intensity pattern as

$$I(\theta) = I_0 \left[\frac{\sin\{\pi b(\sin\theta - \sin\theta_0)/\lambda\}}{\pi b(\sin\theta - \sin\theta_0)/\lambda} \right]^2 \tag{64}$$

The principal maximum will appear at $\theta = \theta_0$ and the zeroes will correspond to

$$(\sin\theta - \sin\theta_0) = \pm m \frac{\lambda}{b} \tag{65}$$

Fresnel Diffraction 11

A Quick Review

11.1 FRESNEL DIFFRACTION INTEGRAL

If $U(x, y, 0)$ represents the amplitude and phase distribution on the plane $z = 0$, then the field as it propagates along the $+z$ direction, is given by

$$U(x, y, z) = \frac{1}{i\lambda z} e^{ikz} \iint U(x', y', 0) \exp\left[\frac{ik}{2z}\{(x - x')^2 + (y - y')^2\}\right] dx'dy' \quad (1)$$

The RHS of the above equation is known as the Fresnel diffraction integral and represents the diffraction pattern in the Fresnel approximation. Often we will write the above integral as

$$U(x, y, z) = \frac{1}{i\lambda z} e^{ikz} \iint U(\xi, \eta, 0) \exp\left[\frac{ik}{2z}\{(x - \xi)^2 + (y - \eta)^2\}\right] d\xi d\eta \quad (2)$$

In both equations, the integral is over the area of the aperture on the plane $z = 0$ (see Fig. 11.1).

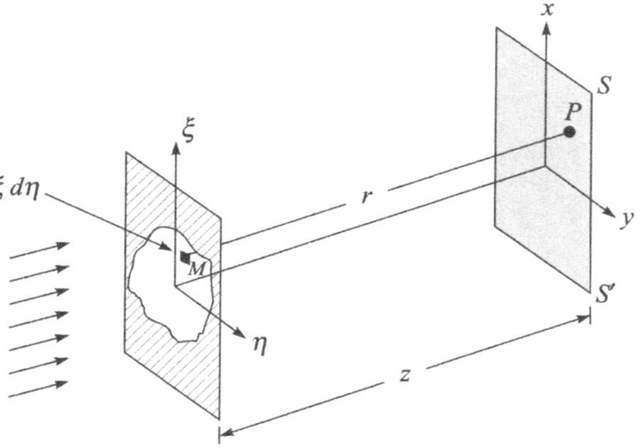

Fig. 11.1 A plane wave incident normally on an aperture. The integration in Eq. (2) is carried out over the area of the aperture.

11.2 DIFFRACTION OF A GAUSSIAN BEAM

Consider a Gaussian beam propagating along the z-direction whose amplitude distribution on the plane $z = 0$ is given by

$$U(x, y, 0) = A \exp\left[-\frac{x^2 + y^2}{w_0^2}\right] \tag{3}$$

implying that the phase front is plane at $z = 0$. From the above equation it follows that at a distance w_0 from the z-axis, the amplitude falls by a factor $1/e$ (i.e., the intensity reduces by a factor $1/e^2$). This quantity w_0 is called the *spot size* of the beam. If we substitute Eq. (3) in Eq. (2) and carry out the integration we would obtain

$$U(x, y, z) \approx \frac{A}{(1 + i\gamma)} \exp\left[-\frac{x^2 + y^2}{w^2(z)}\right] e^{i\Phi} \tag{4}$$

where,

$$\gamma = \frac{\lambda z}{\pi w_0^2} \tag{5}$$

$$w(z) = w_0 \sqrt{1 + \gamma^2} = w_0 \sqrt{1 + \frac{\lambda^2 z^2}{\pi^2 w_0^4}} \tag{6}$$

$$\Phi = kz + \frac{k}{2R(z)}(x^2 + y^2) \tag{7}$$

and

$$R(z) \equiv z\left(1 + \frac{1}{\gamma^2}\right) = z\left[1 + \frac{\pi^2 w_0^4}{\lambda^2 z^2}\right] \tag{8}$$

where $R(z)$ represents the radius of curvature of the phase front (see Solution 11.9).

11.3 FRESNEL INTEGRALS

The Fresnel integrals are defined by the following equations:

$$C(\tau) = \int_0^\tau \cos\left(\frac{1}{2}\pi u^2\right) du \tag{9}$$

and

$$S(\tau) = \int_0^\tau \sin\left(\frac{1}{2}\pi u^2\right) du \tag{10}$$

Since the integrands are even functions of τ, the Fresnel integrals $C(\tau)$ and $S(\tau)$ are odd functions of τ.

$$C(-\tau) = -C(\tau) \quad \text{and} \quad S(-\tau) = -S(\tau) \tag{11}$$

Further,

$$\int_{-\infty}^{+\infty} \exp\left[i\frac{\pi u^2}{2}\right] du = \sqrt{\frac{\pi}{-i\pi/2}} = \sqrt{2e^{i\pi/2}} = \sqrt{2}\, e^{i\pi/4} = 1 + i \tag{12}$$

Also, $$\int_{-\infty}^{+\infty} \exp\left[i\frac{\pi u^2}{2}\right] du = 2\left[\int_0^{\infty} \cos\left(\frac{1}{2}\pi u^2\right) du + i\int_0^{\infty} \sin\left(\frac{1}{2}\pi u^2\right) du\right]$$

$$= 2\,[C(\infty) + S(\infty)]$$

Thus, $$C(\infty) = \frac{1}{2} = S(\infty).$$

To summarise, the Fresnel integrals have the following important properties:

$$C(\infty) = S(\infty) = \frac{1}{2}; \quad C(0) = S(0) = 0 \tag{13}$$

$$C(-\tau) = -C(\tau) \quad \text{and} \quad S(-\tau) = -S(\tau) \tag{14}$$

The values of the Fresnel integrals for typical values of τ are tabulated in Table 11.1.

Table 11.1 Table of Fresnel Integrals (adapted from Ref. Ab1; more accurate values can be found there)

τ	$C(\tau)$	$S(\tau)$	τ	$C(\tau)$	$S(\tau)$
0.0	0.00000	0.00000	2.6	0.38894	0.54999
0.2	0.19992	0.00419	2.8	0.46749	0.39153
0.4	0.39748	0.03336	3.0	0.60572	0.49631
0.6	0.58110	0.11054	3.2	0.46632	0.59335
0.8	0.72284	0.24934	3.4	0.43849	0.42965
1.0	0.77989	0.43826	3.6	0.58795	0.49231
1.2	0.71544	0.62340	3.8	0.44809	0.56562
1.4	0.54310	0.71353	4.0	0.49843	0.42052
1.6	0.36546	0.63889	4.2	0.54172	0.56320
1.8	0.33363	0.45094	4.4	0.43833	0.46227
2.0	0.48825	0.34342	4.6	0.56724	0.51619
2.2	0.63629	0.45570	4.8	0.43380	0.49675
2.4	0.55496	0.61969	5.0	0.56363	0.49919
			∞	0.5	0.5

11.4 THE STRAIGHT EDGE DIFFRACTION PATTERN

Consider a plane wave incident normally on a straight edge as shown in Fig. 11.2; from the figure it is obvious that there will be no variation of intensity along the x-axis and, therefore, without any loss of generality, we may assume the co-ordinates of an arbitrary point P (on the screen) to be $(0, y)$, where the origin has been assumed to be on the edge of the geometrical shadow.

Thus, using Eq. (2), we get

$$U(P) = \frac{A}{i\lambda z} e^{ikz} \int_{-\infty}^{+\infty} d\xi \int_0^{+\infty} d\eta \exp\left[\frac{ik}{2z}\{\xi^2 + (y-\eta)^2\}\right] \tag{15}$$

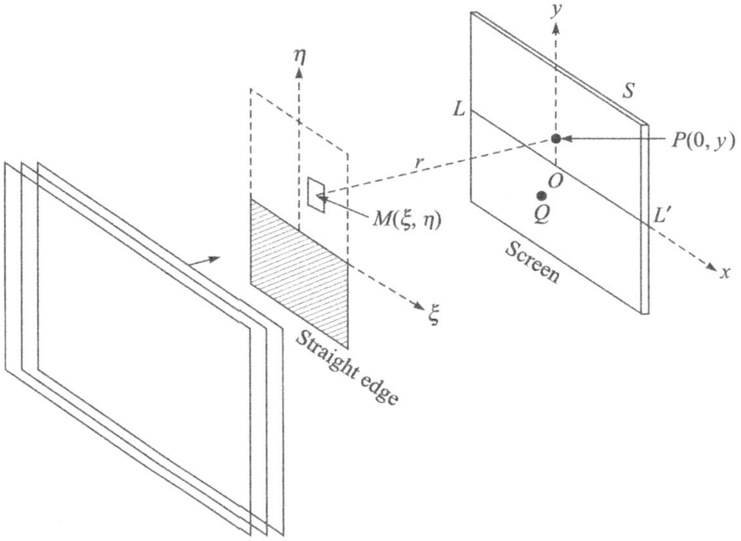

Fig. 11.2 Diffraction of a plane wave incident normally on a straight edge.

The corresponding intensity variation on the screen is given by (see Problem 11.15)

$$I(P) \approx \frac{1}{2} I_0 \left[\left\{ \frac{1}{2} - C(v_0) \right\}^2 + \left\{ \frac{1}{2} - S(v_0) \right\}^2 \right] \tag{16}$$

where,

$$v_0 = -\sqrt{\frac{2}{\lambda z}}\, y \tag{17}$$

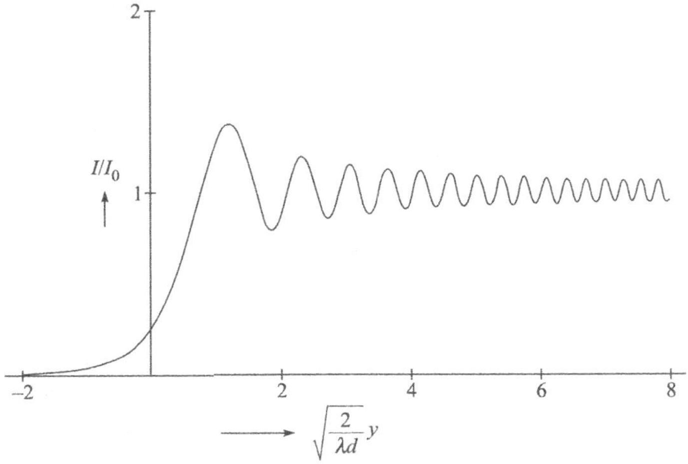

Fig. 11.3 The intensity variation corresponding to the straight edge diffraction pattern.

and $C(v_0)$ and $S(v_0)$ are the Fresnel integrals. The intensity variation is shown in Fig. 11.3 which is an universal curve. The first three maxima occur at $v_0 \approx -1.22$,

−2.34 and −3.08 where $I \approx 1.37\,I_0$, $1.20\,I_0$ and $1.15\,I_0$ respectively. And, the first three minima occur at $v_0 \approx -1.87$, -2.74 and -3.39 where $I \approx 0.778\,I_0$, $0.843\,I_0$ and $0.872\,I_0$ respectively. The actual (computer generated) intensity distribution corresponding to the straight edge diffraction pattern is shown in Fig. 11.4.

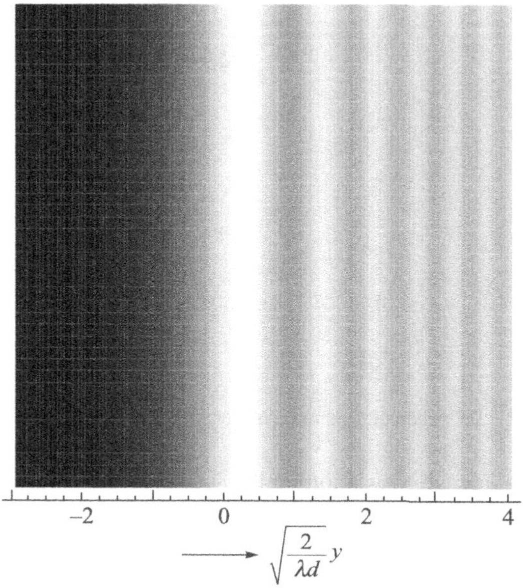

Fig. 11.4 Computer generated intensity distribution corresponding to the straight edge diffraction pattern.

PROBLEMS

11.1 Consider a uniform plane wave

$$U(x, y, 0) = A \text{ for all values of } x \text{ and } y$$

Substitute in Eq. (2) and evaluate the integral to obtain

$$U(x, y, z) = Ae^{ikz} \tag{18}$$

as it indeed should be for a uniform plane wave.

11.2 Consider a plane wave incident normally on a circular aperture of radius a. Using Eq. (2), calculate the intensity variation on the z axis; i.e., for $x = y = 0$. Interpret the result physically in terms of Fresnel half period zones.

11.3 A plane wave ($\lambda = 6 \times 10^{-5}$ cm) is incident normally on a circular aperture of radius a.

(a) Assume $a = 1$ mm. Calculate the values of z (on the axis) for which maximum intensity will occur. Plot the intensity as a function of z and interpret physically. Repeat the calculations for $\lambda = 5 \times 10^{-5}$ cm and discuss chromatic aberration of a zone plate.

(b) Assume $z = 50$ cm. Calculate the values of a for which minimum intensity will occur on the axial point. Plot the intensity variation as a function of a and interpret physically.

11.4 Consider a plane wave incident normally on an opaque circular disc of radius a. Using the results of the Problem 11.2, calculate the intensity variation on the z axis; i.e., for $x = y = 0$.

11.5 (a) Consider a plane wave of wavelength 6×10^{-5} cm incident normally on a circular aperture of radius 0.01 cm. Calculate the positions of the brightest and the darkest points on the axis.

(b) What would happen if the circular aperture is replaced by a circular disc of the same radius?

11.6 Consider a circular aperture of diameter 2 mm illuminated by a plane wave. The most intense point on the axis is at a distance of 200 cm from the aperture. Calculate the wavelength. **[Ans.** 5×10^{-5} cm]

11.7 A plane wave of intensity I_0 is incident normally on a circular aperture as shown in the Fig. 11.5. The point P (at a distance b from the center of the circular aperture) is on the axis and the distance of the point P to the periphery of the circular aperture is $b + \frac{\lambda}{3}$. What will be the intensity on the axial point P?

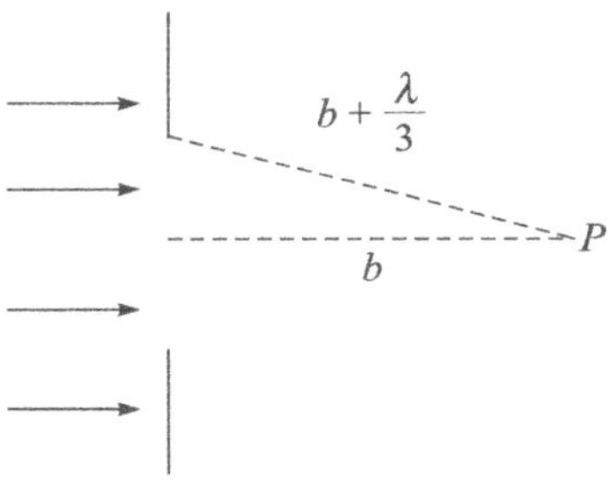

Fig. 11.5

11.8 Consider the propagation of a Gaussian beam as given by Eqs (3)-(8). Show that

(a) The intensity of the propagating Gaussian beam is given by

$$I(x, y, z) = \frac{I_0}{1+\gamma^2} \exp\left[-\frac{2(x^2+y^2)}{w^2(z)}\right] \tag{19}$$

(b) The diffraction angle is given by 2θ, where

$$\tan\theta \approx \frac{\lambda}{\pi w_0} \tag{20}$$

Interpret the result physically.

(c) Evaluate the integral $\int\int_{-\infty}^{+\infty} I(x, y, z)\, dx dy$ and show that it is independent of z. Interpret the result physically.

11.9 In Eq. (8), show that $R(z)$ represents the radius of curvature of the phase front[1].

---

1. Using these results, in Problems 17.3 and 17.4, we will discuss the condition for a Gaussian beam to resonate between two mirrors; this condition is very useful in the design of the resonator in a laser.

11.10 Consider a Gaussian beam with $\lambda = 0.5$ μm. Calculate the spot size of the beam at $z = 10$ m for $w_0 = 1$ mm and for $w_0 = 0.25$ mm; here w_0 represents the spot size at $z = 0$ where the phase front is plane. Interpret the result physically.

11.11 (a) The output of a He-Ne laser ($\lambda = 0.6328$ μm) can be assumed to be Gaussian with plane phase front. For $w_0 = 1$ mm and $w_0 = 0.2$ mm, calculate the beam diameter at $z = 20$ m. (b) Repeat the calculation for $\lambda = 0.5$ μm and interpret the results physically.

11.12 A Gaussian beam is coming out of a laser. Assume $\lambda = 0.6$ μm and that at $z = 0$, the beam width is 1 mm and the phase front is plane. After traversing 10 m through vacuum what will be (a) the beam width and (b) the radius of curvature of the phase front.

11.13 Consider a resonator consisting of a plane mirror and a concave mirror of radius of curvature R (see Fig. 11.6). Assume $\lambda = 1$ μm, $R = 100$ cm and the distance between the 2 mirrors to be 50 cm. Calculate the spot size of the Gaussian beam.

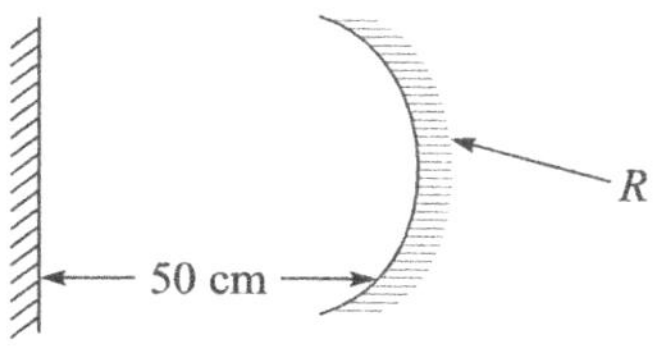

Fig. 11.6

11.14 The output of a semiconductor laser can be approximated described by a Gaussian function with two different widths along the transverse (w_T) and lateral (w_L) directions as

$$\psi(x, y) = A \exp\left(-\frac{x^2}{w_L^2} - \frac{y^2}{w_T^2}\right) \tag{21}$$

where x and y represent axes parallel and perpendicular to the junction plane. Typically $w_T \approx 0.5$ μm and $w_L \approx 2$ μm. Discuss the far field of this beam (see Fig. 11.7 below).

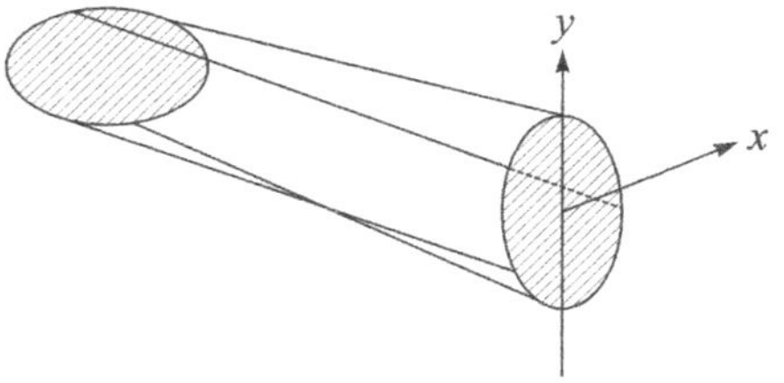

Fig. 11.7

11.15 (a) The field at the screen corresponding to the straight edge diffraction pattern is given by Eq. (15). Show that the corresponding intensity variation on the screen is given by

$$I(P) \approx \frac{1}{2} I_0 \left[\left\{\frac{1}{2} - C(v_0)\right\}^2 + \left\{\frac{1}{2} - S(v_0)\right\}^2\right] \tag{22}$$

(b) Show that for a large positive value of y, the intensity will be I_0 and for a large negative value of y the intensity will tend to zero. Interpret the results physically.

(c) Find the intensity at the edge of the geometrical show ($y = 0$).

(d) Use Table 11.1 to plot the intensity distribution for $-4 < v_0 < +4$ and show that the first three maxima occur at $v_0 \approx -1.22$, -2.34 and -3.08 where $I \approx 1.37 I_0$, $1.20 I_0$ and $1.15 I_0$ respectively. Similarly show that the first three minima occur at $v_0 \approx -1.87$, -2.74 and -3.39 where $I \approx 0.778 I_0$, $0.843 I_0$ and $0.872 I_0$ respectively.

11.16 In the straight edge diffraction pattern (see Fig. 11.8 below), assume $\lambda_0 = 5000$ Å and $d = 100$ cm. Write approximately the values of I/I_0 at the points O, $P(y = 0.5 \text{ mm})$, $Q(y = 1.0 \text{ mm})$ and $R(y = -1.0 \text{ mm})$ where O is at the edge of the geometrical shadow.

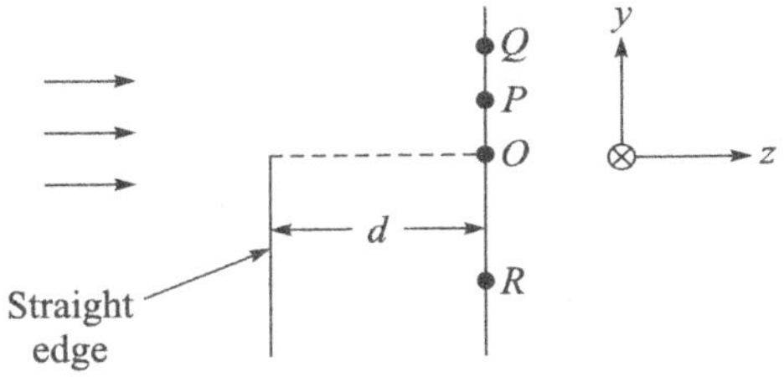

Fig. 11.8

11.17 Consider a straight edge being illuminated by a parallel beam of light with $\lambda = 0.6$ μm. Calculate the positions of the first two maxima and minima on a screen at a distance of 50 cm from the edge.

11.18 In a straight edge diffraction pattern, one observes that the most intense maximum occurs at a distance of 1 mm from the edge of the geometrical shadow. Calculate the wavelength of light, if the distance between the screen and the straight edge is 300 cm. **[Ans.** $\approx$4480 Å]

11.19 In a straight edge diffraction pattern, if the wavelength of the light used is 6000 Å and if the distance between the screen and the straight edge is 100 cm, calculate the distance between the most intense maximum and the next minimum. Find approximately the distance in centimeters inside the geometrical shadow where $I/I_0 = 0.1$. **[Ans.** $y \approx 0.027$ cm]

11.20 The electric field distribution of a light wave is given by (on a plane $z = 0$)

$$E(x) = A \cos\left(\frac{2\pi x}{a}\right)$$

Show that the intensity distribution in the Fresnel diffraction pattern is given by

$$I(x) = A^2 \cos^2\left(\frac{2\pi x}{a}\right)$$

Notice that the intensity distribution in the transverse plane is independent of z.

11.21 A Gaussian beam (at a wavelength of 1 μm) traveling along the z-direction is found to have a *converging* spherical phase front of radius of curvature of

10 m at a certain position and a *diverging* spherical phase front of radius of curvature 10 m at a distance of 2 m from it. Obtain the minimum spot size of the beam and its position.

11.22 Consider a circular aperture of radius 0.4 mm illuminated normally by a plane wave of wavelength 500 nm. Obtain the position of the brightest and darkest point along the axis.

SOLUTIONS

11.1 If we substitute the expression for $U(x, y, 0)$ in Eq. (1), we obtain

$$U(x, y, z) = \frac{A}{i\lambda z} e^{ikz} \int_{-\infty}^{+\infty} e^{\frac{ik}{2z}X^2}\, dX \int_{-\infty}^{+\infty} e^{\frac{ik}{2z}Y^2}\, dY$$

where, $X = x - \xi$ and $Y = y - \eta$. If we now use the integral

$$\int_{-\infty}^{+\infty} e^{-\alpha x^2 + \beta x} dx = \sqrt{\frac{\pi}{\alpha}} \exp\left[\frac{\beta^2}{4\alpha}\right] \tag{23}$$

we would get

$$U(x, y, z) = \frac{A}{i\lambda z} e^{ikz} \left[\sqrt{\frac{\pi 2z}{-ik}}\right]\left[\sqrt{\frac{\pi 2z}{-ik}}\right]$$

or,
$$U(x, y, z) = Ae^{ikz} \tag{24}$$

as it indeed should for a uniform plane wave.

11.2 We assume the circular aperture to be placed on the plane $z = 0$. In Eq (2), we substitute $x = y = 0$ to obtain

$$U(0, 0, z) \approx \frac{A}{i\lambda z} e^{ikz} \iint \exp\left[\frac{ik}{2z}(\xi^2 + \eta^2)\right] d\xi d\eta$$

where we have assumed $u(\xi, \eta, 0) = A$ and the integration is over the aperture; thus in carrying out the above integration, we must have $(\xi^2 + \eta^2) < a^2$. We use polar coordinates

$$\xi = \rho \cos \phi \quad \text{and} \quad \eta = \rho \sin \phi$$

to obtain

$$U(0, 0, z) \approx \frac{A}{i\lambda z} e^{ikz} \int_{\phi=0}^{2\pi} \int_{\rho=0}^{a} \exp\left[\frac{ik}{2z}\rho^2\right] \rho d\rho d\phi$$

$$\approx \frac{A}{i\lambda z} e^{ikz} \left[(e^{2i\alpha} - 1)\frac{z}{ik}\right][2\pi] = U_0(1 - e^{2i\alpha})$$

or,
$$U(0, 0, z) \approx -2iU_0\, e^{i\alpha} \sin \alpha \tag{25}$$

where $\alpha = \dfrac{k}{4z} a^2$ and $U_0 = Ae^{ikz}$ is the field on the z-axis in the absence of the circular aperture. Thus, the intensity variation will be given by

$$I(0, 0, z) \approx 4I_0 \sin^2 \alpha \tag{26}$$

We will have zero intensity when $\alpha = n\pi$; $n = 1, 2, \ldots$ which implies

$$z = \frac{a^2}{2n\lambda}, \quad n = 1, 2, \ldots \quad \text{(Minimum Intensity)} \tag{27}$$

and the circular aperture will contain even number of half period zones. Similarly, we will have maximum intensity when $\alpha = \left(n + \frac{1}{2}\right)\pi$ which implies

$$z = \frac{a^2}{(2n+1)\lambda}, \quad n = 1, 2, \ldots \quad \text{(Maximum Intensity)} \tag{28}$$

and the circular aperture will contain odd number of half period zones.

11.3 (a) As mentioned in Solution 11.2, maxima will occur when

$$z = \frac{a^2}{\lambda(2n+1)} = \frac{10^{-2}}{6 \times 10^{-5} \times (2n+1)} \approx 166.7 \text{ cm, } 55.6 \text{ cm, } 33.3 \text{ cm,} \ldots$$

For $\lambda = 5 \times 10^{-5}$ cm, maxima will occur at $z = 200$ cm, 66.7 cm, 40 cm, …

(b) Minimum intensity will occur when $a = \sqrt{2n\lambda z} \approx 0.0775\sqrt{n}$ cm $\approx$ 0.0775 cm, 0.110 cm, 0.134 cm, …

11.4 If $U_1(P)$ and $U_2(P)$ respectively represent the fields at the point P due to a circular aperture and an opaque disc (of the same radius), then

$$U_1(P) + U_2(P) = U_0(P) \tag{29}$$

where $U_0(P)$ [$= Ae^{ikz}$] represents the field in the absence of any aperture; the above equation is known as the Babinet's principle. Thus,

$$\begin{aligned} U_2(P) &= U_0(P) - U_1(P) \\ &= U_0(P) - U_0(P)[1 - e^{2i\alpha}] = U_0(P)e^{2i\alpha} \end{aligned} \tag{30}$$

where we have used Eq. (25) for $U_1(P)$. Thus,

$$|U_2(P)|^2 = |U_0(P)|^2$$

and therefore the intensity at the point P on the axis of a circular disc, $I_2(P)$, would be given by

$$I_2(P) = I_0 \tag{31}$$

which gives us the remarkable result that the intensity at a point on the axis of an opaque disc is equal to the intensity at the point in the absence of the disc (see Fig. 11.9). This is known as the Poisson spot.

11.5 (a) Maxima will occur when $p = \frac{a^2}{\lambda d} = (2n+1)$; $n = 0, 1, 2, \ldots$ [see Eq. (27)].

$$\text{Thus, } d = \frac{(0.01)^2}{6 \times 10^{-5} \times (2n+1)} \text{ cm} \approx 1.67 \text{ cm, } 0.56 \text{ cm, } 0.33 \text{ cm,} \ldots$$

for $n = 0, 1, 2, \ldots$ The minima will occur when [see Eq. (28)].

$$d = \frac{5}{6n}\ \text{cm} \approx 0.83\ \text{cm},\ 0.42\ \text{cm}, \ldots$$

(b) The central point will always be bright (which is the Poisson spot–see Fig. 11.9).

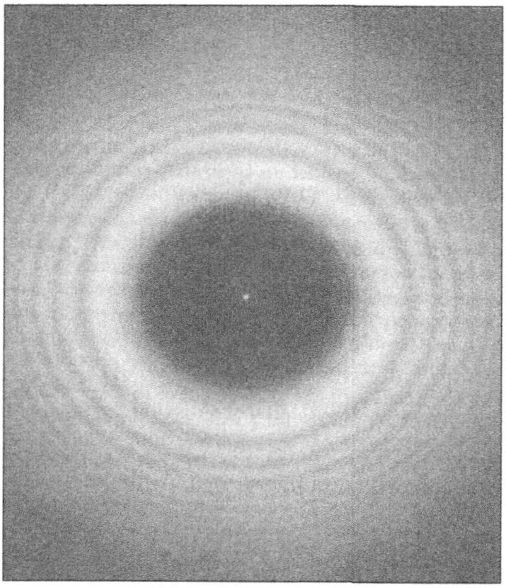

Fig. 11.9 The Poisson spot at the center of the shadow of a one cent coin; the screen is 20 m from the coin and the source of light is also 20 m from the coin. Photograph from Ref. R: 1; used with permission from P. M. Rinard and American association of Physics Teachers.

11.6 The maximum intensity will occur when $\frac{a^2}{\lambda d} = 1$ or when $\lambda = \frac{a^2}{d} = \frac{(0.1)^2}{200}$

$$= 5 \times 10^{-5}\ \text{cm}.$$

11.7 $z = b$ and if a represents the radius of the circular aperture, then

$$\left(b + \frac{\lambda}{3}\right)^2 = b^2 + a^2 \;\Rightarrow\; b + \frac{\lambda}{3} = b\sqrt{1 + \frac{a^2}{b^2}} \approx b + \frac{a^2}{2b}$$

giving $\frac{a^2}{\lambda b} = \frac{2}{3}$. Thus [see Eq. (25)]:

$$\alpha = \frac{k}{4z}a^2 = \frac{\pi}{2\lambda b}a^2 = \frac{\pi}{3}$$

and

$$I(0, 0, z) \approx 4 I_0 \sin^2 \alpha = 4 I_0 \sin^2 \frac{\pi}{3} = 3 I_0$$

11.8 (b) $w(z) = w_0\sqrt{1 + \frac{\lambda^2 z^2}{\pi^2 w_0^4}}$. For large values of $z \left(>> \frac{\pi w_0^2}{\lambda} \right)$, we obtain

$$w(z) \approx w_0 \frac{\lambda z}{\pi w_0^2} = \frac{\lambda z}{\pi w_0} \tag{32}$$

which shows that the width increases linearly with z. We define the diffraction angle as 2θ, where

$$\tan\theta = \frac{w(z)}{z} \approx \frac{\lambda}{\pi w_0} \tag{33}$$

showing that the rate of increase in the width is proportional to the wavelength and inversely proportional to the initial width of the beam.

(c) From Eq. (19), one can readily show that

$$\iint_{-\infty}^{+\infty} I(x, y, z)\, dx\, dy = \frac{\pi w_0^2}{2} I_0 \tag{34}$$

which is independent of z. This is to be expected, as the total energy crossing the entire x-y plane will not change with z.

11.9 For a spherical wave *diverging* from the origin, the field distribution is given by

$$u \sim \frac{1}{r} e^{ikr} \tag{35}$$

Now, on the plane $z = R$ (see Fig. 11.10)

$$r = \sqrt{x^2 + y^2 + R^2} = R\sqrt{1 + \frac{x^2 + y^2}{R^2}} \approx R + \frac{x^2 + y^2}{2R} \tag{36}$$

where we have assumed $|x|, |y| << R$. Thus on the plane $z = R$, the phase distribution (corresponding to a spherical wave of radius R) would be given by

$$e^{ikr} \approx e^{ikR}\, e^{\frac{ik}{2R}(x^2 + y^2)} \tag{37}$$

From the above equation it follows that a phase variation of the type

$$\exp\left[i\frac{k}{2R}(x^2 + y^2)\right] \tag{38}$$

(on the x-y plane) represents a *diverging* spherical wave of radius R. If we compare the above expression with Eqs (4) and (8) we obtain the following approximate expression for the radius of curvature of the phase front:

$$R(z) \approx z\left(1 + \frac{\pi w_0^4}{\lambda^2 z^2}\right) \tag{39}$$

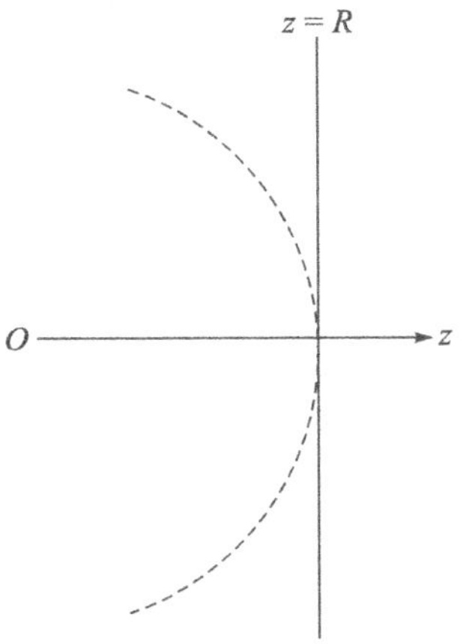

Fig. 11.10 A spherical wave diverging from the point O. The dashed curve represents a section of the spherical wavefront at a distance R from the source.

11.10 For $w_0 = 1$ mm, the diffraction angle is given by

$$\tan\theta \approx \frac{\lambda}{\pi w_0} = \frac{5\times 10^{-4}}{\pi} \text{ implying } 2\theta \approx 0.018°$$

$$\text{and} \quad w(z) = w_0\sqrt{1+\frac{\lambda^2 z^2}{\pi^2 w_0^4}} = 10^{-3}\sqrt{1+\frac{(0.5\times 10^{-6})^2\,10^2}{\pi^2\times 10^{-12}}} \approx 1.88 \text{ mm}$$

at $\quad z = 10$ m

Similarly, for $w_0 = 0.25$ mm

$$2\theta \approx 0.073° \quad \text{and} \quad w \approx 6.35 \text{ mm at } z = 10 \text{ m}$$

(see Fig. 11.11). Thus, smaller the spot size of the beam, greater is the diffraction spreading–as expected from diffraction theory.

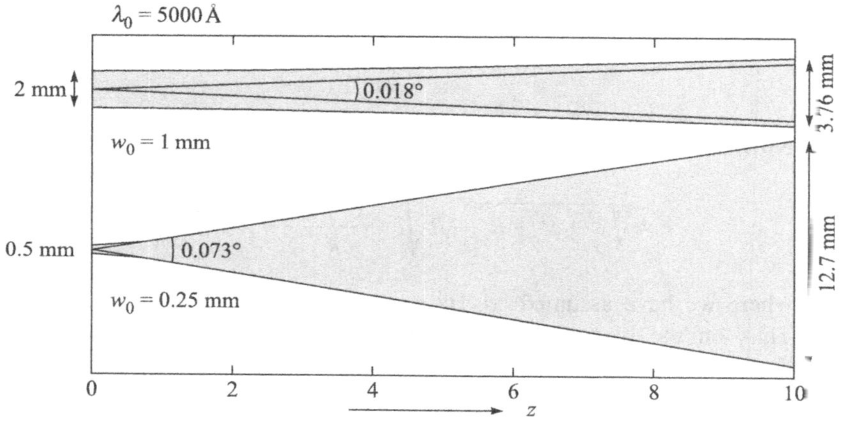

Fig. 11.11 Diffraction divergence of a Gaussian beam whose phase front is plane at $z = 0$. The figure shows the increase in the diffraction divergence as the initial spot size is decreased from 1 mm to 0.25 mm; the wavelength is assumed to be 5000 Å.

11.11 Beam diameter is given by

$$2w = 2w_0\left[1+\frac{\lambda^2 z^2}{\pi^2 w_0^4}\right]^{1/2}$$

For $w_0 = 0.1$ cm, $\lambda = 6.328\times 10^{-5}$ cm, $z = 2000$ cm

$$\frac{\lambda^2 z^2}{\pi^2 w_0^4} \approx 16.23 \Rightarrow 2w \approx 0.83 \text{ cm}$$

For $w_0 = 0.2$ mm $= 0.02$ cm and with same values of λ and z

$$\frac{\lambda^2 z^2}{\pi^2 w_0^4} \approx 10143 \Rightarrow 2w \approx 4.0 \text{ cm}$$

The above results show that the divergence increases as w_0 becomes smaller. At $\lambda = 5 \times 10^{-5}$ cm for $w_0 = 0.1$ cm and $z = 2000$ cm, we have

$$\frac{\lambda^2 z^2}{\pi^2 w_0^4} \approx 10.13 \Rightarrow 2w \approx 0.67 \text{ cm}$$

showing that the divergence decreases as we make the wavelength smaller.

11.12 $\lambda = 6 \times 10^{-5}$ cm, $w_0 = 0.05$ cm and $z = 2000$ cm

$$2w = 2w_0 \left[1 + \frac{\lambda^2 z^2}{\pi^2 w_0^4}\right]^{1/2} = 0.1\,[1 + 58.4]^{1/2} \approx 0.77 \text{ cm}$$

$$R(z) \approx z\left[1 + \frac{\pi^2 w_0^4}{\lambda^2 z^2}\right] \approx 1000\,[1 + 0.017] \approx 1017 \text{ cm}$$

11.13 $\lambda = 10^{-6}$ m, $R = 1$ m, $d = 0.5$ m.

Thus, $$R = d\left[1 + \frac{\pi^2 w_0^4}{\lambda^2 d^2}\right] \Rightarrow w_0 = \sqrt{\frac{\lambda d}{\pi}}\left[\frac{R}{d} - 1\right]^{1/4} \approx 4 \times 10^{-4} \text{ m} = 0.4 \text{ mm}$$

11.14 $$U(x, y, z) = \frac{a}{\sqrt{(1 + i\gamma_T)(1 + i\gamma_L)}} \exp\left[-\frac{x^2}{w_1^2} - \frac{y^2}{w_2^2}\right] e^{i\Phi}$$

Thus, $$I(x, y, z) = \frac{I_0}{\sqrt{(1 + \gamma_T^2)(1 + \gamma_L^2)}} \exp\left[-\frac{2x^2}{w_1^2(z)} - \frac{2y^2}{w_2^2(z)}\right]$$

where, $$w_1^2(z) = w_T(1 + \gamma_T^2)^{1/2} = w_T\left[1 + \frac{\lambda^2 z^2}{\pi^2 w_T^4}\right]^{1/2} \approx \frac{\lambda z}{\pi w_T} \quad \text{(for large } z\text{)}$$

and $$w_2^2(z) = w_L(1 + \gamma_L^2)^{1/2} = w_L\left[1 + \frac{\lambda^2 z^2}{\pi^2 w_L^4}\right]^{1/2} \approx \frac{\lambda z}{\pi w_L} \quad \text{(for large } z\text{)}$$

11.15 In order to carry out the integrations in the equation

$$U(P) = \frac{A}{i\lambda z} e^{ikz} \int_{-\infty}^{+\infty} d\xi \int_{0}^{+\infty} d\eta \exp\left[\frac{ik}{2z}\{\xi^2 + (y - \eta)^2\}\right] \tag{40}$$

we define two parameters u and v such that,

$$\frac{1}{2}\pi u^2 = \frac{k}{2z}\xi^2 \quad \text{and} \quad \frac{1}{2}\pi v^2 = \frac{k}{2z}(\eta - y)^2 = \frac{\pi}{\lambda z}(\eta - y)^2 \tag{41}$$

Thus, $$u = \sqrt{\frac{2}{\lambda z}}\,\xi \quad \text{and} \quad v = \sqrt{\frac{2}{\lambda z}}\,(\eta - y) \tag{42}$$

The integral over ξ gives

$$\int_{-\infty}^{+\infty} d\xi \exp\left[\frac{ik}{2z}\xi^2\right] = \sqrt{\frac{\lambda z}{2}} \int_{-\infty}^{+\infty} du \exp\left[\frac{i\pi u^2}{2}\right] = \sqrt{\frac{\lambda z}{2}}\, 2\,[C(\infty) + iS(\infty)]$$

$$= \sqrt{\frac{\lambda z}{2}}\,(1 + i) \tag{43}$$

The integral over η gives

$$\int_{0}^{+\infty} d\eta \exp\left[\frac{ik}{2z}(\eta - y)^2\right] = \sqrt{\frac{\lambda z}{2}} \int_{v_0}^{+\infty} dv \exp\left(\frac{i\pi v^2}{2}\right)$$

$$= \sqrt{\frac{\lambda z}{2}}\left[\int_{0}^{+\infty} dv \exp\left(\frac{i\pi v^2}{2}\right) - \int_{0}^{v_0} dv \exp\left(\frac{i\pi v^2}{2}\right)\right]$$

$$= \sqrt{\frac{\lambda z}{2}}\,[\{C(\infty) + iS(\infty)\} - \{C(v_0) + iS(v_0)\}]$$

$$= \sqrt{\frac{\lambda z}{2}}\left[\left\{\frac{1}{2} - C(v_0)\right\} + i\left\{\frac{1}{2} - S(v_0)\right\}\right] \tag{44}$$

Substituting in Eq. (40) we readily get

$$U(P) = \frac{(1-i)}{2}\, U_0 \left[\left\{\frac{1}{2} - C(v_0)\right\} + i\left\{\frac{1}{2} - S(v_0)\right\}\right] \tag{45}$$

where $U_0 = Ae^{ikz}$ is the field in the absence of the straight edge.

(b) A large (positive) value of y will correspond to a point which is very far above the edge of the geometrical shadow; for such a point v_0 would tend to $-\infty$ and $C(-\infty) = -\frac{1}{2} = S(-\infty)$ we would get

$$U(P) = \frac{(1-i)}{2}\, U_0 (1 + i) = U_0 \tag{46}$$

Thus, as expected, the amplitude at such a far away point is the same as that in the absence of the edge. On the other hand, when the point P is deep inside the geometrical shadow (i.e., for a large negative value of y), v_0 would tend to $+\infty$ and since $C(\infty) = \frac{1}{2} = S(\infty)$ we would get

$$U(P) \to 0.$$

This is to be expected because deep inside the geometric shadow, the intensity will be very small.

(c) From Eq. (45) it readily follows that the intensity distribution will be given by

$$I(P) = \frac{1}{2} I_0 \left[\left\{\frac{1}{2} - C(v_0)\right\}^2 + \left\{\frac{1}{2} - S(v_0)\right\}^2\right] \tag{47}$$

If the point P is such that it lies on the edge of the geometrical shadow (i.e., on the line LL' (see Fig. 11.2) then $y = 0$ and hence v_0 will be 0 and therefore,

$$I(P) = \frac{1}{2} I_0 \left[\frac{1}{4} + \frac{1}{4} \right] = \frac{1}{4} I_0 \tag{48}$$

Thus, the intensity on the edge of the geometrical shadow is $1/4^{\text{th}}$ of the intensity that would have been in the absence of the edge.

(d) In order to calculate the intensity at an arbitrary value of y, we just calculate the value of the dimensionless parameter v_0 and either use Table 11.1 (or use more accurate values using Ref. Ab1 or a software like Mathematica) to obtain a curve like that shown in Fig. 11.3 which is an universal curve. The corresponding (computer generated) intensity distribution is shown in Fig. 11.4.

11.16 $v_0 = -\sqrt{\frac{2}{\lambda d}}\, y = -\sqrt{\frac{2}{5 \times 10^{-5} \times 100}}\, y = -20\, y$ where y is measured in centimeters. For the points O, P, Q and R (see Fig. 11.8) the values of $v_0 = 0, -1, -2,$ and $+2$.

Now, $C(0) = 0 = S(0)$; $C(-1) = -0.7799$, $S(-1) = -0.4383$
$C(-2) = -0.4883$, $S(-2) = -0.3434$; $C(+2) = +0.4883$,
$S(+2) = +0.3434$

The intensity distribution is given by

$$\frac{I}{I_0} = \frac{1}{2}\left[\left\{ \frac{1}{2} - C(v_0) \right\}^2 + \left\{ \frac{1}{2} - S(v_0) \right\}^2 \right]$$

$$= \frac{1}{4} \quad \text{for} \quad v_0 = 0$$

$$\approx 1.26 \quad \text{for} \quad v_0 = -1\ (y = 0.5 \text{ mm})$$

$$\approx 0.84 \quad \text{for} \quad v_0 = -2\ (y = 1.0 \text{ mm})$$

$$\approx 0.01 \quad \text{for} \quad v_0 = +2\ (y = -1.0 \text{ mm})$$

11.17 $v_0 = -\sqrt{\frac{2}{\lambda d}}\, y = -\sqrt{\frac{2}{6 \times 10^{-5} \times 50}}\, y = -25.82\, y$ where y is measured in centimeters. Now the first two maxima occur at $v_0 \approx -1.22$ and -2.34 giving $y \approx 0.0473$ cm and 0.0906 cm. The first two minima occur at $v_0 \approx -1.87$ and -2.74 giving $y \approx 0.0724$ cm and 0.1061 cm.

11.18 The first maximum occurs at $v_0 \approx -1.22$; thus $v_0 = -\sqrt{\frac{2}{\lambda d}}\, y = -1.22$. Thus,

$$\sqrt{\frac{2}{\lambda \times 300}} \times 0.1 = 1.22 \text{ giving } \lambda \approx 4.479 \times 10^{-5} \text{ cm.}$$

11.19 If the most intense maximum occurs at $y = y_{\text{max}}$ and the next minimum occurs at $y = y_{\text{min}}$, then

$$-\sqrt{\frac{2}{\lambda d}}\, y_{\text{max}} \approx -1.22 \quad \text{and} \quad -\sqrt{\frac{2}{\lambda d}}\, y_{\text{min}} \approx -1.87$$

Thus, $\Delta y = y_{\min} - y_{\max} = (0.65)\sqrt{\dfrac{\lambda d}{2}} = 0.65\sqrt{\dfrac{6 \times 10^{-5} \times 100}{2}} \approx 0.0356$ cm.

Using Table 11.1, we get

At $\quad v_0 = 0, \dfrac{I}{I_0} = 0.25$

At $\quad v_0 = 0.2, \dfrac{I}{I_0} \approx 0.168$

At $\quad v_0 = 0.4, \dfrac{I}{I_0} \approx 0.114$

At $\quad v_0 = 0.6, \dfrac{I}{I_0} \approx 0.079$

Indeed at $v_0 = 0.5$, $C(v_0) = 0.49234$ and $S(v_0) = 0.064732$ giving $\dfrac{I}{I_0} \approx 0.095$. Thus, we may assume very approximately $v_0 \approx 0.5$ giving $y \approx 0.5\sqrt{\dfrac{\lambda d}{2}} \approx 0.027$ cm.

11.20 Substituting the expression for the field in Eq. (2), and using the standard integrals we obtain

$$U(x, y) = Ce^{ikz} \exp\left(i\frac{\pi\lambda z}{a^2}\right) \cos\left(\frac{2\pi x}{a}\right)$$

And hence the intensity distribution at any value of z is given by

$$I(x, z) = C^2 \cos^2\left(\frac{2\pi x}{a}\right)$$

which is the same as the intensity distribution across the object plane. Since $I(x, z)$ is independent of z, we obtain the interesting result that the intensity distribution at all planes normal to the z-axis is the same as that in the object plane.

11.21 From symmetry it is obvious that the waist is at the center between the converging and diverging wavefronts. Thus at a distance of 1 m, the radius of curvature of the wavefront is 10 m. Using the formula for variation of radius of curvature with distance [see Eq. (8)] we obtain

$$w_0 = \sqrt{\frac{3\lambda}{\pi}}$$

Substituting the values, we get $w_0 \sim 0.98$ mm.

11.22 The brightest point on the axis corresponds to having a single Fresnel half period zone within the aperture. This happens when

$$z \approx \frac{a^2}{\lambda} = 32 \text{ cm}$$

The darkest point along the axis will correspond to the aperture having two Fresnel half period zones. This will happen when $z \sim 16$ cm.

Fourier Optics and Holography

12

A Quick Review

12.1 FOURIER OPTICS

The Fraunhoffer diffraction pattern is given by the Fourier transform of the near field distribution. Since a lens focuses plane waves on its focal plane, the Fraunhoffer diffraction pattern of an aperture can be seen on the focal plane of a converging lens. If $f(x, y)$ represents the field distribution on the front focal plane of a converging lens then the field distribution on the back focal plane of the lens is given by

$$g(x, y) = \frac{1}{\lambda f} \iint f(x', y') \exp[-2\pi i(\tilde{u}x' + \tilde{v}y')]\, dx'dy' \qquad (1)$$

where $\tilde{u} = x/\lambda f$, $\tilde{v} = y/\lambda f$ and x and y are measured on the Fourier transform plane. The back focal plane is also called as the spatial frequency plane.

Thus a point with coordinates (x_0, y_0) on the back focal plane would correspond to spatial frequencies $x_0/\lambda f$, $y_0/\lambda f$. Thus points close to the axis correspond to low spatial frequencies and points far away from the axis correspond to large spatial frequencies.

Since any spatial variation of object amplitude distribution can be described by its Fourier transform, the object can as well be described by its spatial frequencies also. The lens provides us the spatial frequency spectrum of the object and thus by placing apertures on the back focal plane of the lens it is possible to alter the spatial frequency spectrum of the object and thus modify the object distribution. A second lens then takes a second Fourier transform and displays the image on its back focal plane. This is the principle behind spatial frequency filtering.

An aperture placed on the axis of the spatial frequency plane would help pass only the low spatial frequencies and a stop placed on the axis will allow only the high spatial frequencies to pass through. These filtering procedures are similar to signal processing of time dependent functions through frequency filters that is common in electronic signal processing.

12.2 HOLOGRAPHY

A hologram is formed by interfering the object beam with a reference wave. In order for the interference pattern to be formed the two waves need to be coherent and hence usually they are derived from the same laser and the path differences are chosen so that they are coherent. The interference pattern contains information about the object wave and is called a hologram.

When the recorded hologram is illuminated by a reference wave which is usually identical to the reference wave, then one of the reconstructed waves reconstructs the original object wave and another wave reconstructs the conjugate of the original object wave. Observing the reconstructed object wave is similar to observing the original object wave and hence the image observed has all the features of the original object and we get a three dimensional reconstruction of the original object.

PROBLEMS

12.1 The field variation on the front focal plane of a lens of focal length 20 cm is given by

$$g(x, y) = A + B\cos 6\pi x + C\cos 12\pi y; \qquad (x, y \text{ in mm}) \qquad (2)$$

(a) What are the spatial frequencies present in the field?

(b) What pattern would you observe at the back focal plane of the lens? Assume a wavelength of 600 nm.

12.2 In a spatial frequency filtering setup (see Fig. 12.1), on plane P_1 lies in an object of the form, $g(x) = A\left[\cos\frac{2\pi x}{a} + \cos\frac{2\pi x}{b}\right]$ which is illuminated by a plane wave. On the plane P_2 is placed a filter having a transmittance,

$$T(x) = \begin{cases} 1 & \text{for } x < 0 \\ 0 & \text{for } x > 0 \end{cases} \qquad (3)$$

Obtain the intensity distribution on the plane P_3.

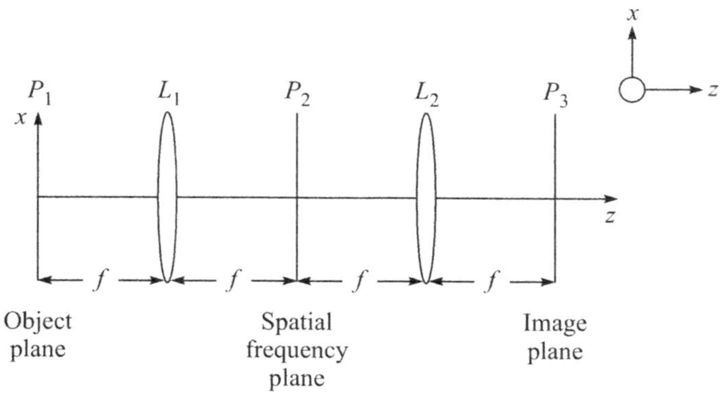

Fig. 12.1 Spatial frequency filtering set up.

12.3 A circular aperture of radius a is placed (with its center on the axis) on the back focal plane of a lens of focal length f. What is the range of spatial frequencies that will be passed by the aperture?

12.4 What is the effect of placing a filter of the form $h(x) = px$, where p is a constant placed in spatial frequency plane of a spatial frequency filtering set up?

12.5 Consider an object distribution of the form,

$$f(x, y) = 1 + 0.2 \cos\left(20x + \frac{\pi}{5}\right) + 0.3 \sin\left(50y + \frac{\pi}{8}\right) \tag{4}$$

where x and y are in millimeters. (a) What are the spatial frequencies present in the object? (b) If the above object is placed in the front focal plane of a lens of focal length 20 cm and illuminated normally by a plane wave of wavelength 1 μm, show schematically what would be observed on the back focal plane of the lens?

12.6 On the front focal plane of a lens of focal length f one finds an amplitude distribution of the form

$$I(x, y) = A \exp\left(-\frac{\pi^2 x^2}{0.04}\right) \exp\left(-\frac{\pi^2 y^2}{0.01}\right) \tag{5}$$

where x and y are measured in mm. Obtain the amplitude distribution on the back focal plane if $\lambda = 1.0$ μm and $f = 20$ cm.

12.7 In a spatial frequency filtering setup a field distribution $g(x)$ is produced on the front focal plane P_1 of the first lens. On the plane P_2 is placed a filter having a transmittance given by $\sin 2\pi px$, where p is a constant. Calculate the amplitude distribution on the amplitude distribution on the plane P_3. (Consider the problem in one dimension only).

12.8 Consider an object having a transmittance given by $f(x) = 1 + \frac{1}{2}\cos(200x)$ placed in the front panel P_1 of a lens of focal length 10 cm and illuminated by a normally incident plane wave (here x is in mm). The focal length of lens L_2 is also 10 cm. (i) what are the spatial frequencies present in the object. (ii) A circular aperture of radius 0.15 cm is placed centered on plane P_2. What intensity distribution would you observe on plane P_3 if, (a) $\lambda = 0.6$ μm (b) $\lambda = 0.4$ μm.

12.9 What are the spatial frequencies present in an object described by

(i) $f(x) = A + B\cos(20\pi x)\sin(50\pi x)$,
(ii) $f(x) = A + B\cos^2(100\pi x)$

where x is measured in millimeters.

12.10 Consider an object with a transmittance given by $g(x, y) = A + B\cos \alpha x$ placed on the front focal plane of a lens and illuminated by a normally incident plane wave in a spatial frequency filtering arrangement. (a) What would you observe on the back focal plane of the lens? (b) If an opaque disc is placed on axis at the back focal plane to obstruct the light at the central spot, what intensity pattern would you observe on the back focal plane of the second lens? (c) if only the light from the central spot is allowed to propagate further, what would you observe on the back focal plane of the second lens?

12.11 The beam coming out of a laser oscillating in the fundamental mode at a wavelength of 600 nm is found to have the following amplitude distribution:

$$f(x, y) = A(1 + 0.1 \sin(20\pi x))\, e^{-(x^2 + y^2)} \tag{6}$$

where x and y are measured in millimeters. (a) Plot the x-variation of the intensity pattern of the laser along the axis $y = 0$ and interpret the result. (b) How would you use a spatial filtering arrangement to remove the fast oscillatory noisy term in the intensity pattern?

12.12 A hologram is formed between a point source placed on the axis at a distance d from the recording plane and a reference plane wave propagating at an angle of θ with the axis, both waves having a wavelength λ. Obtain an expression for the transmittance of the recorded hologram assuming linear recording.

12.13 If the hologram recorded in Problem 12.12 is illuminated by a reconstruction plane wave identical to the recording wave and at the same wavelength, obtain the different waves that will emanate from the reconstruction and interpret the terms as to what they would represent.

12.14 If the reconstruction wave is the conjugate of the recording wave, then obtain the position of the virtual and real images formed by the hologram.

12.15 Consider a hologram formed between a point object and a normally incident plane wave, both having the wavelength λ. Show that the fringes formed have a spatial frequency which increases with the distance from the axis. If the recording medium has a maximum recordable spatial frequency of S_m, obtain the radius of the region where the interference pattern will be recorded by the recording medium. What effect would this have on the reconstruction process?

SOLUTIONS

12.1 (a) Spatial frequencies present are [0, 0], [3, 0] mm^{-1} and [0, 6] mm^{-1}; the first number corresponds to x-direction and the second to the y-direction.

(b) On the back focal plane of the lens we would see five spots. These will be at the following coordinates:

$$x = 0, y = 0$$
$$x = +0.36 \text{ mm}, y = 0$$
$$x = -0.36 \text{ mm}, y = 0$$
$$x = 0, y = +0.72 \text{ mm}$$
$$x = 0, y = -0.72 \text{ mm}$$

12.2 The object contains the spatial frequencies $1/a$ and $1/b$. So there will be four spots lying along the x-axis corresponding to the positions $\pm\lambda f/a$ and $\pm\lambda f/b$. The filter will allow only the spots lying on the positive x-axis to be imaged by the second lens. Corresponding to the spot at $+\lambda f/a$ the field distribution would be $\exp[-2\pi i x/a]$ and corresponding to the spot $+\lambda f/b$ the field distribution would be $\exp[-2\pi i x/b]$. Thus, the total field would be

$$\exp[-2\pi i x/a] + \exp[-2\pi i x/b]$$

and the intensity pattern would be

$$2 + 2\cos[2\pi x(1/a - 1/b)]$$

Note that the intensity pattern does not contain the two spatial frequencies of the object.

12.3 Range of spatial frequencies passed would be 0 to $\lambda f/a$ in the radial direction.

12.4 An object distribution $f(x)$ would produce the following amplitude distribution on the back focal plane

$$g(x, y) = \frac{1}{\lambda f}\iint f(x', y')\exp[-2\pi i(\tilde{u}x' + \tilde{v}y')]\,dx'dy' \qquad (7)$$

where $\tilde{u} = x/\lambda f$, $\tilde{v} = y/\lambda f$ and x and y are measured on the Fourier transform plane. Now, if we place a filter with a transmittance given by px, then the second lens takes the Fourier transform of $g(x, y)px$ and the field pattern as seen on the back focal plane of the second lens would be

$$\begin{aligned} h(x, y) &= \frac{1}{\lambda f}\iint px'g(x', y')\exp[-2\pi i(\tilde{u}x' + \tilde{v}y')]\,dx'dy' \\ &= \left(\frac{1}{\lambda f}\right)^2 p\frac{1}{2\pi i}\frac{\partial}{\partial \tilde{u}}\left[\iint g(x', y')\exp[-2\pi i(\tilde{u}x' + \tilde{v}y')]\,dx'dy'\right] \qquad (8) \\ &= \left(\frac{1}{\lambda f}\right)p\frac{1}{2\pi i}\frac{\partial}{\partial x}[f(-x, -y)] \end{aligned}$$

Thus in the image plane we would obtain an intensity pattern proportional to the x-differential of the object distribution. Since the differential of a function is maximum near edges, such a filtering process leads to edge enhancement.

12.5 (a) Spatial frequencies present: (0, 0), $(10/\pi, 0)$ mm^{-1} and $(0, 25/\pi)$ mm^{-1}.

(b) On the back focal plane we would observe five spots of light at the following points:

$$x = 0, y = 0$$
$$x = \pm 2/\pi \text{ mm}, 0$$
$$x = 0, y = \pm 5/\pi \text{ mm}$$

12.6 On the back focal plane the field distribution will be the Fourier transform of the distribution on the front focal plane and will be given by

$$\begin{aligned} g(x, y) &= \frac{A}{\lambda f}\iint \exp\left(-\frac{\pi^2 x'^2}{0.04}\right)\exp\left(-\frac{\pi^2 y'^2}{0.01}\right)\exp[-2\pi i(\tilde{u}x' + \tilde{v}y')]\,dx'dy' \\ &= 0.02\frac{A}{\pi\lambda f}\exp\left[-\frac{(4x^2 + y^2)}{100\lambda^2 f^2}\right] \qquad (9) \end{aligned}$$

12.7 The amplitude distribution on the back focal plane (apart from a constant factor) is given by

$$g(x) = \int f(x')\,e^{-2\pi i\tilde{u}x'}dx' \qquad (10)$$

The field distribution after the filter would be given by

$$h(x) = g(x)\sin 2\pi px \qquad (11)$$

Thus, on the back focal plane of the second lens, the field distribution would be given by

$$\begin{aligned} p(x) &= \int h(x'')\, e^{-2\pi i \tilde{u} x''}\, dx'' \\ &= \int g(x'') \sin 2\pi p x'' \int f(x')\, e^{-2\pi i x'' x'/\lambda f} dx'\, e^{-2\pi i x x''/\lambda f} \qquad (12) \\ &= \int dx' \int dx'' f(x') \sin 2\pi p x''\, e^{-2\pi i x'' x'/\lambda f} e^{-2\pi i x x''/\lambda f} \end{aligned}$$

Writing the sine function as a sum of two exponentials, it is clear that we obtain

$$\begin{aligned} p(x) &= \frac{1}{2i} \int f(x')\, dx' \int dx'' \left[\exp\left(-2\pi i x'' \left\{ \frac{x+x'}{\lambda f} - p \right\} \right) - \exp\left(-2\pi i x'' \left\{ \frac{x+x'}{\lambda f} + p \right\} \right) \right] \\ &= \frac{1}{2i} \int f(x') \left[\delta\left(\frac{x+x'}{\lambda f} - p \right) - \delta\left(\frac{x+x'}{\lambda f} + p \right) \right] dx' \\ &= \frac{(\lambda f)^2}{2i} [f(-x + p\lambda f) - f(-x - p\lambda f)] \qquad (13) \end{aligned}$$

Thus on the back focal plane of the second lens, we would obtain the same distribution as the object distribution but one centered at $x = p\lambda f$ and the other at $x = -p\lambda f$.

12.8 (a) Spatial frequencies present in the object are 0 and $100/\pi$ mm^{-1} along the x-direction.

(b) For a wavelength of 600 nm, the back focal plane would have spots of light on the axis and at $x \approx 1.91$ mm and $x \approx -1.91$ mm. Similarly for a wavelength of 400 nm, there would be spots of light on the axis and at $x \approx -1.27$ mm and $x \approx 1.27$ mm. Thus, for 600 nm wavelength the circular aperture of radius 1.5 mm placed on the back focal plane would only allow the central spot to contribute to the image on the plane P_3, which will be a uniform intensity pattern. On the other hand for the wavelength of 400 nm, light from all the spots would be able to contribute to the image and thus the image would be the same as the object pattern.

12.9 To obtain the spatial frequencies we need to write the field distribution in terms of sine and cosine functions. Thus,

(i) $f(x) = A + \dfrac{B}{2}[\sin(70\pi x) + \sin(30\pi x)]$

Hence the spatial frequencies are 0, 35 mm^{-1} and 15 mm^{-1}.

(ii) $f(x) = A + \dfrac{B}{2}[1 + \cos 200\pi x]$

Hence the spatial frequencies present are 0 and 100 mm^{-1}.

12.10 (a) We would observe three spots of light, one on the axis and one each at $x = \lambda f \alpha$ and the other at $x = -\lambda f \alpha$ both spots lying on the y-axis.

(b) The opaque disc would stop the zero frequency component to take part in the imaging and hence the amplitude distribution on the image plane would be $g(x, y) = B \cos \alpha x$. Thus the intensity distribution would be proportional to $g^2(x, y) = B^2 \cos^2 \alpha x$.

(c) If only the central spot is allowed to propagate further, then this will result in uniform illumination of the final image plane.

12.11 The amplitude distribution along the x-direction is given in the figure. As can be seen it corresponds to a Gaussian distribution with a sinusoidal noise term riding on the amplitude. This could be for example caused by interference taking place between some reflections on the path of the laser beam. In order to use such beams, we need to first 'clean' the beam and remove the oscillatory amplitude variation. Since the oscillatory amplitude term corresponds to high spatial frequency, the noise term can be removed by using spatial frequency filtering.

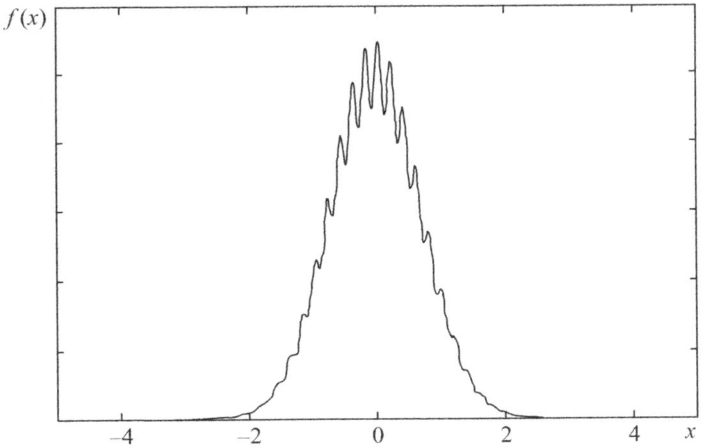

Fig. 12.2

On the spatial frequency plane we would observe the Fourier transform of the amplitude distribution and we get

$$g(x, y) = \frac{1}{\lambda f}\iint A(1 + 0.1 \sin(20\pi x'))\, e^{-(x'^2 + y'^2)} \exp[-2\pi i(\tilde{u}x' + \tilde{v}y')]\, dx'dy'$$

$$= \frac{A\pi}{\lambda f}\Big[\exp\{-\pi^2\{\tilde{u}^2 + \tilde{v}^2\}\} + \frac{0.1}{2i}\exp\{-\pi^2 \{(\tilde{u}+10)^2 + \tilde{v}^2\}\} - \frac{0.1}{2i}\exp\{-\pi^2\{(\tilde{u}-10)^2 + \tilde{v}^2\}\}\Big] \quad (14)$$

Thus on the back focal plane we would get three Gaussian distributions, one centered on the origin, and two others centered at $\tilde{u} = x/\lambda f = \pm 10$. If the three Gaussians are spatially separated, then if we place a pinhole on the axis of the back focal plane to pass only the central Gaussian beam, then on the back focal plane of the second lens, we would obtain a Gaussian without the noise term. This is the principle behind spatial frequency filtering to remove noise from laser beams.

12.12 The amplitude distribution of the spherical wave from the point source is given by

$$f(x, y) = \frac{A}{d} \exp(ikd) \exp\left[i\frac{k}{2d}(x^2 + y^2)\right] \tag{15}$$

where we used the paraxial approximation in writing the equation.

Similarly the plane wave falling on the screen will be given by

$$g(x, y) = B\exp[ikx\sin\theta] \tag{16}$$

where we have assumed the plane of the screen to be $z = 0$ and that the plane wave makes an angle q with the z-axis and lies in the x-z plane.

The transmittance of the hologram would be given by the intensity pattern obtained by the interference of the two waves. This is given (apart from a proportionality constant) by

$$\begin{aligned} T(x, y) &= |f(x, y) + g(x, y)|^2 \\ &= \frac{A^2}{d^2} + B^2 + \frac{AB}{d} \exp\left[+ik\left\{\left(\frac{x^2}{2d} + x\sin\theta\right) + \frac{y^2}{2d}\right\}\right] \\ &\quad + \frac{AB}{d} \exp\left[+ik\left\{\left(\frac{x^2}{2d} - x\sin\theta\right) + \frac{y^2}{2d}\right\}\right] \end{aligned} \tag{17}$$

12.13 The reconstruction wave is given again by $g(x, y)$. Thus the transmitted wave from the hologram would be the product of $g(x, y)$ and $T(x, y)$ which has three terms:

First term:

$$\left(\frac{A^2}{d^2} + B^2\right) Be^{ikx\sin\theta}$$

which corresponds to the same wave as the reconstruction wave except for a change in amplitude.

Second term:

$$\frac{AB^2}{d} \exp\left[ik\left\{\frac{x^2 + y^2}{2d}\right\}\right]$$

which corresponds to a diverging spherical wave and is the same as the object wave that was recorded. This gives the virtual image of the point object.

Third term:

$$\frac{AB}{d} \exp\left[-ik\left\{\left(\frac{x^2}{2d} - 2x\sin\theta\right) + \frac{y^2}{2d}\right\}\right]$$

which represents a converging spherical wave but propagating along a different direction. This wave is responsible for the creation of the real image.

12.14 In this case the reconstruction wave would be given by

$$h(x, y) = B\exp[-ikx\sin\theta] \tag{18}$$

Using the same procedure given in the solution to Problem 12.13 we again obtain three terms corresponding to the emerging waves while the hologram is reconstructed. The reconstructed wave would now be

$$T(x, y) = \left(\frac{A^2}{d^2} + B^2\right)Be^{-ikx\sin\theta} + \frac{AB^2}{d}\exp\left[+ik\left\{\left(\frac{x^2}{2d} - 2x\sin\theta\right) + \frac{y^2}{2d}\right\}\right]$$
$$+ \frac{AB}{d}\exp\left[-ik\left\{+\frac{x^2+y^2}{2d}\right\}\right] \tag{19}$$

The second term gives the virtual image and is now displaced from the axis as contained in the term linearly dependent on x. The last term gives the real image as it corresponds to a converging spherical wavefront and the term gives a wave focusing at a distance d from the hologram on the axis of the arrangement. Thus if the reconstruction wave is the conjugate of the reference wave then the real image is formed on the axis while the virtual image is displaced from the original position.

12.15 The recorded intensity distribution is given by

$$T(x, y) = \frac{A^2}{d^2} + B^2 + \frac{AB}{d}\exp\left[-ik\left\{\frac{x^2+y^2}{2d}\right\}\right] + \frac{AB}{d}\exp\left[+ik\left\{\frac{x^2+y^2}{2d}\right\}\right]$$
$$= \frac{A^2}{d^2} + B^2 + \frac{2AB}{d}\cos\left[2\pi\left\{\frac{x^2+y^2}{2\lambda d}\right\}\right] \tag{20}$$

where we have replaced k by $2\pi/\lambda$. So we obtain oscillatory intensity distribution. In order to get the spatial frequency, we note that a variation of the form $\cos(2\pi\alpha x)$ has a spatial frequency α, which can be obtained by using the following formula:

$$s = \frac{1}{2\pi}\frac{d}{dx}(2\pi\alpha x) = \alpha \tag{21}$$

Hence a linear x-variation in the amplitude corresponds to a constant spatial frequency. If the argument of the cosine term is some function of x, say $\phi(x)$, then we obtain the local spatial frequency by taking the x-derivative of $\phi(x)$ and dividing by 2π.

In the present case since the expression is symmetric in x and y, we consider the variation along x. Hence in the present case, the local spatial frequency is obtained as

$$s(x) = \frac{1}{2\pi}\frac{d}{dx}\left(2\pi\frac{x^2}{2\lambda d}\right) = \frac{x}{\lambda d} \tag{22}$$

which shows that the spatial frequency increases with x.

If the recording medium has a maximum recordable spatial frequency of S_m, then the radius of the recorded hologram would be the value of x where the spatial frequency becomes equal to S_m. This is given by

$$R = \lambda d S_m \tag{23}$$

The finite size of the recorded hologram would impact the resolution of the images formed by the hologram. The hologram would act as if it is limited by a circular aperture of radius R which would then restrict the resolution of the image formed by the hologram.

Polarisation I: Basics and Double Refraction

13

A Quick Review

13.1 LINEARLY POLARISED WAVES

The electric field associated with an x-polarised wave is described by the following equation:

$$\mathbf{E} = \hat{\mathbf{x}} E_0 \cos(kz - \omega t + \phi_0) \quad (1)$$

where E_0 (always assumed to be positive and measured in V/m) is known as the amplitude of the wave, z is the direction of propagation, ϕ_0 is the initial phase, $\hat{\mathbf{x}}$ represents the unit vector along the x-axis and

$$k = \frac{2\pi}{\lambda_0} n \quad (2)$$

n being the refractive index of the medium and λ_0 free space wavelength. Similarly, the electric field associated with a y-polarised wave is described by the equation:

$$\mathbf{E} = \hat{\mathbf{y}} E_0 \cos(kz - \omega t + \phi_0) \quad (3)$$

where the various symbols have the same meaning and $\hat{\mathbf{y}}$ represents the unit vector along the y-axis. For a linearly polarised wave with its $\mathbf{E}$ vector making an angle α with the y-axis (see Fig. 13.1), we will have

$$\left.\begin{aligned} E_x &= E_0 \sin\alpha \cos(kz - \omega t + \phi_0) \\ E_y &= E_0 \cos\alpha \cos(kz - \omega t + \phi_0) \end{aligned}\right\} \quad (4)$$

and

Notice that the x and y components of the electric field are in phase. Perhaps the easiest way for producing linearly polarised light waves is to pass an unpolarised beam through a Polaroid which consists of long chain polymer molecules that contain atoms (like iodine) which provide high conductivity along the length of the chain. These long chain molecules are aligned so that they are almost parallel to each other.

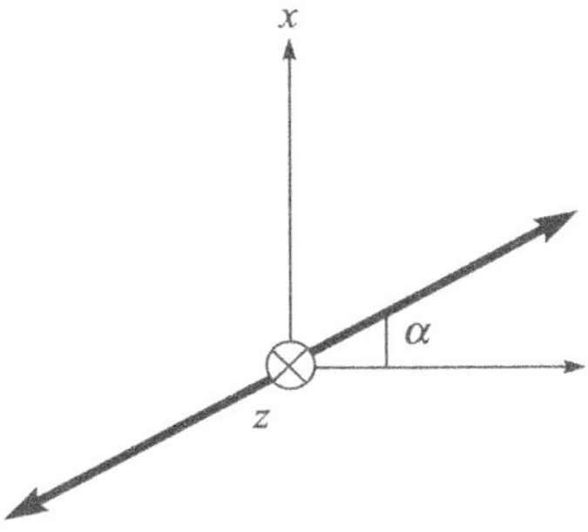

Fig. 13.1 A linearly polarised wave with its **E** vector making an angle α with the y-axis. The propagation of the wave is in the $+z$ direction.

When a light beam is incident on such a polaroid, the molecules (aligned parallel to each other) absorb the component of electric field which is parallel to the direction of alignment because of the high conductivity provided by the iodine atoms; the component perpendicular to it passes through. Thus, linearly polarised light waves are produced.

13.2 MALUS' LAW

If an x-polarised beam is passed through a Polaroid P_2 whose pass axis makes an angle θ with the x-axis, then the intensity of the emerging beam will be given by (see Fig. 13.2)

$$I = I_0 \cos^2 \theta \tag{5}$$

where I_0 represents the intensity of the emergent beam when the pass axis of P_2 is also along the x-axis (i.e., when $\theta = 0$). Equation (5) represents the Malus' Law.

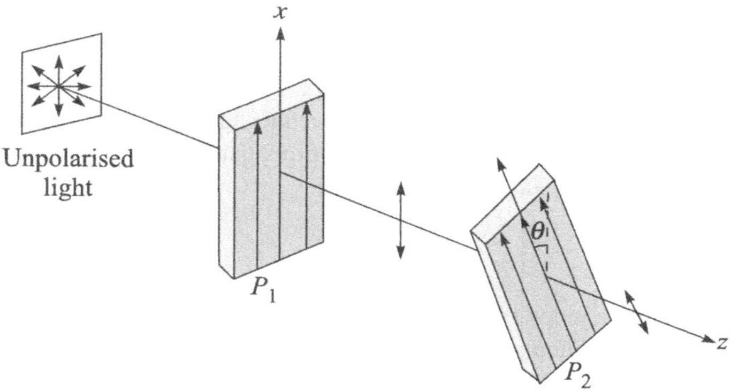

Fig. 13.2 An unpolarised light beam gets x-polarised after passing through the polaroid P_1, the pass axis of the second polaroid P_2 makes an angle θ with the x-axis. The intensity of the emerging beam will vary as $\cos^2 \theta$.

13.3 CIRCULARLY POLARISED WAVES

For a right circularly polarised wave (usually abbreviated as RCP), E_x and E_y will be given by

RCP: $E_x = E_0 \cos (kz - \omega t)$

and $E_y = -E_0 \sin (kz - \omega t) = E_0 \cos \left(kz - \omega t + \dfrac{\pi}{2} \right)$ (6)

The y component (of the electric field) is ahead in phase of the x component by $\dfrac{\pi}{2}$. At $z = 0$, we will have

$$E_x = E_0 \cos \omega t$$

and

$$E_y = E_0 \sin \omega t$$

and the electric field vector will rotate on the circumference of a circle in the clockwise direction as shown in Fig. 13.3. Similarly, for a left circularly polarised wave (usually abbreviated as LCP), E_x and E_y will be given by

$$\text{LCP:} \quad \left.\begin{aligned} E_x &= E_0 \cos(kz - \omega t) \\ \text{and} \quad E_y &= E_0 \sin(kz - \omega t) = E_0 \cos\left(kz - \omega t - \frac{\pi}{2}\right) \end{aligned}\right\} \tag{7}$$

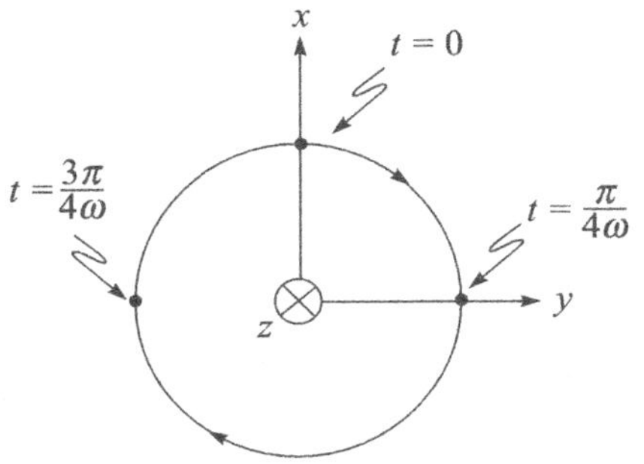

Fig. 13.3 A RCP (right circularly polarised) wave propagating in the $+z$ direction.

The y component now lags behind in phase of the x component by $\frac{\pi}{2}$. The electric field vector rotates on the circumference of a circle in the anti-clockwise direction (see Fig. 13.4)

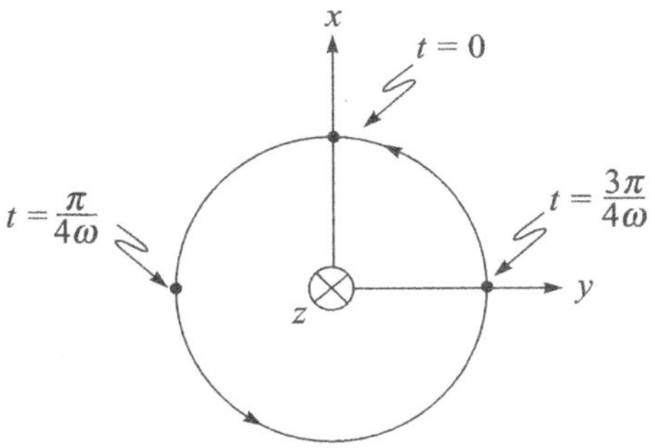

Fig. 13.4 A LCP (left circularly polarised) wave propagating in the $+z$ direction.

13.4 ELLIPTICALLY POLARISED WAVES

For a right elliptically polarised wave (usually abbreviated as REP) with its major and minor axes along x and y axes, E_x and E_y will be given by

$$\text{REP:} \quad \left.\begin{aligned} E_x &= E_1 \cos(kz - \omega t) \\ \text{and} \quad E_y &= -E_2 \sin(kz - \omega t) \end{aligned}\right\} \tag{8}$$

with $E_1 \neq E_2$; if $E_1 = E_2$, we will have a circularly polarised wave. Similarly, for a left elliptically polarised wave (usually abbreviated as LEP) with its major and minor axes along x and y axes, E_x and E_y will be given by

$$\text{LEP:}\quad E_x = E_1 \cos (kz - \omega t) \quad \text{and} \quad E_y = E_2 \sin (kz - \omega t) \tag{9}$$

13.5 POLARISATION BY REFLECTION: BREWSTER'S LAW

If an unpolarised plane wave is incident on a dielectric at an angle of incidence (θ) such that

$$\theta = \theta_p = \tan^{-1}\left(\frac{n_2}{n_1}\right) \tag{10}$$

then the reflected beam will be linearly polarised with its electric vector perpendicular to the plane of incidence [see Fig. 13.5]. The above equation is known as the *Brewster's law* and the angle θ_p is known as the polarising angle (also known as the Brewster angle).

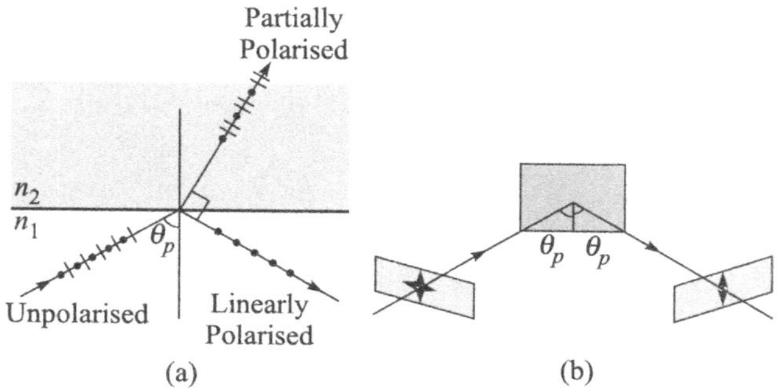

Fig. 13.5 When an unpolarised beam of light is incident on a dielectric at the polarising angle [i.e., the angle of incidence is equal to $\tan^{-1}(n_2/n_1)$] then the reflected beam is plane-polarised with its **E**-vector perpendicular to the plane of incidence. The transmitted beam is partially polarised. The dashed line in (b) is normal to the reflecting surface.

13.6 ANISOTROPIC MEDIA: BIAXIAL MEDIA

1. Electric displacement vector and the applied electric field are related through the following equation:

$$\mathbf{D} = \varepsilon \mathbf{E} \tag{11}$$

In isotropic media the displacement and the electric field are parallel to each other and ε is a scalar quantity. In anisotropic media, the two vectors are not parallel to each other and ε is a tensor and we write Eq. (11) in the form

$$\mathbf{D} = \bar{\varepsilon}\, \mathbf{E} \tag{12}$$

where we have put a bar on ε to indicate that it is not a scalar.

2. In the principal axis system of the medium, $\bar{\varepsilon}$ can be represented by a diagonal matrix:

$$\bar{\varepsilon} = \begin{pmatrix} \varepsilon_{xx} & 0 & 0 \\ 0 & \varepsilon_{yy} & 0 \\ 0 & 0 & \varepsilon_{zz} \end{pmatrix} \tag{13}$$

and the three diagonal terms give the principal dielectric permittivities of the medium.

3. For isotropic media,

$$\varepsilon_{xx} = \varepsilon_{yy} = \varepsilon_{zz} = \varepsilon \tag{14}$$

For uniaxial media,

$$\varepsilon_{xx} = \varepsilon_{yy} \neq \varepsilon_{zz} \tag{15}$$

and for biaxial media,

$$\varepsilon_{xx} \neq \varepsilon_{yy} \neq \varepsilon_{zz} \tag{16}$$

4. We can also define the principal dielectric constants and the principal refractive indices through the following equations:

$$K_{ij} = \frac{\varepsilon_{ij}}{\varepsilon_0}; \quad n_{ij}^2 = \sqrt{K_{ij}} \tag{17}$$

5. Since in the principal axis system ε is diagonal, the principal refractive indices are also sometimes referred to as n_x, n_y and n_z.
6. In anisotropic media along any given direction of propagation there are two linearly polarised eigenmodes which propagate, in general, with different phase velocities.
7. When a wave crosses a boundary, according to Snell's law the component of the **k** parallel to the interface for the incident wave, the reflected wave and the refracted wave are equal. On the other hand, the Poynting vector **S** of the waves do not satisfy such a condition.
8. The index ellipsoid equation for an anisotropic medium is defined by

$$\frac{x^2}{n_{xx}^2} + \frac{y^2}{n_{yy}^2} + \frac{z^2}{n_{zz}^2} = 1 \tag{18}$$

13.7 ANISOTROPIC MEDIA: UNIAXIAL MEDIA

1. For uniaxial media, the principal refractive indices are referred to as

$$n_o = \sqrt{K_{xx}} = \sqrt{K_{yy}} \quad \text{and} \quad n_e = \sqrt{K_{zz}} \tag{19}$$

2. In uniaxial media, there is one optic axis also called the z-axis along which the two eigenmodes have the same velocity.
3. Both ordinary and extra-ordinary waves are linearly polarised:

$$\mathbf{D} \cdot \mathbf{k} = 0 \quad \text{for both } o\text{- and } e\text{-waves} \tag{20}$$

Thus **D** is always at right angles to **k** and for this reason the direction of **D** is chosen as the direction of 'vibrations'.

4. For the o- wave, the **D** vector is at right angles to the optic axis as well as to **k** (see Fig. 13.6).

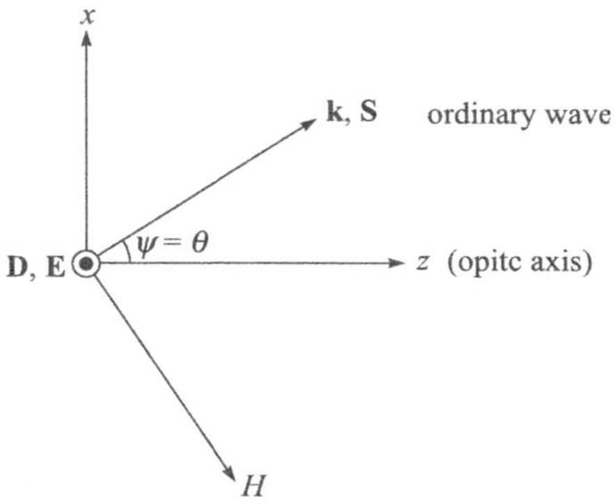

Fig. 13.6 For the ordinary wave (in uniaxial crystals), **D** and **E** vectors are in the *y* direction; **k** and **S** are in the same direction in the *x-z* plane and **H** also lies in the *x-z* plane.

5. On the other hand, for the e-wave, **D** lies in the plane containing **k** and the optic axis (and of course, $\mathbf{D} \cdot \mathbf{k} = 0$) (see Fig. 13.7).

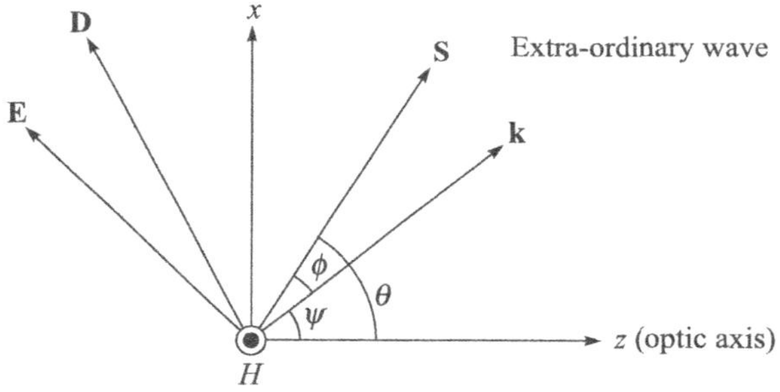

Fig. 13.7 For the extraordinary wave (in uniaxial crystals), **E**, **D**, **S** and **k** vectors would lie in the *x-z* plane and **H** will be in the *y* direction. **S** is at right angles to **E** and **H**; **D** is at right angles to **k** and **H**.

6. Referring to Fig. 13.7, we have

$$\theta = \phi + \psi = \tan^{-1}\left(\frac{n_o^2}{n_e^2}\tan\psi\right) \tag{21}$$

13.8 VELOCITY OF ORDINARY AND EXTRAORDINARY WAVES

We write the electric vector as

$$\mathbf{E} = \mathbf{E}_0 e^{i(\mathbf{k}\cdot\mathbf{r} - \omega t)}$$

where the vector $\mathbf{E}_0$ is independent of space and time; **k** represents the propagation vector of the wave and ω the angular frequency. The wave velocity v_w (also known as the phase velocity) and the wave refractive index n_w are defined through the following equation:

$$v_w = \frac{\omega}{k} = \frac{c}{n_w} \tag{22}$$

Thus,

$$|\mathbf{k}| = k = \frac{\omega}{c} n_w \tag{23}$$

For the ordinary and extra-ordinary waves, we have

$$n_w = n_{wo} = n_o \quad \Rightarrow \quad v_w = v_{wo} = \frac{c}{n_o} \tag{24}$$

$$\frac{1}{n_w^2} = \frac{1}{n_{we}^2} = \frac{\cos^2 \psi}{n_o^2} + \frac{\sin^2 \psi}{n_e^2} \tag{25}$$

where ψ is the angle that the **k** vector makes with the optic axis (see Figs 13.6 and 13.7). Thus

$$v_{we}^2 = \frac{c^2}{n_{we}^2} = \frac{c^2}{n_o^2} \cos^2 \psi + \frac{c^2}{n_e^2} \sin^2 \psi \tag{26}$$

13.9 VELOCITY OF ORDINARY AND EXTRAORDINARY RAYS

The ray propagates along the direction of the Poynting vector $\mathbf{S} (\equiv \mathbf{E} \times \mathbf{H})$

$$\frac{1}{v_r^2} = \frac{1}{v_{re}^2} = \frac{n_{re}^2}{c^2} = \frac{\cos^2 \theta}{c^2 / n_o^2} + \frac{\sin^2 \theta}{c^2 / n_e^2} \tag{27}$$

where we have chosen the y-axis in such a way that the ray propagates in the x-z plane making an angle θ with the z-axis (see Figs 13.6 and 13.7).

1. For a uniaxial medium the index ellipsoid equation becomes

$$\frac{x^2 + y^2}{n_o^2} + \frac{z^2}{n_e^2} = 1 \tag{28}$$

PROBLEMS

13.1 For a right circularly polarised wave described by Eq. (6), plot the electric vector at $t = 0$, $t = \frac{\pi}{4\omega}$, $t = \frac{\pi}{2\omega}$, $t = \frac{3\pi}{4\omega}$, $t = \frac{\pi}{\omega}$, $t = \frac{3\pi}{2\omega}$, and at $t = \frac{2\pi}{\omega}$.

13.2 For a left circularly polarised wave described by Eq. (7), plot the electric vector at $t = 0$, $t = \frac{\pi}{4\omega}$, $t = \frac{\pi}{2\omega}$, $t = \frac{3\pi}{4\omega}$, $t = \frac{\pi}{\omega}$, $t = \frac{3\pi}{2\omega}$ and at $t = \frac{2\pi}{\omega}$.

13.3 Discuss the state of polarisation when the x and y components of the electric field are given by the following equations:

(a)
$$\left.\begin{aligned} E_x &= E_0 \cos(\omega t + kz) \\ E_y &= \frac{1}{\sqrt{2}} E_0 \cos(\omega t + kz + \pi) \end{aligned}\right\}$$

(b) $\left.\begin{array}{l} E_x = E_0 \sin(\omega t + kz) \\ E_y = E_0 \cos(\omega t + kz) \end{array}\right\}$

(c) $\left.\begin{array}{l} E_x = E_0 \sin\left(kz - \omega t + \dfrac{\pi}{3}\right) \\ E_y = E_0 \sin\left(kz - \omega t - \dfrac{\pi}{6}\right) \end{array}\right\}$

(d) $\left.\begin{array}{l} E_x = E_0 \sin\left(kz - \omega t + \dfrac{\pi}{4}\right) \\ E_y = \dfrac{1}{\sqrt{2}} E_0 \sin(kz - \omega t) \end{array}\right\}$

13.4 The electric field components of a plane electromagnetic wave are

$$E_x = 2E_0 \cos(\omega t - kz + \phi); \quad E_y = 2E_0 \sin(\omega t - kz)$$

Draw the diagram showing the state of polarisation (i.e., circular, plane, elliptical or unpolarised) when

(a) $\phi = 0$
(b) $\phi = \pi/2$
(c) $\phi = \pi/4$

13.5 Calculate the polarising angle for the air-glass interface, $n_1 \approx 1$ and $n_2 \approx 1.5$ and $\theta_p \approx 53°$ and also for the air-water interface, $n_1 \approx 1$ and $n_2 \approx 1.33$.

13.6 At $z = 0$, the field associated with an RCP beam (propagating in the $+z$ direction) is described by the following equation

$$E_x^r = E_0 \cos \omega t; \quad E_y^r = E_0 \sin \omega t \tag{29}$$

where the superscript r signifies that we are considering an RCP beam. Similarly, the field at $z = 0$, associated with an LCP beam (propagating in the $+z$ direction) is described by the following equation

$$E_x^l = E_0 \cos(\omega t - \phi); \quad E_y^l = -E_0 \sin(\omega t - \phi) \tag{30}$$

where the superscript l signifies that we are considering an LCP beam. Determine the SOP (state of polarisation) of the superposed beam (at $z = 0$).

13.7 Discuss the superposition of an RCP with an LCP of the same amplitude and both propagating in the $+z$ direction but with slightly different phase velocities:

$$E_x^r = E_0 \cos(k_r z - \omega t); \quad E_y^r = -E_0 \sin(k_r z - \omega t)$$
$$E_x^l = E_0 \cos(k_l z - \omega t); \quad E_y^l = E_0 \sin(k_l z - \omega t)$$

where $k_r = \dfrac{\omega}{c} n_r$ and $k_l = \dfrac{\omega}{c} n_l$; once again, the superscripts (and the subscripts) r and l signify that we are considering an RCP and LCP respectively; n_r and n_l are the corresponding refractive indices.

13.8 Consider a linearly polarised beam (propagating in the $+z$ direction) incident normally on a calcite (or quartz) crystal whose optic axis is along the y direction and the oscillating electric field is assumed to make an angle ϕ with the y-axis (in Fig. 13.8 we have shown $\phi = \frac{\pi}{4}$). Find the value of the thickness of the crystal (d) for which it will act as a QWP (quarter wave plate) or as a HWP (half wave plate).

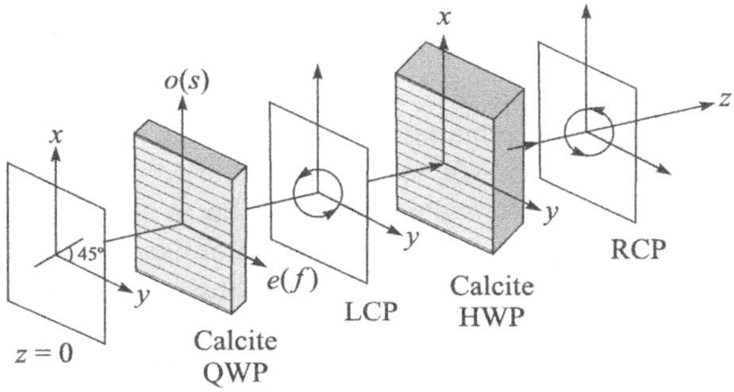

Fig. 13.8 A linearly polarised beam making an angle 45° with the y-axis gets converted to a LCP after propagating through a calcite QWP whose optic axis is along the y-axis as shown by lines parallel to the y-axis. Further, an LCP gets converted to a RCP after propagating through a calcite HWP; the optic axis of the HWP is also assumed to be along the y-direction.

13.9 (a) For calcite, $n_o = 1.65836$, $n_e = 1.48641$ (at 18°C and for $\lambda_0 = 5893$ Å). Calculate the thickness of the quarter wave plate. (b) In the above problem if we assume $\phi = \pi/4$ obtain the output SOP. (c) What would happen if the calcite QWP is replaced by a quartz QWP with its optic axis again along the y-axis.

13.10 What would be the output SOP, if an RCP is incident normally on a calcite QWP. What would happen if the calcite QWP is replaced by a calcite HWP.

13.11 A left circularly polarised beam ($\lambda_0 = 5893$ Å) is incident on a quartz crystal (with its optic axis cut parallel to the surface) of thickness 0.025 mm. Determine the state of polarisation of the emergent beam. Assume $n_o = 1.54425$ and $n_e = 1.55336$.

13.12 Consider the propagation of an extra-ordinary wave through a KDP crystal. If the wave vector is at an angle of 45° to the optic axis, calculate the angle between **S** and **k**. Repeat the calculation for $LiNbO_3$. For KDP $n_o = 1.5074$, $n_e = 1.4669$ and for $LiNbO_3$, $n_o = 2.2967$, $n_e = 2.2082$.

13.13 Prove that when the angle of incidence corresponds to the Brewster angle, the reflected and refracted rays are at right angles to each other.

13.14 (a) Consider two crossed polaroids placed in the path of an unpolarised beam of intensity I_0. If we place a third polaroid in between the two then, in general, some light will be transmitted through. Explain this phenomenon.

(b) Assuming the pass axis of the third polaroid to be at 45° to the pass axis of either of the polaroids, calculate the intensity of the transmitted beam. Assume that all the polaroids are perfect.

13.15 A quarter-wave plate is rotated between two crossed polaroids. If an unpolarised beam is incident on the first polaroid, discuss the variation of intensity of the emergent beam as the quarter-wave plate is rotated. What will happen if we have a half-wave instead of a quarter-wave plate?

13.16 In the previous problem, if the optic axis of the quarter-wave plate makes an angle of 45° with the pass axis of either polaroid, show that only a quarter of the incident intensity will be transmitted. If the quarter-wave plate is replaced by a half-wave plate, show that half of the incident intensity will be transmitted through.

13.17 For calcite, the values of n_o and n_e for $\lambda_0 = 4046$ Å are 1.68134 and 1.49694 respectively; corresponding to $\lambda_0 = 7065$ Å the values are 1.65207 and 1.48359 respectively. We have a calcite quarter-wave plate corresponding to $\lambda_0 = 4046$ Å. A left-circularly polarised beam of $\lambda_0 = 7065$ Å is incident on this plate. Obtain the state of polarisation of the emergent beam.

13.18 A HWP (half wave plate) is introduced between two crossed polaroids P_1 and P_2. The optic axis makes an angle 15° with the pass axis of P_1 as shown in Fig. 13.9. If an unpolarised beam of intensity I_0 is normally incident on P_1 and if I_1, I_2, and I_3 are the intensities after P_1, after HWP and after P_2 respectively, then calculate I_1/I_0, I_2/I_0 and I_3/I_0.

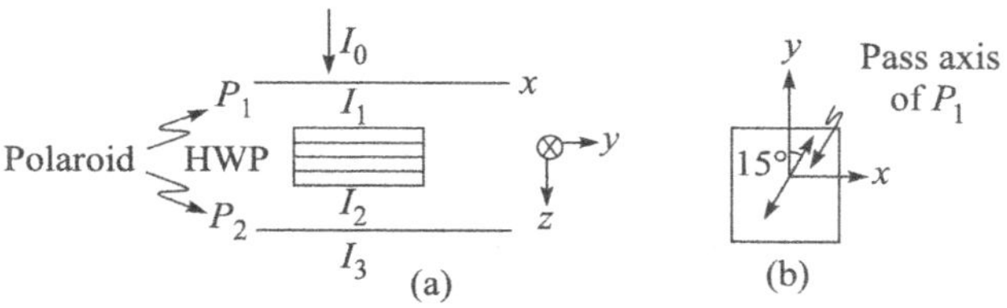

Fig. 13.9

13.19 Two prisms of calcite ($n_o > n_e$) are cemented together as shown in Fig. 13.10. Lines and dots show the direction of the optic axis. A beam of unpolarised light is incident normally from region *I*. Assume the angle of the prism to be 12°. Determine the path of rays in regions II, III and IV indicating the direction of vibrations (i.e., the direction of **D**).

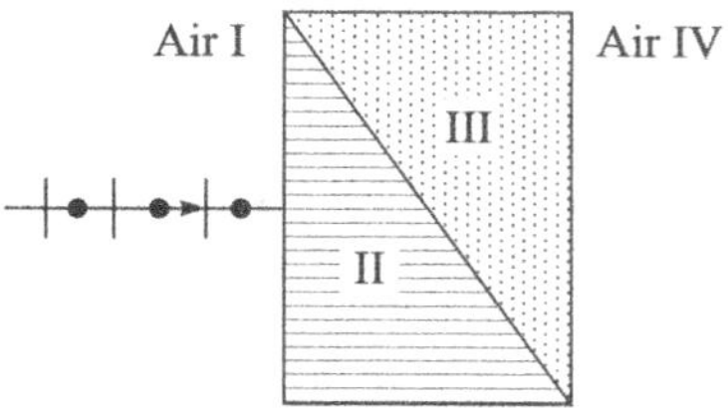

Fig. 13.10

13.20 A $\lambda/6$ plate is introduced in between the two crossed polarisers in such a way that the optic axis of the $\lambda/6$ plate makes an angle of 45° with the pass axis of the first polariser (see Fig. 13.11). Consider an unpolarised beam of intensity I_0 to be incident normally on the polariser. Assume the optic axis to be along the z-axis and the propagation along the x-axis. Write the y and z components of the electric fields (and the corresponding total intensities) after passing through (i) P_1 (ii) $\lambda/6$ plate and (iii) P_2.

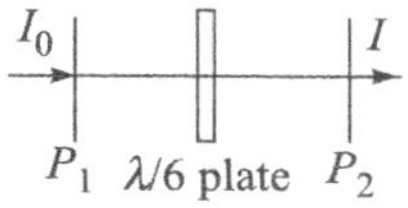

Fig. 13.11

13.21 A beam of light is passed through a polariser. If the polariser is rotated with the beam as an axis, the intensity I of the emergent beam does not vary. What are the possible states of polarisation of the incident beam? How to ascertain its state of polarisation with the help of the given polariser and a QWP?

13.22 Give the correct choice: A linearly polarised wave is incident normally on a $\lambda/4$ plate. The light coming out of the plate is (a) always CP, (b) always LP, (c) can never be LP, (d) can be LP, CP or even EP. (L: Linearly; C: Circularly; E: Elliptically; P: Polarised).

13.23 Consider a plane electromagnetic wave propagating in KDP with its **k** lying in the y-z plane and making an angle of 45° with the optic axis. Assuming $n_o = 1.507$ and $n_e = 1.467$, obtain the phase velocities of the ordinary and extraordinary waves. What is the angle between **E** and **D** vector for the two waves?

13.24 An anisotropic medium is characterised by $n_o = 2.6$ and $n_e = 2.9$. Calculate the refractive index of an extraordinary wave propagating with its **k** at 60° to the optic axis.

13.25 Calculate the thickness of a $\lambda/4$ and a $\lambda/2$ plate required for an operating wavelength of 0.589 μm when it is made of

(a) Calcite with $n_o = 1.66584$ and $n_e = 1.4864$

(b) Quartz with $n_o = 1.5442$ and $n_e = 1.5574$

What would be the difference in the output SOP when linearly polarised wave at 45° to the axes of the wave plate is incident on a $\lambda/4$ plate made of quartz and calcite?

13.26 Consider a 2 mm thick rectangular block of lithium niobate ($n_o = 2.26$, $n_e = 2.20$) with its optic axis at 45° to the surface as shown in the Fig. 13.12. A lightwave at a wavelength of 0.6 μm is incident normally on the block.

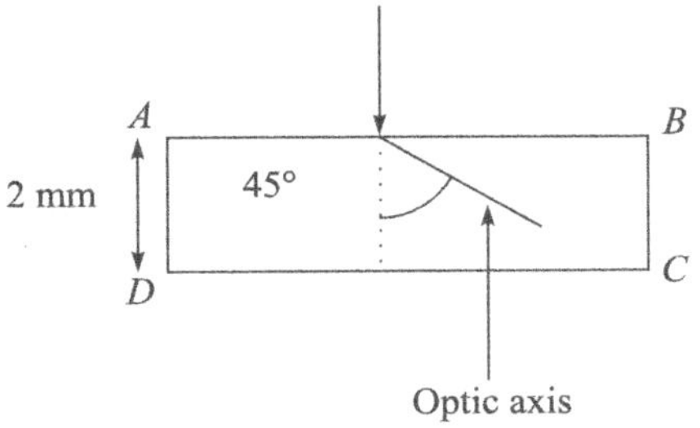

Fig. 13.12

(a) What should be the polarisation state of the incident wave so that it propagates as an e-wave?

(b) Calculate the phase difference accumulated between the o-wave and the e-wave in propagating from face AB to face CD.

(c) If the face CD is made reflecting, giving reasons, state the angle at which the reflected e-wave will re-emerge from the face AB.

13.27 A plane polarised wave polarised at 30° to the x-direction propagates along the z-direction. If you wish to convert this to a circularly polarised wave.

(a) What wave plate would you use to achieve this?

(b) What should be the orientation of the axes of the wave plate with respect to the x-direction to achieve this?

13.28 What should be the direction of propagation in a uniaxial medium to accumulate maximum phase difference between the ordinary and extraordinary waves?

13.29 Consider a right angled isosceles prism made from calcite with n_o = 1.658 and n_e = 1.486 with its optic axis perpendicular to the side AC (see Fig. 13.13). Light is incident normally on the face AB. What will happen to the waves as they strike the surface AC?

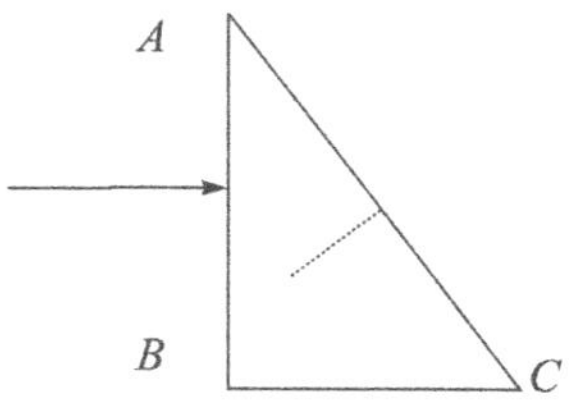

Fig. 13.13

13.30 Consider propagation of the extraordinary wave (in an uniaxial crystal) with its **k** making an angle of 45° with the optic axis and lying in the x-z plane (see Fig. 13.14). Show schematically **D, E, H, B** and **S** corresponding to the wave specifying values of all angles. What will be the wave velocity of the wave?

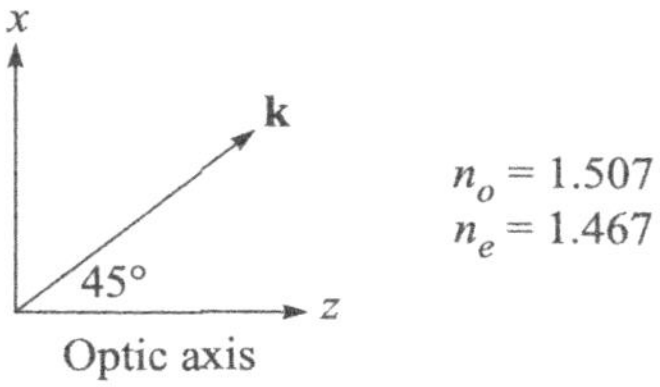

Fig. 13.14

13.31 Consider a Wollaston prism consisting of two similar prisms of calcite (n_o = 1.66 and n_e = 1.49) as shown in Fig. 13.15, with angle of prism equal to 25°. Calculate the angular divergence of the two emerging beams.

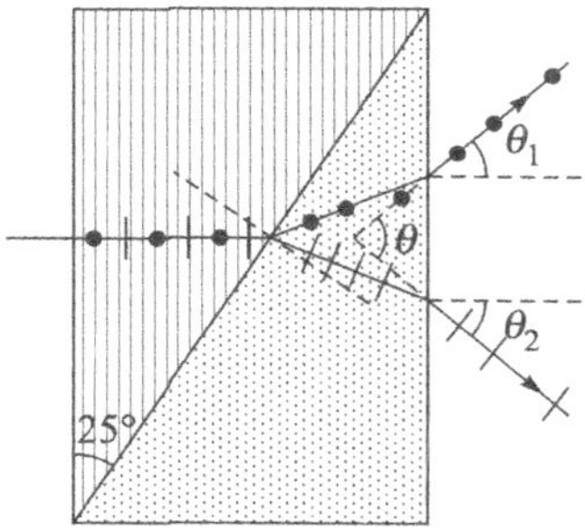

Fig. 13.15 A Wollaston prism. The lines and dots show the direction of the optic axis.

13.32 (a) Consider a plane wave incident normally on a calcite crystal with its optic axis making an angle of 20° with the normal. Thus ψ = 20°. Calculate the angle that the Poynting vector will make with the normal to the surface. Assume $n_o \approx 1.66$ and $n_e \approx 1.49$.

(b) In the above problem assume the crystal to be quartz with $n_o \approx 1.544$ and $n_e \approx 1.553$.

13.33 (a) In an anisotropic dielectric, we may assume $\mathbf{B} = \mu_0 \mathbf{H}$. Substitute plane wave solutions

$$\mathbf{E} = \mathbf{E}_0 e^{i(\mathbf{k}\cdot\mathbf{r} - \omega t)}, \ \mathbf{H} = \mathbf{H}_0 e^{i(\mathbf{k}\cdot\mathbf{r} - \omega t)}, \ \mathbf{D} = \mathbf{D}_0 e^{i(\mathbf{k}\cdot\mathbf{r} - \omega t)} \tag{31}$$

in Maxwell's equations. Show that **D** is always at right angles to **k** and **H** is at right angles to **k**, **E** and **D** implying **k**, **E** and **D** will always be in the same plane.

(b) Using the equations so derived, show that

$$\mathbf{D} = \frac{n_w^2}{c^2 \mu_0} [\mathbf{E} - (\hat{\kappa} \cdot \mathbf{E})\hat{\kappa}] \tag{32}$$

where $\hat{\kappa}$ is the unit vector along the direction of propagation of the wave and

$$|\mathbf{k}| = k = \frac{\omega}{c} n_w \tag{33}$$

13.34 In an anisotropic medium let the x, y and z axes are chosen along the principal axes so that

$$D_x = \varepsilon_x E_x; \quad D_y = \varepsilon_y E_y \quad \text{and} \quad D_z = \varepsilon_z E_z \tag{34}$$

where ε_x, ε_y and ε_z are known as the principal dielectric constants. In an uniaxial crystal we choose the z-axis along the optic axis and the x and y directions can be arbitrarily chosen as long as they are perpendicular to the z-axis. Also,

$$D_x = \varepsilon_x E_x; \quad D_y = \varepsilon_y E_y \quad \text{and} \quad D_z = \varepsilon_z E_z \tag{35}$$

where,

$$\varepsilon_x = \varepsilon_y = \varepsilon_0 n_o^2 \quad \text{and} \quad \varepsilon_z = \varepsilon_0 n_e^2 \tag{36}$$

and n_o and n_e are known as the ordinary and extra-ordinary refractive indices and at a given temperature (for a given wavelength) they are constants of the crystal. Without any loss of generality, we may assume $\kappa_y = 0$. Using the result derived in the previous problem, derive expressions for n_w for the ordinary and extra-ordinary waves.

13.35 If we dissolve cane sugar in water, then because of the spiral like structure of sugar molecules, the relation between **D** and **E** is given by the following relation

$$\mathbf{D} = \varepsilon_0 n^2 \mathbf{E} + ig\,\hat{\kappa} \times \mathbf{E} \tag{37}$$

Without any loss of the generality, we may assume propagation along the z-axis so that $\kappa_x = \kappa_y = 0$ and $\kappa_z = 1$. Solve the equation

$$\frac{n_w^2}{c^2 \mu_0}\left[\mathbf{E} - (\hat{\kappa}\cdot\mathbf{E})\,\hat{\kappa}\right] = \mathbf{D} \tag{38}$$

and show that modes are RCP and LCP. Calculate their velocities.

13.36 Consider the incidence of the following REP beam on a sugar solution at $z = 0$:

$$E_x = 5\cos\omega t; \quad E_y = 4\sin\omega t$$

with $\lambda = 6328$ Å. Assume $n_l - n_r \approx 10^{-5}$ and $n_l = 4/3$. Study the evolution of the SOP of the beam.

13.37 Consider a biaxial crystal with $n_x = 1.619$, $n_y = 1.620$ and $n_z = 1.626$. A circularly polarised plane wave at $\lambda_0 = 600$ nm propagates along the x-axis. After how much distance will the wave become linearly polarised?

13.38 Consider a biaxial crystal with $n_x = 1.56$, $n_y = 1.59$ and $n_z = 1.60$. Along which direction should a circularly polarised wave propagate so that it does not change its state of polarisation as it propagates?

SOLUTIONS

13.1 At $z = 0$, we will have

$$E_x = E_0 \cos\omega t \quad \text{and} \quad E_y = E_0 \sin\omega t$$

Thus,

at $\quad t = 0: \quad E_x = E_0 \quad \text{and} \quad E_y = 0$

at $t = \dfrac{\pi}{4\omega}$: $E_x = \dfrac{E_0}{\sqrt{2}}$ and $E_y = \dfrac{E_0}{\sqrt{2}}$

at $t = \dfrac{\pi}{2\omega}$: $E_x = 0$ and $E_y = E_0$ (see Fig. 13.3).

13.2 See Fig. 13.4.

13.3 (a) Propagation along the $-z$ direction (into the page). At $z = 0$

$$E_x = E_0 \cos \omega t; \quad E_y = -\frac{1}{\sqrt{2}} E_0 \cos \omega t$$

$$\frac{E_x}{E_y} = -\sqrt{2} = \tan(125.3°)$$

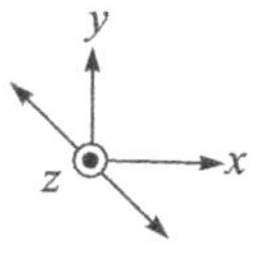

Fig. 13.16

LP (Linear Polarised) along the direction shown in the figure; the propagation is along the $-z$ axis; i.e., into the page (see Fig. 13.16).

(b) Propagation along the $-z$ direction (into the page). At $z = 0$

$$E_x = E_0 \sin \omega t; \quad E_y = E_0 \cos \omega t$$

$$\Rightarrow \quad E_x^2 + E_y^2 = E_0^2 \Rightarrow \text{Circularly polarised}$$

Since propagation is along the $-z$ axis; i.e., into the page, we have a RCP wave (see Fig. 13.17).

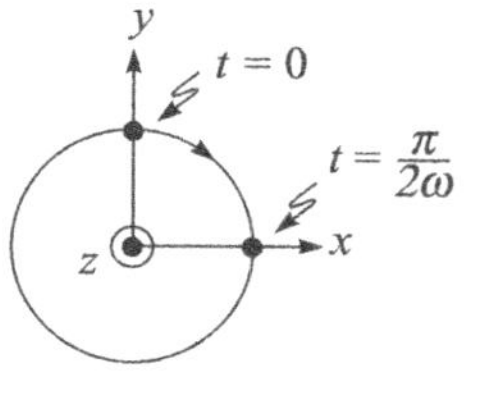

Fig. 13.17

(c) Propagation is along the $+z$ direction (into the page). At $z = \dfrac{\pi}{6k}$

$$E_x = E_0 \sin\left(\frac{\pi}{6} - \omega t + \frac{\pi}{3}\right) = E_0 \cos \omega t$$

$$E_y = -E_0 \sin \omega t$$

$$\Rightarrow \quad E_x^2 + E_y^2 = E_0^2 \Rightarrow \text{Circularly polarised wave}$$

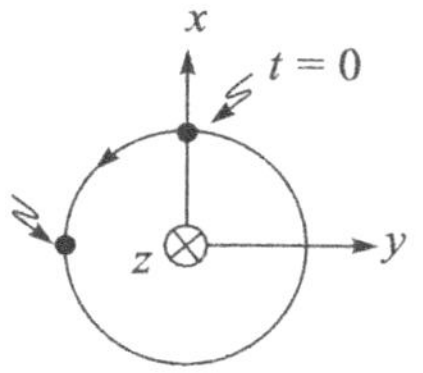

Fig. 13.18

Since propagation is along the $+z$ axis; i.e., into the page, we have a LCP wave.

(d) Propagation is along the $+z$ axis (into the page). Now, at $z = 0$

$$E_x = E_0 \sin\left(\frac{\pi}{4} - \omega t\right); \quad E_y = -\frac{E_0}{\sqrt{2}} \sin \omega t$$

Thus,

at $t = 0$; $E_x = \dfrac{E_0}{\sqrt{2}}$ and $E_y = 0$

$t = \dfrac{\pi}{4\omega}$; $E_x = 0$ and $E_y = -\dfrac{E_0}{2}$

$t = \dfrac{\pi}{2\omega}$; $E_x = -\dfrac{E_0}{\sqrt{2}}$ and $E_y = -\dfrac{E_0}{\sqrt{2}}$

$t = \dfrac{3\pi}{4\omega}$; $E_x = -E_0$ and $E_y = -\dfrac{E_0}{2}$

$t = \dfrac{\pi}{\omega}$; $E_x = -\dfrac{E_0}{\sqrt{2}}$ and $E_y = 0$

$$t = \frac{5\pi}{4\omega}; \qquad E_x = 0 \qquad \text{and} \qquad E_y = +\frac{E_0}{2}$$

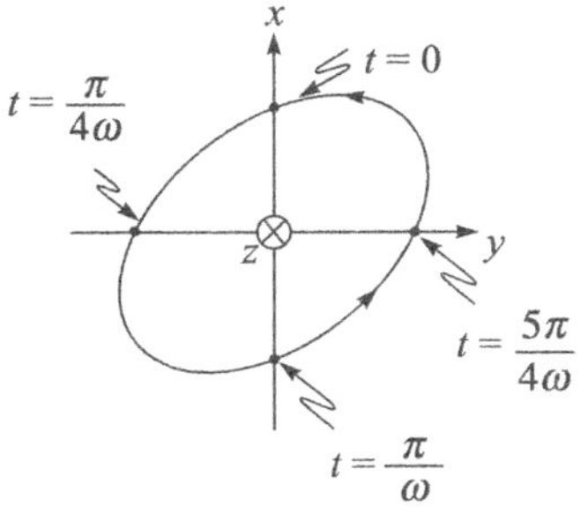

Fig. 13.19

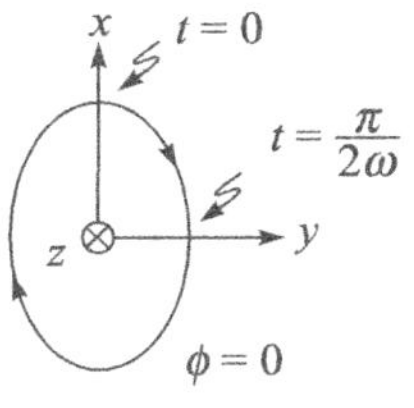

Fig. 13.20

The wave is LEP (Left Elliptically Polarised) as shown in Fig. 13.19. Further

$$E_x = \frac{E_0}{\sqrt{2}} \cos \omega t - \frac{E_0}{\sqrt{2}} \sin \omega t$$

or, $$E_x - E_y = \frac{E_0}{\sqrt{2}} \cos \omega t \quad \text{and} \quad E_y = -\frac{E_0}{\sqrt{2}} \sin \omega t$$

Thus $E_x^2 + 2E_y^2 - 2E_x E_y = \frac{E_0^2}{2}$ which represents ellipse in the $E_x - E_y$ plane.

13.4 Propagation is in the $+z$ direction. Now, at $z = 0$

$$E_x = 2E_0 \cos(\omega t + \phi) \quad \text{and} \quad E_y = E_0 \sin \omega t$$

(a) For $\phi = 0$: $E_x = 2E_0 \cos \omega t$, $E_y = E_0 \sin \omega t$

$$\left(\frac{E_x}{2E_0}\right)^2 + \left(\frac{E_y}{E_0}\right)^2 = 1$$

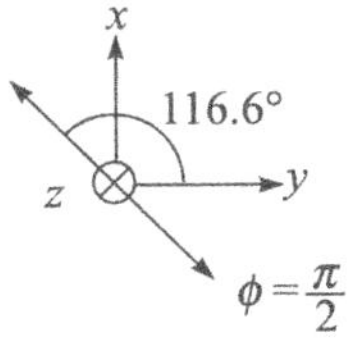

Fig. 13.21

As can be seen form Fig. 13.20, the wave is REP (Right Elliptically Polarised) with major axis along the x-direction.

(b) For $\phi = \frac{\pi}{2}$: $E_x = -2E_0 \sin \omega t$ and $E_y = E_0 \sin \omega t$

$$\frac{E_x}{E_y} = -2 \approx \tan(116.56°)$$

Thus we have a linearly polarised wave (see Fig. 13.21).

(c) For $\phi = \frac{\pi}{2}$: $E_x = -2E_0 \cos\left(\omega t + \frac{\pi}{4}\right)$ and $E_y = E_0 \sin \omega t$

$$t = 0; \qquad E_x = \sqrt{2} E_0, \qquad E_y = 0$$

$$t = \frac{\pi}{4\omega}; \qquad E_x = 0, \qquad E_y = \frac{E_0}{\sqrt{2}}$$

$$t = \frac{\pi}{2\omega}; \quad E_x = -\sqrt{2}E_0, \quad E_y = E_0$$

$$t = \frac{3\pi}{4\omega}; \quad E_x = -2E_0, \quad E_y = \frac{E_0}{\sqrt{2}}$$

$$t = \frac{\pi}{\omega}; \quad E_x = -\sqrt{2}E_0, \quad E_y = 0$$

etc. The wave is REP (Right Circularly Polarised)—see Fig. 13.22.

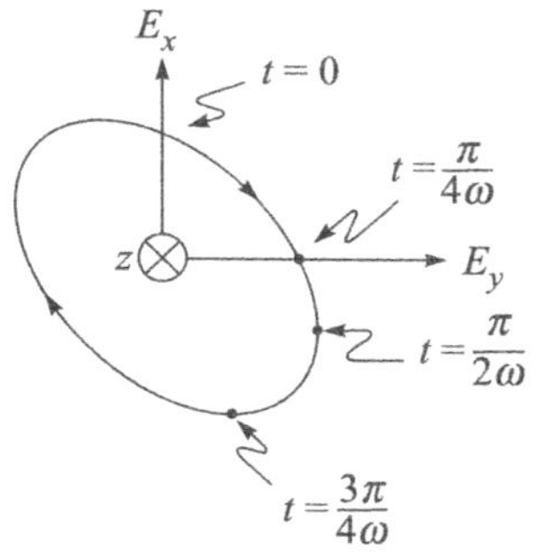

Fig. 13.22

13.5 (a) For the air-glass interface, $n_1 \approx 1$ and $n_2 \approx 1.5$ and

$$\theta_p = \tan^{-1}(1.5) \approx 56.3°$$

For the air-water interface, $n_1 \approx 1$ and $n_2 \approx 1.33$ and

$$\theta_p = \tan^{-1}(1.33) \approx 53.1°$$

If we assume the simultaneous propagation of the two beams then the x and y components of the resultant fields would be given by the following equations:

13.6 At $z = 0$, the superposed field will be given by

$$E_x = E_0[\cos \omega t + \cos(\omega t - \phi)] = 2E_0 \cos\frac{\phi}{2} \cos\left(\omega t - \frac{\phi}{2}\right)$$

$$E_y = E_0[\sin \omega t - \sin(\omega t - \phi)] = 2E_0 \sin\frac{\phi}{2} \cos\left(\omega t - \frac{\phi}{2}\right)$$

Thus, the resultant wave is linearly polarised with the direction of the oscillating electric vector making an angle $\phi/2$ with the x-axis.

13.7 If we superpose the two fields we would obtain

$$E_x = 2E_0 \cos\left[\frac{\phi(z)}{2}\right] \cos\left[\omega t - \frac{\theta(z)}{2}\right]$$

$$E_y = 2E_0 \sin\left[\frac{\phi(z)}{2}\right] \cos\left[\omega t - \frac{\theta(z)}{2}\right]$$

where, $\phi(z) = (k_l - k_r)z$ and $\theta(z) = (k_l + k_r)z$

Thus the resultant wave is *always* linearly polarised with the direction of the oscillating electric vector making an angle $\phi(z)/2$ with the x-axis; thus the SOP rotates as it propagates through the optically active medium. Further,

$$\phi(z) = (k_l - k_r)z = \frac{2\pi}{\lambda_0}(n_l - n_r)z$$

where λ_0 is the free space wavelength. Now, if

$n_l > n_r \Leftrightarrow$ the optically active substance is said to be right-handed or dextro-rotatory

$n_r > n_l \Leftrightarrow$ the optically active substance is said to be left-handed or laevo-rotatory.

13.8 Let the plane $z = 0$ represent the surface of the crystal on which the beam is incident. The x- and y-components of the incident beam can be written in the form

$$E_x = E_0 \sin\phi \cos(kz - \omega t); \quad E_y = E_0 \cos\phi \cos(kz - \omega t)$$

where $k (= \omega/c)$ represents the free-space wave number. Inside the crystal, the x and y components propagate as ordinary and extraordinary waves:

$$E_x = E_0 \sin\phi \cos(n_o kz - \omega t) \quad \text{(ordinary wave)}$$
$$E_y = E_0 \cos\phi \cos(n_e kz - \omega t) \quad \text{(extra-ordinary wave)} \qquad (39)$$

If the thickness of the crystal is d, then at the emerging surface, we will have

$$E_x = E_0 \sin\phi \cos(\omega t - \theta_o); \quad E_y = E_0 \cos\phi \cos(\omega t - \theta_e)$$

where, $\theta_o = n_o kd$ and $\theta_e = n_e kd$. By appropriately choosing the instant $t = 0$, the components may be rewritten as

$$E_x = E_0 \sin\phi \cos(\omega t - \theta); \quad E_y = E_0 \cos\phi \cos\omega t$$

where, $$\theta = \theta_o - \theta_e = kd(n_o - n_e) = \frac{2\pi}{\lambda_0}(n_o - n_e)d \qquad (40)$$

represents the phase difference between the ordinary and the extra-ordinary beams. Clearly, if the thickness of the crystal is such, that $\theta = 2\pi, 4\pi, 6\pi\ldots$ the emergent beam will have the same state of polarisation as the incident beam. Now, if the thickness d of the crystal is such that $\theta = \pi/2$, the crystal is said to be a quarter wave plate (usually abbreviated as QWP)—a phase difference of $\pi/2$ implies a path difference of a quarter of a wavelength—and if $\phi = \frac{\pi}{4}$ (see Fig. 13.8) the output will be LCP. On the other hand, if the thickness of the crystal is such that $\theta = \pi$, the crystal is said to be a half wave plate (usually abbreviated as HWP); an LCP incident normally on a HWP will get converted to RCP (see Fig. 13.8).

13.9 (a) $d = \frac{1}{4}\frac{\lambda_0}{(n_o - n_e)} = \frac{5893 \times 10^{-8}}{4 \times 0.17195} = 0.0000857$ cm.

(b) The output SOP will be LCP

(c) If the calcite QWP is replaced by a quartz QWP (with its optic-axis along the y-direction) the output will be RCP.

13.10 We assume propagation along the $+z$ axis (see Fig. 13.23). For a RCP, we may assume

$$E_x = E_0 \cos(\omega t - kz); \quad E_y = E_0 \sin(\omega t - kz)$$

Thus at $z = 0$, $E_x = E_0 \cos\omega t$ and $E_y = E_0 \sin\omega t$ which represents a RCP. The x-polarised wave propagates as an o-wave and the y-polarised wave propagates as an e-wave. Thus

$$E_x = E_0 \cos(\omega t - \theta); \quad E_y = E_0 \sin\omega t$$

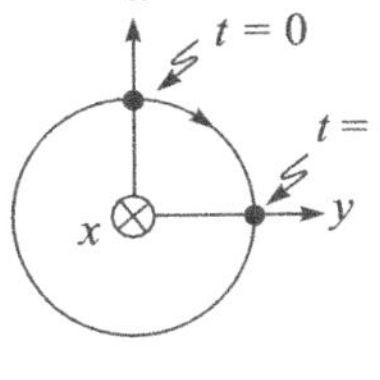

Fig. 13.23

where $\theta = (k_o - k_e)z = \frac{2\pi}{\lambda_0}(n_o - n_e)z$. For a calcite HWP, $n_o > n_e$ and $\theta = \pi$; thus after emerging from the half wave plate, it will be

$$E_x = -E_0 \cos \omega t,\ E_y = E_0 \sin \omega t$$

which represents a LCP.

13.11 The electric field for the incident beam at $z = 0$ would be

$$E_x = \frac{E_0}{\sqrt{2}} \sin \omega t; \quad E_y = \frac{E_0}{\sqrt{2}} \cos \omega t \tag{41}$$

The x-polarised wave propagates as an o-wave and the y-polarised wave propagates as an e-wave. Further,

$$\theta' = (n_e - n_o)\frac{2\pi}{\lambda_0} d = 2\pi \frac{0.00911 \times 0.025}{5893 \times 10^{-7}} \approx 0.77\pi$$

Thus, the emergent beam will be

$$E_x = \frac{E_0}{\sqrt{2}} \sin(\omega t + 0.77\pi); \quad E_y = \frac{E_0}{\sqrt{2}} \cos(\omega t)$$

which will represent a right elliptically polarised light.

13.12 We refer to Fig. 13.7; thus $\psi = 45°$. Now the angle ϕ between **S** and **k** is given by

$$\phi = \tan^{-1}\left[\frac{n_o^2}{n_e^2} \tan \psi\right] - \psi$$

For KDP,

$$\phi = \tan^{-1}\left[\left(\frac{1.5074}{1.4669}\right)^2\right] - 45° \approx 1.56°$$

For $LiNbO_3$,

$$\phi = \tan^{-1}\left[\left(\frac{2.2967}{2.2082}\right)^2\right] - 45° \approx 2.25°$$

13.13 When the angle of incidence is the Brewster angle θ_p,

$$\tan \theta_p = \frac{n_2}{n_1} \tag{42}$$

Now, using Snell's law (see Fig. 13.24)

$$n_1 \sin \theta_p = n_2 \sin r \Rightarrow \sin \theta_p = \frac{n_2}{n_1} \sin r = \tan \theta_p \sin r$$

$$\Rightarrow \quad \cos \theta_p = \sin r \Rightarrow r = \frac{\pi}{2} - \theta_p \Rightarrow r + \theta_p = \frac{\pi}{2}.$$

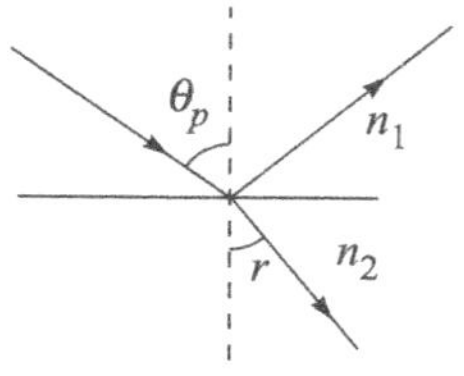

Fig. 13.24

13.14

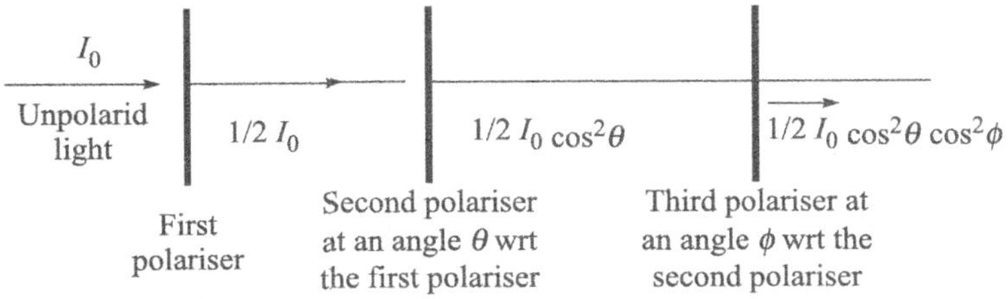

Fig. 13.25

(a) The first polaroid produces linearly polarised light of electric filed strength E_0. When it encounters the next polariser which is placed at an angle $\neq 90°$ to the first polaroid, then some component is passed through, the amplitude of resulting electric field being $E_0 \cos\theta$, where θ is the angle between the first two polaroids (see Fig. 13.25).

(b) Now when this encounters a third polaroid, a part of the electric field $E_0 \cos\theta \cos\phi$ is passed through. If $\theta = \phi = \frac{\pi}{4}$, then

$$I = I_0 \cdot \frac{1}{2}\left(\cos^2\frac{\pi}{4}\right)\left(\cos^2\frac{\pi}{4}\right) = \frac{1}{8} I_0.$$

13.15

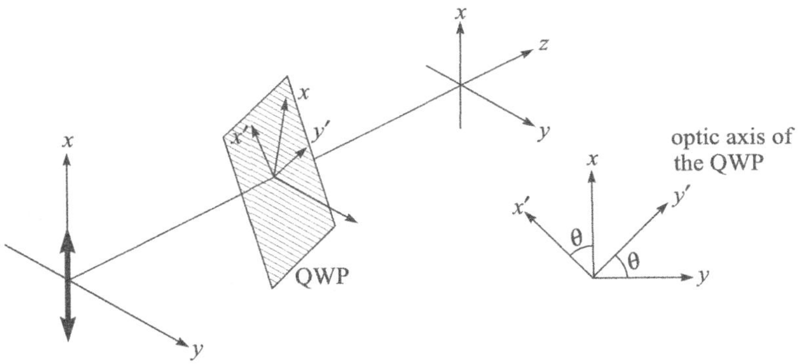

Fig. 13.26

Let the optic axis of the QWP (shown as y' axis in Fig. 13.26) make an angle θ with the y-axis. Then, for a x polarised beam, just before the QWP

$$E_{x'} = E_0 \cos\theta \cos\omega t; \quad E_{y'} = E_0 \sin\theta \cos\omega t$$

After passing through the (calcite) QWP, we will have (see Solution 13.10):

$$E_{x'} = E_0 \cos\theta \cos\left(\omega t - \frac{\pi}{2}\right) = E_0 \cos\theta \sin\omega t$$

$$E_{y'} = E_0 \sin\theta \cos\omega t$$

[If $\theta = \frac{\pi}{4}$, we will have a LCP]. Thus, component along the y-axis that is transmitted by the second Polaroid is given by,

$$E_y = -E_0 \cos\theta \sin\theta \sin\omega t + E_0 \sin\theta \cos\theta \cos\omega t$$

Thus, $$I = K\langle E_y^2\rangle = KE_0^2\left[\frac{1}{2}\sin^2\theta\cos^2\theta + \frac{1}{2}\sin^2\theta\cos^2\theta\right] = \frac{1}{2}I_0\sin^2 2\theta$$

where $I_0\left(=\frac{1}{2}KE_0^2\right)$ is the intensity incident on the QWP. This follows from the fact that when $\theta = \frac{\pi}{4}$, the SOP after the QWP is LCP and the intensity (after the analyser) must be $\frac{1}{2}I_0$. Further when $\theta = 0$, the x-polarised wave travels as on o-wave and the SOP (after traversing through the QWP) remains unchanged. Thus, the intensity after the analyser will be zero. Similarly, when $\theta = \frac{\pi}{2}$, the x-polarised wave travels as on e-wave and the SOP remains unchanged giving zero intensity after the analyser.

If instead of a QWP, we have a (calcite) HWP, then after passing through the HWP, we will have,

$$E_{x'} = E_0\cos\theta\cos(\omega t - \pi) = -E_0\cos\theta\cos\omega t;\ E_{y'} = E_0\sin\theta\cos\omega t$$

Thus the component along the y-axis that is transmitted by the second polaroid is given by,

$$E_y = +2E_0\cos\theta\sin\theta\cos\omega t \quad \Rightarrow \quad I = I_0\sin^2 2\theta$$

Obviously, when $\theta = \frac{\pi}{4}$, after the HWP the SOP of the beam will be rotated by $\frac{\pi}{2}$ and the (y-polarised) beam will pass through the analyser.

13.17 For $\lambda_0 = 4046$ Å, the thickness of the QWP is given by

$$d = \frac{\lambda_0}{4(n_o - n_e)} = \frac{4.046\times10^{-5}}{4(1.68134 - 1.49694)} \approx 5.49\times10^{-5}\ \text{cm}$$

At $\lambda_0 = 7065$ Å, the phase difference introduced is given by:

$$\theta = \frac{2\pi}{\lambda_0}(n_o - n_e)d = \frac{2\pi}{7.065\times10^{-5}} \times \frac{4.046\times10^{-5}}{4} \approx \frac{\pi}{3.49} \approx \frac{\pi}{4}\ \text{(say)}$$

The LCP is incident on the QWP is given by

$$E_x = E_0\sin\omega t;\quad E_y = E_0\cos\omega t$$

The output beam will be

$$E_x = E_0\sin\left(\omega t - \frac{\pi}{4}\right);\quad E_y = E_0\cos\omega t$$

which is a LEP beam (see Fig. 13.27).

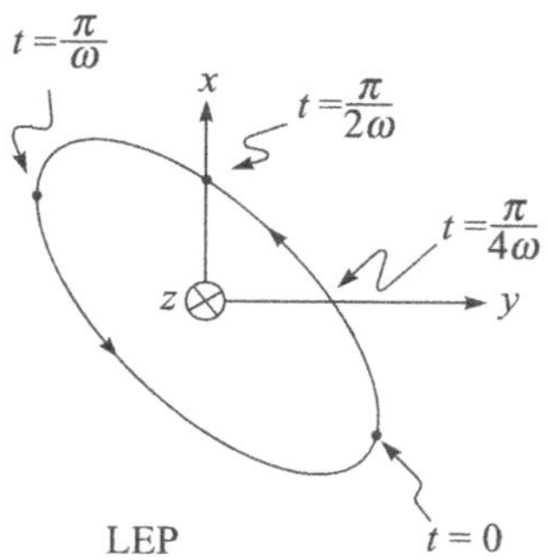

Fig. 13.27

13.18 $I_1 = \frac{1}{2}I_0$. After the HWP, the intensity remains the same $I_2 = I_1 = \frac{1}{2}I_0$. I_3 can be obtained using Problem 13.14:

$$I_3 = \frac{1}{2}I_0\sin^2 2\theta = \frac{1}{2}I_0\sin^2 30° = \frac{1}{8}I_0.$$

13.19 The polarisation in the plane of the paper will pass through undeviated. The polarisation normal to the plane of the paper will pass through as an ordinary wave in the first crystal (see Fig. 13.6) and will pass through an extra-ordinary wave in the second crystal; thus,

$$n_o \sin 12° = n_e \sin r \Rightarrow r = \sin^{-1}\left[\frac{n_o}{n_e}\sin 12°\right] = \sin^{-1}\left[\frac{1.658}{1.486}\sin 12°\right] \approx 13.41°$$

Therefore the angle of incidence at the second surface will be 13.41° – 12° = 1.41°. The emerging angle will be

$$\sin\theta = n_e \sin(1.41°) = 1.486 \times 0.0246 \Rightarrow \theta \approx 2.1°.$$

13.20

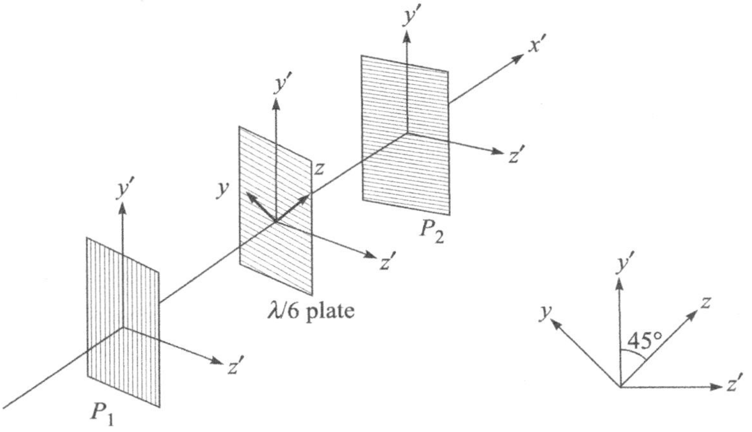

Fig. 13.28

After passing through P_1, the beam is polarised along the y' axis (see Fig. 13.28). Obviously, if I_0 is the intensity of the incident beam then the intensity of the beam coming out from P_1 is given by

$$I_1 = \frac{1}{2} I_0$$

Let the amplitude of the beam emerging from P_1 be E_1, then

$$E_y = \frac{1}{\sqrt{2}} E_1 \cos\omega t; \quad E_z = \frac{1}{\sqrt{2}} E_1 \cos\omega t$$

The optic axis of the $\lambda/6$ plate is along the z-axis; thus, after passing through the $\lambda/6$ plate

$$E_y = \frac{1}{\sqrt{2}} E_1 \cos\left(\omega t - \frac{\pi}{3}\right); \quad E_z = \frac{1}{\sqrt{2}} E_1 \cos\omega t$$

[The y-polarised beam propagates as an o-wave and the z-polarised wave propagates as an e-wave and we have assumed the crystal to be negative like calcite]. The intensity is still $1/2\,I_1$. Now, only z' component passes through the Polaroid P_2; thus,

$$E_{z'} = -\frac{1}{\sqrt{2}}E_y + \frac{1}{\sqrt{2}}E_z = \frac{1}{2}E_1\left[-\cos\left(\omega t - \frac{\pi}{3}\right) + \cos\omega t\right]$$

$$= -E_1 \sin\left(\omega t - \frac{\pi}{6}\right)\cdot \sin\frac{\pi}{6} = -\frac{1}{2}E_1 \sin\left(\omega t - \frac{\pi}{6}\right)$$

Thus, $I_2 = \frac{1}{4}I_1 = \frac{1}{8}I_0.$

13.21 The incident beam may be either (a) unpolarised or (b) circularly polarised or (c) a mixture of both unpolarised and circularly polarised beam. If we now place a QWP after the beam then the circularly polarised beam will become linearly polarised which would give complete extinction at two positions if we put a polariser after the QWP and rotate this polariser.

On the after hand, if there is no intensity variation as the second polariser is rotated then the incident beam is unpolarised light and if there is some intensity variation then the incident beam will be a mixture of unpolarised light and circularly polarised light.

13.22 (d)

13.23 The ordinary wave will have an index of 1.507 and hence its phase velocity will be $c/1.507 \approx 1.9907 \times 10^8$ m/s. The refractive index of the extraordinary wave will be given by Eq. (25) with $\psi = 45°$. This gives us an index of 1.4866 for the refractive index of the extraordinary wave. Thus, the phase velocity of the extraordinary wave will be $c/1.4866 \approx 2.018 \times 10^8$ m/s.

For the ordinary wave **E** and **D** will be along the same direction.

For the extraordinary wave, since the direction of propagation is at 45° to the optic axis, the displacement vector of the extraordinary wave will also subtend an angle of 135° with the z-axis since the displacement vector is always perpendicular to the propagation direction. Hence,

$$\frac{D_x}{D_z} = -1 = \frac{\varepsilon_0 n_o^2 E_x}{\varepsilon_0 n_e^2 E_z} = \frac{n_o^2 E_x}{n_e^2 E_z}$$

Hence if ϕ is the angle between **E** and **D** then

$$\tan(\psi + \phi) = -\frac{E_z}{E_x} = \frac{n_o^2}{n_e^2} = 1.5528 \qquad (43)$$

Since $\psi = 45°$, we get $\phi \approx 1.54°$.

13.24 Substitute for n_o and n_e in Eq. (25) and assume $\psi = 60°$

13.25 (a) 0.81 μm and 1.62 μm (b) 11.16 μm and 22.32 μm. Calcite is negative uniaxial while quartz is positive uniaxial. Hence in one case the output will be right circularly polarised and in the other case it would be left circularly polarised.

13.26 (a) Since the polarisation state of an e-wave is perpendicular to the propagation direction and lies in the plane containing the optic axis and the propagation direction, the incident polarisation must be in the plane of the figure so that it propagates as an e-wave.

(b) Since the incident wave is perpendicular to the interface, the propagation vectors of both the ordinary and the extra ordinary refracted waves would be perpendicular to the interface. Hence the phase difference between the two waves in propagating from the face AB to the face CD would be

$$\Delta\phi = \frac{2\pi}{\lambda_0}(n_o - n_e)L = 400\pi$$

(c) The propagation vector of the wave reflected from the face CD would again be perpendicular to the interface and as it emerges from the face AB it will again emerge perpendicular to the interface. This is due to the conservation of the tangential component of the propagation vector at the interfaces.

13.27 (a) We would use a quarter wave plate.

(b) The fast and slow axes of the quarter wave plate should make an angle of 45° with the orientation of the input polarisation state.

13.28 The wave should propagate perpendicular to the optic axis of the uniaxial medium.

13.29 The ordinary wave will be polarised perpendicular to the plane of the figure while the extra ordinary wave will be polarised in the plane of the figure. Since the angle of incidence on the face AC will be 45°, both the waves will be incident at an angle greater than the critical angle at the interface and will get total internally reflected.

13.30

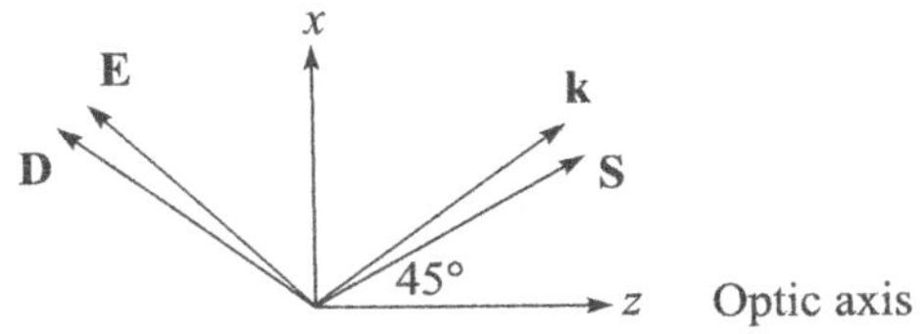

Fig. 13.29

13.31 Corresponding to the perpendicular polarisation [see Fig. 13.15], the angle of refraction is given by

$$n_o \sin 25° = n_e \sin r_1 \Rightarrow \sin r_1 = \frac{1.66}{1.49} \times \sin 25° \Rightarrow r_1 \approx 28.1°$$

Thus the angle of incidence at the second surface will be $i_1 \approx 28.1° - 25° = 3.1°$. The output angle θ_1 will be given by

$$n_e \sin 3.1° = \sin \theta_1 \Rightarrow \theta_1 \approx 4.62°$$

For the y-polarised beam,

$$n_e \sin 25° = n_o \sin r_2 \Rightarrow \sin r_2 = \frac{1.49}{1.66} \times \sin 25° \Rightarrow r_2 \approx 22.3°$$

Thus, $\quad i_2 \approx 25° - 22.3° = 2.7°.$

The output of angle θ_2 will be given by

$$n_o \sin 2.7° = \sin \theta_2 \Rightarrow \theta_2 \approx 4.5°$$

Thus, $\quad \theta = \theta_1 + \theta_2 \approx 7.2°.$

13.32 (a) $\psi = 20°$, $n_o \approx 1.66$ and $n_e \approx 1.49$. Thus,

$$\phi = \tan^{-1}\left[\frac{n_o^2}{n_e^2}\tan\psi\right] - \psi \approx 24.31° - 20° = 4.31°$$

(b) If $n_o \approx 1.544$, $n_e \approx 1.553$ then for $\psi = 20°$

$$\phi \approx 19.79° - 20° = -0.21°$$

13.33 (a) In a (charge free) dielectric div $\mathbf{D} = 0$. If we substitute the plane wave solutions, we would obtain

$$i(k_x D_x + k_y D_y + k_z D_z) = 0 \Rightarrow \mathbf{D}\cdot\mathbf{k} = 0 \tag{44}$$

Thus $\mathbf{D}$ is always at right angles to $\mathbf{k}$. Similarly since in a nonmagnetic medium div $\mathbf{H} = 0$ giving $\mathbf{H}\cdot\mathbf{k} = 0$. Thus,

$$\mathbf{D} \text{ and } \mathbf{H} \text{ will be right angles to } \mathbf{k} \tag{45}$$

Now, in the absence of any currents (i.e., $J = 0$) Maxwell's curl equations become

$$\nabla\times\mathbf{E} = -\frac{\partial B}{\partial t} = i\omega\mathbf{B} = i\omega\mu_0\mathbf{H} \tag{46}$$

and

$$\nabla\times\mathbf{H} = \frac{\partial D}{\partial t} = -i\omega\mathbf{D} \tag{47}$$

Substituting the plane wave solutions, we would obtain

$$(\nabla\times\mathbf{E})_x = \frac{\partial E_z}{\partial y} - \frac{\partial E_y}{\partial z} = (ik_y E_{0z} - ik_z E_{0y})\,e^{i(\mathbf{k}\cdot\mathbf{r}-\omega t)}$$

$$= i(k_y E_z - k_z E_y) = i(\mathbf{k}\times\mathbf{E})_x$$

Thus,

$$\nabla\times\mathbf{E} = i(\mathbf{k}\times\mathbf{E}) = i\omega\mu_0\mathbf{H} \Rightarrow \mathbf{H} = \frac{1}{\omega\mu_0}(\mathbf{k}\times\mathbf{E}) \tag{48}$$

and

$$\nabla\times\mathbf{H} = i(\mathbf{k}\times\mathbf{H}) = -i\omega\mathbf{D} \Rightarrow \mathbf{D} = \frac{1}{\omega}(\mathbf{H}\times\mathbf{k}) \tag{49}$$

The above equations show that

$$\mathbf{H} \text{ is at right angles to } \mathbf{k}, \mathbf{E} \text{ and } \mathbf{D} \tag{50}$$

implying

k, **E** and **D** will always be in the same plane.

(b) Substituting for $\mathbf{H}$ in Eq. (49), we get

$$\mathbf{D} = \frac{1}{\omega^2\mu_0}[(\mathbf{k}\times\mathbf{E})\times\mathbf{k}]$$

$$= \frac{1}{\omega^2\mu_0}[(\mathbf{k}\cdot\mathbf{k})\mathbf{E} - (\mathbf{k}\cdot\mathbf{E})\mathbf{k}] \tag{51}$$

where we have used the vector identity

$$(\mathbf{A}\times\mathbf{B})\times\mathbf{C} = (\mathbf{A}\cdot\mathbf{C})\mathbf{B} - (\mathbf{B}\cdot\mathbf{C})\mathbf{A} \tag{52}$$

Thus, $$\mathbf{D} = \frac{k^2}{\omega^2 \mu_0}[\mathbf{E} - (\hat{\kappa}\cdot\mathbf{E})\,\hat{\kappa}] = \frac{n_w^2}{c^2 \mu_0}[\mathbf{E} - (\hat{\kappa}\cdot\mathbf{E})\,\hat{\kappa}] \tag{53}$$

where $$\hat{\kappa} = \frac{\mathbf{k}}{k} \tag{54}$$

represents the unit vector along k.

13.34 Since $D_x = \varepsilon_x E_x = \varepsilon_0 n_x^2 E_x$, we have for the x-component of Eq. (53)

$$\frac{\varepsilon_0 \mu_0 c^2 n_x^2}{n_w^2} E_x = E_x - \kappa_x(\kappa_x E_x + \kappa_y E_y + \kappa_z E_z)$$

Since $c^2 = \dfrac{1}{\varepsilon_0 \mu_0}$, we have

$$\left(\frac{n_x^2}{n_w^2} - \kappa_y^2 - \kappa_z^2\right) E_x + \kappa_x \kappa_y E_y + \kappa_x \kappa_z E_z = 0 \tag{55}$$

where we have used the relation

$$\kappa_x^2 + \kappa_y^2 + \kappa_z^2 = 1 \quad \text{(since } \hat{\kappa} \text{ is a unit vector)} \tag{56}$$

Similarly,

$$\kappa_x \kappa_y E_x + \left(\frac{n_y^2}{n_w^2} - \kappa_x^2 - \kappa_z^2\right) E_y + \kappa_y \kappa_z E_z = 0 \tag{57}$$

$$\kappa_x \kappa_z E_x + \kappa_y \kappa_z E_y + \left(\frac{n_z^2}{n_w^2} - \kappa_x^2 - \kappa_y^2\right) E_z = 0 \tag{58}$$

Without any loss of generality, we assume that the y-axis is at right angles to k; thus $\kappa_y = 0$ and we may write

$$\kappa_x = \sin\psi,\ \kappa_y = 0 \quad \text{and} \quad \kappa_z = \cos\psi \tag{59}$$

where ψ is the angle that the k vector makes with the optic axis (see Fig. 13.7). Equations (55) – (57) therefore become

$$\left(\frac{n_o^2}{n_w^2} - \cos^2\psi\right) E_x + \sin\psi\cos\psi\, E_z = 0 \tag{60}$$

$$\left(\frac{n_o^2}{n_w^2} - 1\right) E_y = 0 \tag{61}$$

and $$\sin\psi\cos\psi\, E_x + \left(\frac{n_e^2}{n_w^2} - \sin^2\psi\right) E_z = 0 \tag{62}$$

Since two equations involve only E_x and E_z and one equation involves only E_y we have the following two independent solutions:

First Solution: We assume $E_y \neq 0$ then $E_x = 0 = E_z$. From Eq. (61) one obtains the solution

$$n_w = n_{wo} = n_o \qquad \text{(ordinary wave)} \tag{63}$$

The corresponding wave velocity is

$$v_w = v_{wo} = \frac{c}{n_o} \qquad (y\text{-polarised } o\text{-wave}) \tag{64}$$

Since the wave velocity is independent of the direction of the wave, it is referred to as the ordinary wave (usually abbreviated as the o-wave) and hence the subscript 'o' on n_w and v_w.

Second Solution: The second solution of Eqs (60) – (62) will correspond to

$$E_y = 0 \quad \text{and} \quad E_x, E_z \neq 0 \tag{65}$$

We use Eqs (60) – (62) to obtain

$$\frac{E_z}{E_x} = -\frac{\dfrac{n_o^2}{n_w^2} - \cos^2\psi}{\sin\psi\cos\psi} = -\frac{\sin\psi\cos\psi}{\dfrac{n_e^2}{n_w^2} - \sin^2\psi}$$

Simple manipulations would give us

$$\frac{1}{n_w^2} = \frac{1}{n_{we}^2} = \frac{\cos^2\psi}{n_o^2} + \frac{\sin^2\psi}{n_e^2} \tag{66}$$

where the subscript e refers to the fact that the wave refractive index corresponds to the extra ordinary wave. The corresponding wave velocity would be given by

$$v_{we}^2 = \frac{c^2}{n_{we}^2} = \frac{c^2}{n_o^2}\cos^2\psi + \frac{c^2}{n_e^2}\sin^2\psi \tag{67}$$

Since the wave velocity is dependent on the direction of the wave, it is referred to as the extra ordinary wave and hence the subscript e. Of course, for the extra-ordinary wave, we must have

$$D_y = \varepsilon_y E_y = 0$$

Thus the displacement vector **D** of the wave is normal to the y-axis and also to **k** implying that

> *the displacement vector* **D** *associated with the extraordinary wave lies in the plane containing the propagation vector* **k** *and the optic axis and is normal to* **k**.

Let ϕ and θ represent the angles that the **S** vector makes with the **k** vector and the optic axis respectively (see Fig. 13.7). In order to determine the angle ϕ we note that

$$\frac{\varepsilon_z E_z}{\varepsilon_x E_x} = \frac{D_z}{D_x} = -\tan\psi$$

and since $$\frac{E_z}{E_x} = -\tan(\phi + \psi) \tag{68}$$

we get $$\frac{n_e^2}{n_o^2}\tan(\phi + \psi) = \tan\psi \tag{69}$$

or, $$\phi = \tan^{-1}\left[\frac{n_o^2}{n_e^2}\tan\psi\right] - \psi \tag{70}$$

Obviously, for negative crystals $n_o > n_e$ and ϕ will be positive implying that ray direction is further away from the optic axis as shown in Fig. 13.7.

13.35 We write

$$\mathbf{D} = \varepsilon_0 n^2 \mathbf{E} + ig\,\hat{\kappa} \times \mathbf{E}$$

$$= \varepsilon_0 n^2[\mathbf{E} + i\alpha\,\hat{\kappa} \times \mathbf{E}] \tag{71}$$

where, $$\alpha = \frac{g}{\varepsilon_0 n^2} \tag{72}$$

and $\hat{\kappa}$ is the unit vector along the direction of propagation of the wave. The parameter α can be positive or negative but it is usually an extremely small number ($<< 1$). Without any loss of the generality, we may assume propagation along the z-axis so that $\kappa_x = \kappa_y = 0$ and $\kappa_z = 1$ giving

$$\hat{\kappa} \times \mathbf{E} = \begin{pmatrix} \hat{\mathbf{x}} & \hat{\mathbf{y}} & \hat{\mathbf{z}} \\ 0 & 0 & 1 \\ E_x & E_y & E_z \end{pmatrix} = -\hat{\mathbf{x}}E_y + \hat{\mathbf{y}}E_x$$

Thus, $$\begin{pmatrix} D_x \\ D_y \\ D_z \end{pmatrix} = \begin{pmatrix} \varepsilon_0 n^2 & -ig & 0 \\ ig & \varepsilon_0 n^2 & 0 \\ 0 & 0 & \varepsilon_0 n^2 \end{pmatrix} \begin{pmatrix} E_x \\ E_y \\ E_z \end{pmatrix} \tag{73}$$

The ε matrix is still Hermitian but there is a 'small' off diagonal imaginary element. The presence of these off diagonal terms give rise to optical activity. Now,

$$\frac{n_w^2}{c^2\mu_0}[\mathbf{E} - (\hat{\kappa}\cdot\mathbf{E})\hat{\kappa}] = \mathbf{D}$$

We write the x and y components of the above equation and since $\kappa_x = 0 = \kappa_y$ and $\kappa_z = 1$, we get

$$\frac{n_w^2}{c^2\mu_0}E_x = D_x = \varepsilon_0 n^2 E_x - igE_y$$

and $$\frac{n_w^2}{c^2\mu_0}E_y = D_y = igE_x + \varepsilon_0 n^2 E_y$$

Thus, $$\left(\frac{n_w^2}{n^2} - 1\right)E_x = -i\alpha E_y \tag{74}$$

and $$\left(\frac{n_w^2}{n^2} - 1\right) E_y = i\alpha E_x \tag{75}$$

where we have used the fact that $c = 1/\sqrt{\varepsilon_0 \mu_0}$. For nontrival solutions,

$$\left(\frac{n_w^2}{n^2} - 1\right)^2 = \alpha^2$$

giving $$n_w = n\sqrt{1 \pm \alpha} \tag{76}$$

and $$E_y = \pm i E_x \tag{77}$$

We write the two solutions as $n_r\,(= n\sqrt{1+\alpha})$ and $n_l\,(= n\sqrt{1-\alpha})$; the corresponding propagation constants will be given by

$$k = k_r = \frac{\omega}{c} n_r = \frac{\omega}{c} n\sqrt{1+\alpha} \tag{78}$$

and $$k = k_l = \frac{\omega}{c} n_l = \frac{\omega}{c} n\sqrt{1-\alpha} \tag{79}$$

For $n_w = n_r$, if

$$E_x = E_0 e^{i(k_r z - \omega t)}$$

then $$E_y = +iE_x = E_0 e^{i\left(k_r z - \omega t + \frac{\pi}{2}\right)}$$

which would represent an RCP (Right Circularly Polarised) wave and hence the subscript r. Similarly, For $n_w = n_l$, if

$$E_x = E_0 e^{i(k_l z - \omega t)}$$

then $$E_y = -iE_x = E_0 e^{i\left(k_l z - \omega t - \frac{\pi}{2}\right)}$$

which would represent an LCP (Left Circularly Polarised) wave and hence the subscript l. The RCP and LCP waves are the two 'modes' of the 'optically active' substance and for an arbitrary incident state of polarisation, we must write it as a superposition of the two modes and study the independent propagation of the two modes. Now, for $\alpha << 1$

$$n_r - n_l = n[\sqrt{1+\alpha} - \sqrt{1-\alpha}]$$
$$\approx n\alpha \tag{80}$$

13.36 Since $\frac{5+4}{2} = 4.5$ and $\frac{5-4}{2} = 0.5$, we write the incident beam as superposition of the following two circularly polarised beams

$$\left\{\begin{matrix} E_{1x} = 4.5 \cos \omega t \\ E_{1y} = 4.5 \sin \omega t \end{matrix}\right\} \textbf{(RCP)} + \left\{\begin{matrix} E_{2x} = 0.5 \cos \omega t \\ E_{2y} = -0.5 \sin \omega t \end{matrix}\right\} \textbf{(LCP)}$$

As the beam propagates, we will have

$$E_x = 4.5 \cos(\omega t - \phi_1) + 0.5 \cos(\omega t - \phi_2)$$

$$= (4.5 \cos \phi_1 + 0.5 \cos \phi_2) \cos \omega t + (4.5 \sin \phi_1 + 0.5 \sin \phi_2) \sin \omega t$$
$$= a_1 \cos (\omega t - \theta_1)$$

where, $\phi_1 = k_r z = \dfrac{\omega}{c} n_r z,\ \phi_2 = k_l z = \dfrac{\omega}{c} n_l z$

$$a_1 \cos \theta_1 = 4.5 \cos \phi_1 + 0.5 \cos \phi_2 \quad \text{and} \quad a_1 \sin \theta_1 = 4.5 \sin \phi_1 + 0.5 \sin \phi_2$$

from which one can easily calculate a_1 and θ_1. Similarly,

$$E_y = 4.5 \sin (\omega t - \phi_1) - 0.5 \sin (\omega t - \phi_2)$$
$$= (4.5 \cos \phi_1 - 0.5 \cos \phi_2) \sin \omega t - (4.5 \sin \phi_1 - 0.5 \sin \phi_2) \cos \omega t$$
$$= a_2 \sin (\omega t - \theta_2)$$

where, $a_2 \cos \theta_2 = 4.5 \cos \phi_1 - 0.5 \cos \phi_2$

and $a_2 \sin \theta_2 = 4.5 \sin \phi_1 - 0.5 \sin \phi_2$

13.37 Since the wave propagates along the x-direction, the two modes will be polarised along y and z. The one polarised along y will have an index 1.620 and the one polarised along z will have an index 1.626. For converting a circularly polarised wave to linearly polarised wave, the phase difference between the two components must increase by $\pi/2$. Hence,

$$\frac{2\pi}{\lambda_0} (n_z - n_y) d = \frac{\pi}{2} \tag{81}$$

which gives $d = 25$ μm.

13.38 The given medium is biaxial. Hence it must propagate along one of the optic axes of the biaxial medium. Now since as given $n_x < n_y < n_z$ the wave must propagate along the x-z plane so that one of the indices would be n_y and the other index would lie between n_x and n_z which can be made equal to n_y. If the direction of propagation which lies in the x-z plane makes an angle θ with the z-axis then the refractive index of the wave would be

$$\frac{1}{n^2(\theta)} = \frac{\cos^2 \theta}{n_x^2} + \frac{\sin^2 \theta}{n_z^2} \tag{82}$$

In our case, we need to obtain θ such that $n(\theta) = n_y$. Solving we get $\theta = 60.24°$.

Polarisation II: Jones Vectors and Jones Matrices 14

A Quick Review

Through Jones calculus, it becomes quite straightforward to determine the polarisation state of the beam emerging from a polariser or a phase retarder (like a QWP or a HWP). An x-polarised plane wave (propagating in the $+z$-direction) is described by

$$\mathbf{E} = \hat{\mathbf{x}} E_0 \cos(kz - \omega t) = \hat{\mathbf{x}} E_0 \,\mathrm{Re}\,[e^{i(kz - \omega t)}] \tag{1}$$

Such a wave is written as $E_0|x>$, where

$$|x> = \begin{pmatrix} 1 \\ 0 \end{pmatrix} \tag{2}$$

is the normalised Jones vector representing a x-polarised wave. Similarly, a y-polarised wave

$$\mathbf{E} = \hat{\mathbf{y}} E_0 \cos(kz - \omega t) = \hat{\mathbf{y}} E_0 \,\mathrm{Re}\,[e^{i(kz - \omega t)}] \tag{3}$$

is written as $E_0|y>$, where

$$|y> = \begin{pmatrix} 0 \\ 1 \end{pmatrix} \tag{4}$$

is the normalised Jones vector representing a y-polarised wave; in writing Eqs. (2) and (3), we have neglected the common phase factor $e^{i(kz-\omega t)}$ which is implicitly assumed. A linearly polarised wave whose **E** vector makes an angle α with the y-axis (see Fig. 14.1) is represented by the normalised Jones vector

$$|\mathrm{LP}\,\alpha> = \sin\alpha\,|x> + \cos\alpha\,|y> = \begin{pmatrix} \sin\alpha \\ \cos\alpha \end{pmatrix} \tag{5}$$

For an RCP (propagating in the z-direction) we may write (see Fig. 14.2):

$$\mathbf{E} = \hat{\mathbf{x}} e^{i(kz - \omega t)} + \hat{\mathbf{y}} e^{i(kz - \omega t + \pi/2)} \tag{6}$$

Thus, neglecting the (unimportant) phase factor, the normalised Jones vector representing an RCP will be

$$|\mathrm{RCP}> = \frac{1}{\sqrt{2}}\begin{pmatrix} 1 \\ e^{i\pi/2} \end{pmatrix} = \frac{1}{\sqrt{2}}\begin{pmatrix} 1 \\ i \end{pmatrix} \tag{7}$$

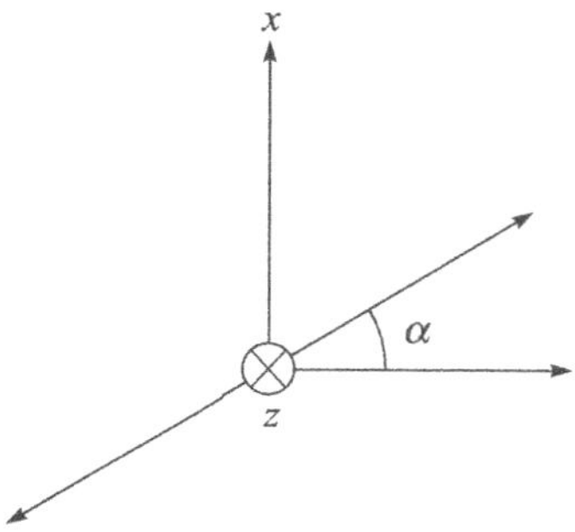

Fig. 14.1 A linearly polarised wave with its **E** vector making an angle α with the y-axis. The propagation of the wave is in the $+z$ direction.

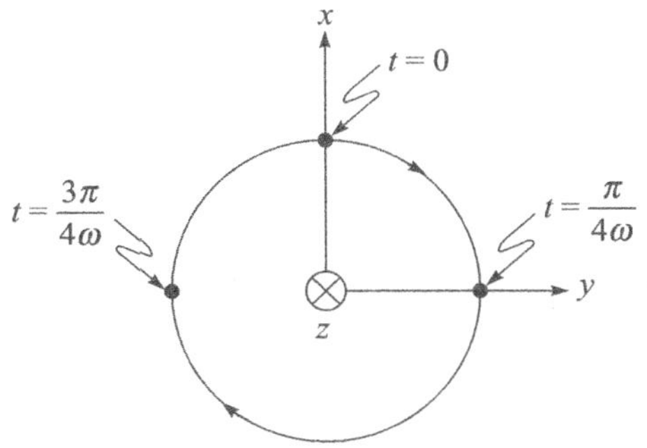

Fig. 14.2 A right circularly polarised wave propagating in the +*z* direction.

Similarly, the normalised Jones vector representing an LCP will be

$$|\text{LCP}> = \frac{1}{\sqrt{2}}\begin{pmatrix} 1 \\ -i \end{pmatrix} \tag{8}$$

Let us next consider a calcite (or a quartz) phase retarder like a QWP or a HWP; we assume its optic axis to be along the y-axis (see Fig. 14.3). The 'modes' of such a device are linearly polarised along the x and y-directions; the x-polarised wave will be the ordinary wave and the y-polarised wave will be the extra-ordinary wave. Thus, if E_x' and E_y' are the x and y **components of the electric field after propagating through** the retardation plates (of thickness d), then

$$E_x' = e^{ik_o d}E_x$$
$$E_y' = e^{ik_e d}E_y$$

where, $$k_o = \frac{2\pi}{\lambda_0}n_o \quad \text{and} \quad k_e = \frac{2\pi}{\lambda_0}n_e \tag{9}$$

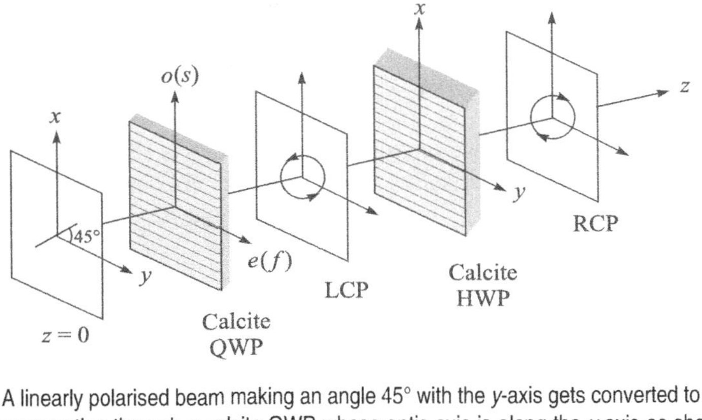

Fig. 14.3 A linearly polarised beam making an angle 45° with the *y*-axis gets converted to a LCP after propagating through a calcite QWP whose optic axis is along the *y*-axis as shown by lines parallel to the *y*-axis. Further, an LCP gets converted to a RCP after propagating through a calcite HWP; the optic axis of the HWP is also assumed to be along the *y*-direction.

Since, only the relative phase difference is of interest, we may write

$$E'_x = e^{i\Phi} E_x$$
$$E'_y = E_y$$

where, $$\Phi = (k_o - k_e)d = \frac{2\pi}{\lambda_0}(n_o - n_e)d \tag{10}$$

is the phase difference introduced by the phase retarder. Thus, we may write

$$\begin{pmatrix} E'_x \\ E'_y \end{pmatrix} = \begin{pmatrix} e^{i\Phi} & 0 \\ 0 & 1 \end{pmatrix} \begin{pmatrix} E_x \\ E_y \end{pmatrix} = T_{PR} \begin{pmatrix} E_x \\ E_y \end{pmatrix} \tag{11}$$

where T_{PR} is Jones matrix for the phase retarder and is given by

$$T_{PR} = \begin{pmatrix} e^{i\Phi} & 0 \\ 0 & 1 \end{pmatrix} \tag{12}$$

For calcite (which is a negative crystal), n_o = 1.65836 and n_e = 1.48641 at λ_0 = 5893 Å. Since $n_o > n_e$, Φ will be positive; the y-polarised extra-ordinary wave will travel faster than the x-polarised ordinary wave $\left(\frac{c}{n_o} < \frac{c}{n_e}\right)$. Thus, for a calcite QWP (with its optic axis along the y-direction), $\Phi = +\frac{\pi}{2}$ and

$$T_{\text{QWP}} = \begin{pmatrix} i & 0 \\ 0 & 1 \end{pmatrix} \text{ (fast axis along the } y\text{-direction)} \tag{13}$$

For a quartz QWP, $n_o < n_e$ and with its optic axis along the y-direction, $\Phi = -\frac{\pi}{2}$ and

$$T_{\text{QWP}} = \begin{pmatrix} -i & 0 \\ 0 & 1 \end{pmatrix} \text{ (slow axis along the } y\text{-direction)} \tag{14}$$

On the other hand, for a HWP, $\Phi = +\pi$ for calcite and $\Phi = -\pi$ for quartz. Thus, for both cases

$$T_{\text{HWP}} = \begin{pmatrix} -1 & 0 \\ 0 & 1 \end{pmatrix} \tag{15}$$

The Jones matrix for a phase retarder (like a QWP or a HWP) whose optic axis makes an angle α with the horisontal axis (y-axis) is discussed in Problem 14.13. The Jones matrix for a linear polariser making an angle α with the horizontal axis (y-axis) is given by (see Problem 14.8)

$$T_{LP}(\alpha) = \begin{pmatrix} \sin^2\alpha & \sin\alpha\cos\alpha \\ \sin\alpha\cos\alpha & \cos^2\alpha \end{pmatrix} \tag{16}$$

An elliptically polarised wave (with its major axis either along the x or y-direction) is given by

$$E_x = a\cos(kz - \omega t) = a \operatorname{Re} e^{i(kz - \omega t)}$$
$$E_y = \mp b\sin(kz - \omega t) = b \operatorname{Re} e^{i\left(kz - \omega t \pm \frac{\pi}{2}\right)}$$

where the upper (and lower) signs correspond to the REP and LEP respectively. Thus, the corresponding normalised Jones vectors can be written as

$$|\text{REP}> = \frac{1}{\sqrt{a^2+b^2}}\begin{pmatrix} a \\ be^{+i\pi/2}\end{pmatrix} = \begin{pmatrix} \cos\varepsilon \\ i\sin\varepsilon\end{pmatrix};\ \varepsilon = \tan^{-1}\left(\frac{b}{a}\right) \text{ with } 0 \le \varepsilon \le \frac{\pi}{2} \quad (17)$$

where a and b are assumed to be real and positive; and

$$|\text{LEP}> = \frac{1}{\sqrt{a^2+b^2}}\begin{pmatrix} a \\ be^{-i\pi/2}\end{pmatrix} = \begin{pmatrix} \cos\varepsilon \\ -i\sin\varepsilon\end{pmatrix};\ \varepsilon = \tan^{-1}\left(\frac{b}{a}\right) \text{ with } 0 \le \varepsilon \le \frac{\pi}{2} \quad (18)$$

The parameter ε is known as the ellipticity. For the REP shown in Fig. 14.4

$$\frac{b}{a} \approx 3 \quad \Rightarrow \quad \varepsilon \approx \frac{\pi}{2.5}$$

and the normalised Jones vector will be

$$|\text{REP}> = \begin{pmatrix} 0.31 \\ 0.95i\end{pmatrix}$$

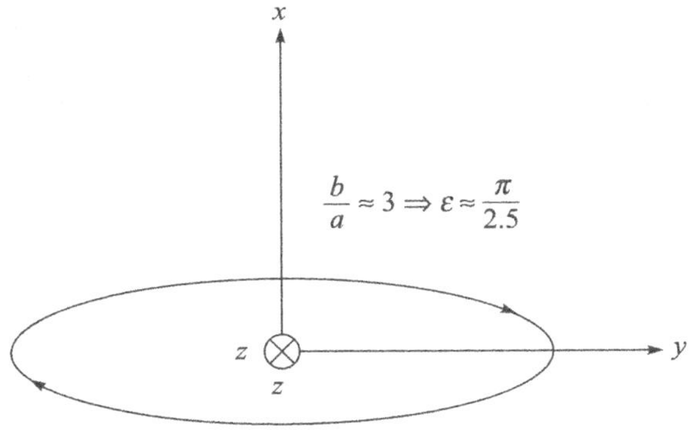

Fig. 14.4 A right elliptically polarised wave propagating in the +z direction.

Obviously $\varepsilon = 0$ represents the x-polarised wave, $\varepsilon = \frac{\pi}{4}$ represents a circularly polarised wave and $\varepsilon = \frac{\pi}{2}$ represents the y-polarised wave.

The use of Jones matrices makes it very straightforward to consider more complicated cases like two QWP with their axes at an angle.

PROBLEMS

14.1 The state of polarisation is described by the following normalised Jones vector

$$\begin{pmatrix} 0.8 \\ -0.6i\end{pmatrix}$$

Determine the x and y components of correspond electric field and the sense of rotation and ellipticity.

14.2 Write the normalised Jones vector for the following values of E_x and E_y

$$E_x = a\cos(kz - \omega t)$$
$$E_y = -b\cos(kz - \omega t - \phi)$$

14.3 Consider a calcite QWP whose optic axis is along the y-axis (see Fig. 14.3). By using Jones matrices, obtain the output state of polarisation when the incident beam is

(a) x polarised
(b) y polarised
(c) Left Circularly Polarised (LCP)
(d) Linearly Polarised with its **E** making an angle of 45° with the y-axis
(e) Linearly Polarised with its **E** making an angle of 30° with the y-axis
(f) Left Elliptically Polarised (LEP) with its **E** given by

$$E_x = \frac{1}{2}E_0\cos(kz - \omega t)$$
$$E_y = \frac{\sqrt{3}}{2}E_0\sin(kz - \omega t) \qquad (19)$$

14.4 Consider a calcite HWP whose optic axis is along the y-axis (see Fig. 14.3). By using Jones matrices, obtain the output state of polarisation when the incident beam is

(a) x-polarised
(b) y polarised
(c) Left Circularly Polarised (LCP)
(d) Linearly Polarised with its **E** making an angle of 45° with the y-axis
(e) Linearly Polarised with its **E** making an angle of 30° with the y-axis
(f) Left Elliptically Polarised (LEP) with its **E** given by Eq. (19).

14.5 (a) Consider a calcite QWP followed by a calcite HWP; in both of them the optic axis is along the y-axis (see Fig. 14.3). Find the Jones matrix for the combination and obtain the output state of polarisation when the incident beam is linearly polarised with its **E** making an angle of 45° with the y-axis,

(b) Show that if we put the calcite HWP first and then the QWP, the effect will be the same.

14.6 Consider a quartz QWP whose optic axis is along the y-axis (see Fig. 14.3). By using Jones matrices, obtain the output state of polarisation when the incident beam is

(a) x polarised
(b) y polarised
(c) Left Circularly Polarised (LCP)
(d) Linearly Polarised with its **E** making an angle of 45° with the y-axis
(e) Linearly Polarised with its **E** making an angle of 30° with the y-axis
(f) Left Elliptically Polarised (LEP) with its **E** given by Eq. (19).

14.7 Consider now a calcite QWP followed by a quartz QWP; in both of them the optic axis is along the y-axis. Find the Jones matrix for the combination.

14.8 (a) Show that the Jones matrix for a linear polariser making an angle α with the horisontal axis (y-axis) is given by Eq. (16). (b) Write the Jones matrix for the x-polariser, for the y-polariser and for a polariser which polarises at +45° angle and at 135° angle with the y-axis (see Figs 14.1 and 14.5).

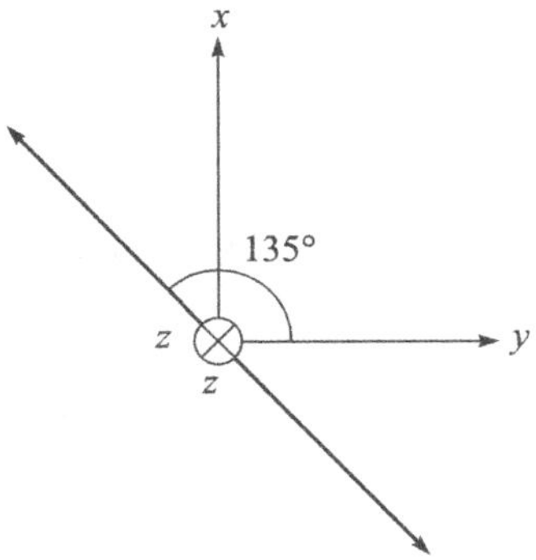

Fig. 14.5 A linearly polarised wave making an angle of 135° with the y-axis; the propagation is in the $+z$ direction.

14.9 (a) Consider a calcite QWP (with its optic axis along the y-axis) followed by a Polaroid with its pass axis making angle α with the y-axis (a) Find the Jones matrix for the combination. (b) For an incident x-polarised beam, find the state of polarisation after it comes out of the QWP.

14.10 Consider a Polaroid (with its pass axis making angle α with the y-axis) followed by a calcite QWP with its optic axis along the y-axis. (a) Find the Jones matrix for the combination; (b) For an incident x-polarised beam, find the state of polarisation after it comes out of the QWP; (c) what will be the output SOP if $\alpha = \pi/4$.

14.11 Consider a REP with its major axis along the η direction and described by the following Jones vector (see Fig. 14.6).

$$|\text{REP}> = \begin{pmatrix} E_\xi \\ E_\eta \end{pmatrix} = \begin{pmatrix} \cos\varepsilon \\ -i\sin\varepsilon \end{pmatrix} \tag{20}$$

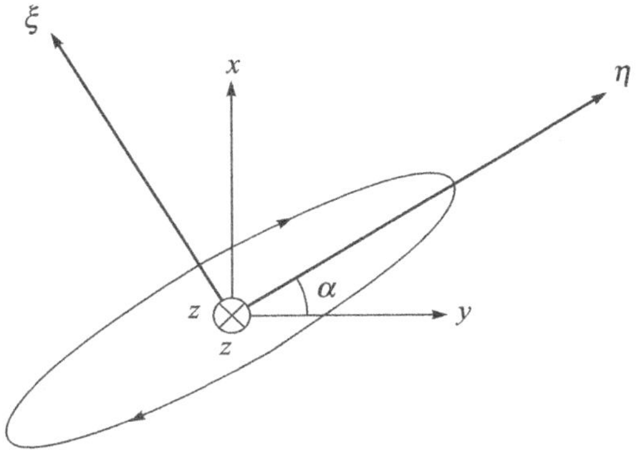

Fig. 14.6 A right elliptically polarised wave with its major axis making an angle α with the y-axis; the wave is propagating in the $+z$ direction.

The η axis makes an angle α with the y-axis as shown in Fig. 14.6. Calculate the Jones vector

$$\begin{pmatrix} E_x \\ E_y \end{pmatrix}$$

14.12 For calcite, the values of n_o and n_e for $\lambda_0 = 4046$ Å are 1.68134 and 1.49694 respectively; corresponding to $\lambda_0 = 7065$ Å the values are 1.65207 and 1.48359 respectively. At $\lambda_0 = 4046$ Å the calcite plate is a QWP (a) Write the Jones matrix for the calcite plate for $\lambda_0 = 4046$ Å and for $\lambda_0 = 7065$ Å. (b) A left-circularly polarised beam of $\lambda_0 = 7065$ Å is incident on this calcite plate. Obtain the state of polarisation of the emergent beam.

14.13 Consider a calcite (or quartz) phase retarder whose optic axis makes an angle α with the y-axis; we choose the η axis along this direction and the ξ axis perpendicular to that (see Fig. 14.7). If E_x and E_y are the x and y components of the electric field that is incident on the phase retarder and if E_x' and E_y' are the x and y components of the electric field after propagating through the phase retarder then show that

$$\begin{pmatrix} E_x' \\ E_y' \end{pmatrix} = T_{PR}(\alpha) \begin{pmatrix} E_x \\ E_y \end{pmatrix} \tag{21}$$

where, $$T_{PR}(\alpha) = \begin{pmatrix} e^{i\Phi}\cos^2\alpha + \sin^2\alpha & (1 - e^{i\Phi})\sin\alpha\cos\alpha \\ (1 - e^{i\Phi})\sin\alpha\cos\alpha & e^{i\Phi}\sin^2\alpha + \cos^2\alpha \end{pmatrix} \tag{22}$$

represents the Jones matrix of a calcite (or quartz) phase retarder whose optic axis makes an angle α with the y-axis.

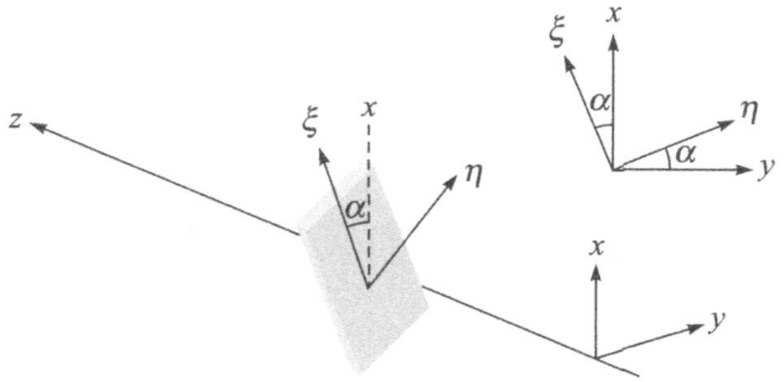

Fig. 14.7 A calcite (or quartz) phase retarder whose optic axis is along the η axis which makes an angle α with the y-axis; the wave is propagating in the $+z$ direction.

14.14 Use the results of the previous problem to calculate the output state of polarisation for an x-polarised light incident on a (a) calcite QWP (b) calcite HWP; in each case, the optic axis makes an angle $\pi/4$ with the y-axis.

SOLUTIONS

14.1 $|\text{LEP}> = \begin{pmatrix} 0.8 \\ -0.6i \end{pmatrix} = \begin{pmatrix} 0.8 \\ 0.6e^{-i\pi/2} \end{pmatrix}$

Thus,

$$E_x = 0.8E_0 \text{ Re } e^{i(kz - \omega t)} = 0.8E_0 \cos(kz - \omega t)$$

$$E_y = 0.6E_0 \text{ Re } e^{i\left(kz - \omega t - \frac{\pi}{2}\right)} = 0.6E_0 \sin(kz - \omega t)$$

At $z = 0$, we will have

$$E_x = 0.8\,E_0 \cos \omega t$$

$$E_y = -0.6\,E_0 \sin \omega t$$

which will represent an LEP with ellipticity given by

$$\varepsilon = \tan^{-1}\left(\frac{0.6}{0.8}\right) \approx \frac{\pi}{4.9}$$

14.2 $E_x = a\cos(kz - \omega t) = a \text{ Re } e^{i(kz - \omega t)}$

$E_y = -b\cos(kz - \omega t - \phi) = -b \text{ Re } e^{i(kz - \omega t - \phi)}$

Thus, the normalised Jones vector will be

$$\frac{1}{\sqrt{a^2 + b^2}}\begin{pmatrix} a \\ -be^{-i\phi} \end{pmatrix}$$

14.3 The Jones matrix for a calcite QWP with its fast axis along the y-direction is given by

$$\begin{pmatrix} i & 0 \\ 0 & 1 \end{pmatrix}$$

(a) The x polarised beam will remain x polarised.

(b) The y polarised beam will remain y polarised

(c) The normalised Jones vector for the LCP is given by $|\text{LCP}> = \frac{1}{\sqrt{2}}\begin{pmatrix} 1 \\ -i \end{pmatrix}$. Thus,

$$|\text{output}> = \begin{pmatrix} i & 0 \\ 0 & 1 \end{pmatrix} \frac{1}{\sqrt{2}}\begin{pmatrix} 1 \\ -i \end{pmatrix} = \frac{i}{\sqrt{2}}\begin{pmatrix} 1 \\ -1 \end{pmatrix}$$

which is a linearly polarised wave with its **E** making an angle of 135° with the x-axis (see Fig. 14.5).

(d) The Jones matrix for linearly polarised wave with its **E** making an angle of 45° with the y-axis is given by

$$|45°> = \frac{1}{\sqrt{2}}\begin{pmatrix} 1 \\ 1 \end{pmatrix}$$

Thus,

$$|\text{output}> = \begin{pmatrix} i & 0 \\ 0 & 1 \end{pmatrix} \frac{1}{\sqrt{2}} \begin{pmatrix} 1 \\ 1 \end{pmatrix} E_0 = \frac{1}{\sqrt{2}} \begin{pmatrix} i \\ 1 \end{pmatrix} E_0 = \frac{i}{\sqrt{2}} \begin{pmatrix} 1 \\ -i \end{pmatrix} E_0$$

which is a left circularly polarised wave.

(e) The Jones vector for a linearly polarised wave with its **E** making an angle of 30° with the y-axis is given by

$$|60° > = \begin{pmatrix} \sin 30° \\ \cos 30° \end{pmatrix} E_0 = \frac{1}{2} \begin{pmatrix} 1 \\ \sqrt{3} \end{pmatrix} E_0$$

Thus, $$|\text{output}> = \begin{pmatrix} i & 0 \\ 0 & 1 \end{pmatrix} \frac{1}{2} \begin{pmatrix} 1 \\ \sqrt{3} \end{pmatrix} E_0 = \frac{1}{2} \begin{pmatrix} i \\ \sqrt{3} \end{pmatrix} E_0 = \frac{i}{2} \begin{pmatrix} 1 \\ -i\sqrt{3} \end{pmatrix} E_0$$

which is a left elliptically polarised wave with its minor axis along the x-axis.

(f) The incident left elliptically polarised wave is given by

$$E_x = \frac{1}{2} E_0 \cos(kz - \omega t) = \frac{1}{2} E_0 \,\text{Re}\, e^{i(kz - \omega t)}$$

$$E_y = \frac{\sqrt{3}}{2} E_0 \sin(kz - \omega t) = \frac{\sqrt{3}}{2} E_0 \,\text{Re}\, e^{i\left(kz - \omega t - \frac{\pi}{2}\right)}$$

Thus, the Jones vector for the incident LEP will be

$$|\text{input}> = \frac{1}{2} \begin{pmatrix} 1 \\ -i\sqrt{3} \end{pmatrix} E_0$$

and the beam coming out of the QWP will be given by

$$|\text{output}> = \begin{pmatrix} E'_x \\ E'_y \end{pmatrix} = \begin{pmatrix} i & 0 \\ 0 & 1 \end{pmatrix} \frac{1}{2} \begin{pmatrix} 1 \\ -i\sqrt{3} \end{pmatrix} E_0 = \frac{1}{2} \begin{pmatrix} i \\ -i\sqrt{3} \end{pmatrix} E_0$$

$$= \frac{1}{2} \begin{pmatrix} i \\ -i\sqrt{3} \end{pmatrix} E_0$$

which is LP with its **E** making an angle of 150° with the y-axis.

14.4 In continuation of the previous problem we just have to replace everywhere

$$T_{\text{QWP}} = \begin{pmatrix} i & 0 \\ 0 & 1 \end{pmatrix} \quad \text{by} \quad T_{\text{HWP}} = \begin{pmatrix} -1 & 0 \\ 0 & 1 \end{pmatrix}$$

(c) For a LCP incident on the HWP, the beam coming out will be given by

$$|\text{output}> = \begin{pmatrix} -1 & 0 \\ 0 & 1 \end{pmatrix} \frac{1}{\sqrt{2}} \begin{pmatrix} 1 \\ -i \end{pmatrix} = \frac{-i}{\sqrt{2}} \begin{pmatrix} 1 \\ i \end{pmatrix}$$

which is a RCP.

(d) For a linearly polarised wave (with its **E** making an angle of 45° with the y-axis), the beam coming out of the HWP will be given by

$$|\text{output}> = \begin{pmatrix} -1 & 0 \\ 0 & 1 \end{pmatrix} \frac{1}{\sqrt{2}} \begin{pmatrix} 1 \\ 1 \end{pmatrix} E_0 = \frac{1}{\sqrt{2}} \begin{pmatrix} -1 \\ 1 \end{pmatrix} E_0$$

which is a linearly polarised wave with **E** making an angle of 135° with the y-axis.

14.5 The Jones matrix for the combination will be

$$T = T_{\text{HWP}}\, T_{\text{QWP}} = \begin{pmatrix} -1 & 0 \\ 0 & 1 \end{pmatrix} \begin{pmatrix} i & 0 \\ 0 & 1 \end{pmatrix} = \begin{pmatrix} -i & 0 \\ 0 & 1 \end{pmatrix}$$

This should have been obvious because a QWP followed by a HWP is equivalent to phase retarder with

$$\Phi = \frac{\pi}{2} + \pi = \frac{3\pi}{2} \quad \Rightarrow \quad e^{i\Phi} = -i$$

Since the Jones matrix for a linearly polarised wave with its **E** making an angle of 45° with the y-axis is given by

$$|45° > = \frac{1}{\sqrt{2}} \begin{pmatrix} 1 \\ 1 \end{pmatrix}$$

the beam coming out of the HWP will be given by

$$|\text{output}> = \begin{pmatrix} -i & 0 \\ 0 & 1 \end{pmatrix} \frac{1}{\sqrt{2}} \begin{pmatrix} 1 \\ 1 \end{pmatrix} = \frac{-i}{\sqrt{2}} \begin{pmatrix} 1 \\ i \end{pmatrix}$$

which represents a RCP as shown in Fig. 14.3.

(b) When the HWP is followed by a QWP we will have

$$T = T_{\text{QWP}}\, T_{\text{HWP}} = \begin{pmatrix} i & 0 \\ 0 & 1 \end{pmatrix} \begin{pmatrix} -1 & 0 \\ 0 & 1 \end{pmatrix} = \begin{pmatrix} -i & 0 \\ 0 & 1 \end{pmatrix}$$

which is the same as in (a). This is a consequence of the fact that if we have a QWP and an HWP, then it does not matter which one is put first as long as the optic axes are in the same direction.

14.6 In continuation of Problem 14.5 we just have to replace everywhere

$$T_{\text{QWP}}^{\text{calcite}} = \begin{pmatrix} i & 0 \\ 0 & 1 \end{pmatrix} \quad \text{by} \quad T_{\text{QWP}}^{\text{quartz}} = \begin{pmatrix} -i & 0 \\ 0 & 1 \end{pmatrix} \tag{23}$$

14.7 The Jones matrix of the combination will be

$$T = T_{\text{QWP}}^{\text{quartz}}\, T_{\text{QWP}}^{\text{calcite}} = \begin{pmatrix} i & 0 \\ 0 & 1 \end{pmatrix} \begin{pmatrix} -i & 0 \\ 0 & 1 \end{pmatrix} = \begin{pmatrix} 1 & 0 \\ 0 & 1 \end{pmatrix} \tag{24}$$

14.8 Consider a polaroid whose pass axis makes an angle α with the y-axis; we choose the η axis along this direction and the ξ axis perpendicular to that (see Fig. 14.8). If E_x and E_y are the x and y components of the electric field that is incident on the polaroid, then for the light coming out of the polaroid, the ξ and η components of the field will be given by

$$E'_\xi = 0$$

and
$$E'_\eta = E_x \sin\alpha + E_y \cos\alpha$$

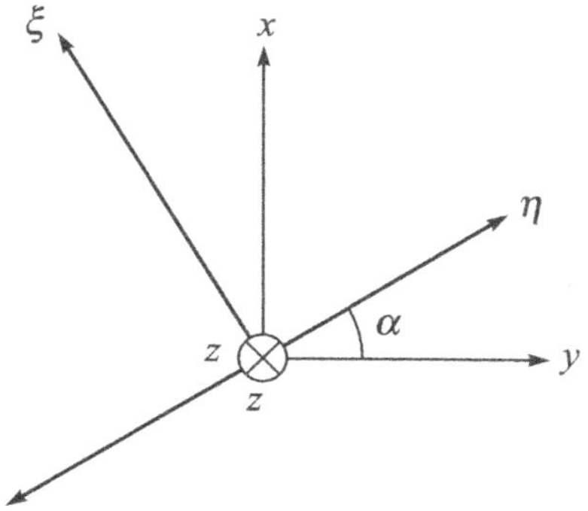

Fig. 14.8 A linearly polarised wave whose electric vector oscillates along a direction making an angle α with the y-axis; the wave is propagating in the $+z$ direction.

Thus, the corresponding x and y components of the field will be given by

$$E'_x = E'_\eta \sin\alpha + E'_\xi \cos\alpha = E_x \sin^2\alpha + E_y \sin\alpha\cos\alpha$$

and
$$E'_y = E'_\eta \cos\alpha - E'_\xi \sin\alpha = E_x \sin\alpha\cos\alpha + E_y \cos^2\alpha$$

Thus
$$\begin{pmatrix} E'_x \\ E'_y \end{pmatrix} = T_{LP}(\alpha)\begin{pmatrix} E_x \\ E_y \end{pmatrix} \tag{25}$$

where,
$$T_{LP}(\alpha) = \begin{pmatrix} \sin^2\alpha & \sin\alpha\cos\alpha \\ \sin\alpha\cos\alpha & \cos^2\alpha \end{pmatrix} \tag{26}$$

(b) For the x-polariser, $\alpha = \dfrac{\pi}{2}$ and we will have

$$T_{LP}\left(\alpha = \frac{\pi}{2}\right) = \begin{pmatrix} 1 & 0 \\ 0 & 0 \end{pmatrix} \quad (x\text{-polariser}) \tag{27}$$

For the y-polariser, $\alpha = 0$ and we will have

$$T_{LP}(\alpha = 0) = \begin{pmatrix} 0 & 0 \\ 0 & 1 \end{pmatrix} \quad (y\text{-polariser}) \tag{28}$$

For a polariser which polariser at $+45°$ angle with the y-axis.

$$T_{LP}\left(\alpha = \frac{\pi}{4}\right) = \frac{1}{2}\begin{pmatrix} 1 & 1 \\ 1 & 1 \end{pmatrix} \quad (45°\text{-polariser}) \tag{29}$$

For a polariser which polarises at +135° angle with the y-axis.

$$T_{LP}\left(\alpha = \frac{3\pi}{4}\right) = \frac{1}{2}\begin{pmatrix} 1 & -1 \\ -1 & 1 \end{pmatrix} \qquad \text{(135°-polariser)} \qquad (30)$$

14.9 A calcite QWP (with its optic axis along the y-axis) is followed by a polaroid with its pass axis making angle α with the y-axis

(a) The Jones matrix for the combination will be

$$T = T_{LP}(\alpha)\,T_{\text{QWP}} = \begin{pmatrix} \sin^2\alpha & \sin\alpha\cos\alpha \\ \sin\alpha\cos\alpha & \cos^2\alpha \end{pmatrix}\begin{pmatrix} i & 0 \\ 0 & 1 \end{pmatrix}$$

$$= \begin{pmatrix} i\sin^2\alpha & \sin\alpha\cos\alpha \\ i\sin\alpha\cos\alpha & \cos^2\alpha \end{pmatrix} \qquad (31)$$

(b) The Jones vector for the incident LEP is [see Solution 14.3(f)]

$$|\,\text{input}> = \frac{1}{2}\begin{pmatrix} 1 \\ -i\sqrt{3} \end{pmatrix} E_0$$

Thus, the state of polarisation after it comes out of the polaroid will be

$$|\,\text{output}> = \begin{pmatrix} i\sin^2\alpha & \sin\alpha\cos\alpha \\ i\sin\alpha\cos\alpha & \cos^2\alpha \end{pmatrix}\frac{1}{2}\begin{pmatrix} 1 \\ -i\sqrt{3} \end{pmatrix} E_0$$

$$= \frac{(i\sin\alpha - \sqrt{3}\cos\alpha)}{2}\begin{pmatrix} \sin\alpha \\ \cos\alpha \end{pmatrix} E_0$$

which is a linearly polarised wave making an angle α with the horizontal axis. This should have been obvious because we have a polaroid at the end so the output has to be linearly polarised along the pass axis of the polaroid.

14.10 A polaroid (with its pass axis making angle α with the y-axis) is followed by a calcite QWP with its optic axis along the y-axis.

(a) The Jones matrix for the combination will be

$$T = T_{\text{QWP}}\,T_{LP}(\alpha) = \begin{pmatrix} i & 0 \\ 0 & 1 \end{pmatrix}\begin{pmatrix} \sin^2\alpha & \sin\alpha\cos\alpha \\ \sin\alpha\cos\alpha & \cos^2\alpha \end{pmatrix}$$

$$= \begin{pmatrix} i\sin^2\alpha & i\sin\alpha\cos\alpha \\ \sin\alpha\cos\alpha & \cos^2\alpha \end{pmatrix}$$

(b) The SOP of the emergent wave will be

$$|\text{output}> = \begin{pmatrix} i\sin^2\alpha & i\sin\alpha\cos\alpha \\ \sin\alpha\cos\alpha & \cos^2\alpha \end{pmatrix}\begin{pmatrix} 1 \\ 0 \end{pmatrix}$$

$$= \sin\alpha\begin{pmatrix} i\sin\alpha \\ \cos\alpha \end{pmatrix} = \sin\alpha\begin{pmatrix} \sin\alpha\, e^{i\pi/2} \\ \cos\alpha \end{pmatrix}$$

Thus, $E_x = E_0\sin^2\alpha \operatorname{Re} e^{i\left(kz - \omega t + \frac{\pi}{2}\right)} = -E_0\sin^2\alpha\sin(kz - \omega t)$

$$E_y = E_0\sin\alpha\cos\alpha \operatorname{Re} e^{i(kz-\omega t)} = E_0\sin\alpha\cos\alpha\cos(kz - \omega t)$$

which is a left elliptically polarised light with major and minor axes along the x and y directions.

(c) For $\alpha = \pi/4$, the output will be a LCP.

14.11 $|\text{REP}> = \begin{pmatrix} E_\xi \\ E_\eta \end{pmatrix} = \begin{pmatrix} \cos\varepsilon \\ -i\sin\varepsilon \end{pmatrix}$

Now, $E_x = E_\xi\cos\alpha + E_\eta\sin\alpha$

and $E_y = -E_\xi\sin\alpha + E_\eta\cos\alpha$

Thus, $$\begin{pmatrix} E_x \\ E_y \end{pmatrix} = \begin{pmatrix} \cos\alpha & \sin\alpha \\ -\sin\alpha & \cos\alpha \end{pmatrix}\begin{pmatrix} E_\xi \\ E_\eta \end{pmatrix}$$

$$= \begin{pmatrix} \cos\alpha & \sin\alpha \\ -\sin\alpha & \cos\alpha \end{pmatrix}\begin{pmatrix} \cos\varepsilon \\ -i\sin\varepsilon \end{pmatrix}$$

$$= \begin{pmatrix} \cos\alpha\cos\varepsilon - i\sin\alpha\sin\varepsilon \\ -\sin\alpha\cos\varepsilon - i\cos\alpha\sin\varepsilon \end{pmatrix} \quad (32)$$

14.12 For the QWP at $\lambda_0 = 4046$ Å

$$\Phi = \frac{2\pi}{\lambda_0}(n_o - n_e)d = \frac{\pi}{2}$$

Thus, $$d = \frac{\lambda_0}{4(n_o - n_e)} = \frac{4.046\times 10^{-5}}{4(1.68134 - 1.49694)} \approx 5.49\times 10^{-5}\text{ cm}$$

At $\lambda_0 = 7065$ Å, the phase difference introduced is given by:

$$\Phi = \frac{2\pi}{\lambda_0}(n_0 - n_e)d = \frac{2\pi}{7.065\times 10^{-5}}\times(1.65207 - 1.48359) \times 5.49\times 10^{-5}\text{ cm}$$

$$\approx \frac{\pi}{3.82}$$

Thus, $$T(\lambda_0 = 4046\text{ Å}) = \begin{pmatrix} e^{i\pi/2} & 0 \\ 0 & 1 \end{pmatrix} = \begin{pmatrix} i & 0 \\ 0 & 1 \end{pmatrix}$$

$$T(\lambda_0 = 7065\ \text{Å}) = \begin{pmatrix} e^{i\pi/3.82} & 0 \\ 0 & 1 \end{pmatrix}$$

Thus, if a left-circularly polarised beam of $\lambda_0 = 7065$ Å is incident on this calcite plate, the state of polarisation of the emergent beam will be given by

$$|\text{output}\rangle = \begin{pmatrix} e^{i\pi/3.82} & 0 \\ 0 & 1 \end{pmatrix} \frac{1}{\sqrt{2}} \begin{pmatrix} 1 \\ -i \end{pmatrix} = \frac{1}{\sqrt{2}} \begin{pmatrix} e^{i\pi/3.82} \\ e^{-i\pi/2} \end{pmatrix}$$

Thus, $\quad E_x = \dfrac{E_0}{\sqrt{2}} \text{Re}\, e^{i\left(kz - \omega t + \frac{\pi}{3.82}\right)}$ and $E_y = \dfrac{E_0}{\sqrt{2}} \text{Re}\, e^{i\left(kz - \omega t - \frac{\pi}{2}\right)}$

At $z = 0$, we will have

$$E_x = \frac{E_0}{\sqrt{2}} \cos\left(\omega t - \frac{\pi}{3.82}\right) \quad \text{and} \quad E_y = -\frac{E_0}{\sqrt{2}} \sin \omega t$$

14.13 Consider a calcite phase retarder whose optic axis makes an angle α with the y-axis; we choose the η axis along this direction and the ξ axis perpendicular to that (see Fig. 14.8). If E_x and E_y are the x and y components of the electric field that incident on the phase retarder, then,

$$\begin{aligned} E_\xi &= E_x \cos \alpha - E_y \sin \alpha \\ E_\eta &= E_x \sin \alpha + E_y \cos \alpha \end{aligned} \tag{33}$$

Thus,
$$\begin{pmatrix} E_\xi \\ E_\eta \end{pmatrix} = \begin{pmatrix} \cos \alpha & -\sin \alpha \\ \sin \alpha & \cos \alpha \end{pmatrix} \begin{pmatrix} E_x \\ E_y \end{pmatrix} \tag{34}$$

Now, if E'_ξ and E'_η are the ξ and η components of the electric field that comes out of the phase retarder, then

$$\begin{pmatrix} E'_\xi \\ E'_\eta \end{pmatrix} = T_{PR} \begin{pmatrix} E_\xi \\ E_\eta \end{pmatrix} = \begin{pmatrix} e^{i\Phi} & 0 \\ 0 & 1 \end{pmatrix} \begin{pmatrix} E_\xi \\ E_\eta \end{pmatrix} \tag{35}$$

where T_{PR} is Jones matrix for the phase retarder [see Eq. (12)]. Thus, if E'_x and E'_y are the x and y components of the electric field after propagating through the phase retarder then

$$\begin{aligned} \begin{pmatrix} E'_x \\ E'_y \end{pmatrix} &= \begin{pmatrix} \cos \alpha & \sin \alpha \\ -\sin \alpha & \cos \alpha \end{pmatrix} \begin{pmatrix} E'_\xi \\ E'_\eta \end{pmatrix} \\ &= \begin{pmatrix} \cos \alpha & \sin \alpha \\ -\sin \alpha & \cos \alpha \end{pmatrix} \begin{pmatrix} e^{i\Phi} & 0 \\ 0 & 1 \end{pmatrix} \begin{pmatrix} E_\xi \\ E_\eta \end{pmatrix} \\ &= \begin{pmatrix} \cos \alpha & \sin \alpha \\ -\sin \alpha & \cos \alpha \end{pmatrix} \begin{pmatrix} e^{i\Phi} & 0 \\ 0 & 1 \end{pmatrix} \begin{pmatrix} \cos \alpha & -\sin \alpha \\ \sin \alpha & \cos \alpha \end{pmatrix} \begin{pmatrix} E_x \\ E_y \end{pmatrix} \end{aligned} \tag{36}$$

Simple matrix multiplication would give the desired result.

14.14 For a QWP $e^{i\Phi} = i$ and for $\alpha = \dfrac{\pi}{4}$ we will have

$$T_{PR}\left(\frac{\pi}{4}\right) = \frac{1}{2}\begin{pmatrix} 1+i & 1-i \\ 1-i & 1+i \end{pmatrix}$$

Thus, for an incident x-polarised light,

$$\begin{pmatrix} E_x' \\ E_y' \end{pmatrix} = \frac{1}{2}\begin{pmatrix} 1+i & 1-i \\ 1-i & 1+i \end{pmatrix}\begin{pmatrix} 1 \\ 0 \end{pmatrix} = \frac{1}{2}\begin{pmatrix} 1+i \\ 1-i \end{pmatrix} = \frac{1}{\sqrt{2}}\begin{pmatrix} e^{i\pi/2} \\ e^{-i\pi/2} \end{pmatrix}$$

Maxwell's Equations and The Wave Equation

15

A Quick Review

The equation

$$\frac{\partial^2 \Psi}{\partial x^2} = \frac{1}{v^2}\frac{\partial^2 \Psi}{\partial t^2} \tag{1}$$

is known as the *one-dimensional wave equation*. This is because of the fact that the most general solution of the above equation is given by (see Problem 15.7)

$$\Psi(x, t) = f(x - vt) + g(x + vt) \tag{2}$$

where f and g are arbitrary functions of their arguments. The term $f(x - vt)$ represents a wave propagating in the $+x$ direction with speed v and the term $g(x + vt)$ represents a wave propagating in the $-x$ direction with speed v. For example, the function

$$\Psi(x, t) = A\cos(kx - \omega t) \tag{3}$$

can be written as

$$\Psi(x, t) = A\cos[k(x - vt)] \tag{4}$$

Since x and t appear as $(x - vt)$, the above form of $\Psi(x, t)$ would satisfy Eq. (1) and would represent a wave propagating in the $+x$ direction with speed v given by

$$v = \frac{\omega}{k} = \nu\lambda \tag{5}$$

where,

$$\nu = \frac{\omega}{2\pi} \quad \text{and} \quad \lambda = \frac{2\pi}{k} \tag{6}$$

represent respectively the frequency and wavelength associated with the wave. Similarly, the function

$$\Psi(x, t) = A\sin(kx + \omega t) \tag{7}$$

can be written as

$$\Psi(x, t) = A\sin[k(x + vt)] \tag{8}$$

and would represent a wave propagating in the $-x$ direction with speed $v = \omega/k$. Thus whenever, from physical considerations, we are able to derive an equation of the type given by Eq. (1), we can predict the existence of waves and calculate the velocity of propagation of these waves. For example, if we apply Newton's laws of motion to a vibrating string under tension T, we can derive the following equation (see, e.g., Ref. Gh1)

$$\frac{\partial^2 y}{\partial x^2} = \frac{1}{T/\rho} \frac{\partial^2 y}{\partial t^2} \tag{9}$$

where ρ is the mass per unit length of the string (which is assumed to be along the x direction). The above equation shows the existence of transverse waves on a string and that the velocity of these waves will be given by:

$$v = \sqrt{\frac{T}{\rho}} \tag{10}$$

Similarly for sound waves propagating in a gas, one can derive from physical considerations (see, e.g., Chapter 11 of Ref. Gh1)

$$\frac{\partial^2 \xi}{\partial x^2} = \frac{\rho}{\gamma P} \frac{\partial^2 \xi}{\partial t^2} \tag{11}$$

where $\xi(x)$ represents the (longitudinal) displacement of the gas, ρ the density of the gas and $\gamma = C_p/C_v$ **represents the ratio of specific heats. Thus, the velocity of the** (longitudinal) sound waves in a gas will be given by:

$$v = \sqrt{\frac{\gamma P}{\rho}} \tag{12}$$

The equation

$$\nabla^2 \Psi = \frac{1}{v^2} \frac{\partial^2 \Psi}{\partial t^2} \tag{13}$$

is known as the *three-dimensional wave equation.* In the above equation

$$\nabla^2 \Psi = \text{div grad } \Psi \tag{14}$$

and in the Cartesian system of coordinates

$$\nabla^2 \Psi = \frac{\partial^2 \Psi}{\partial x^2} + \frac{\partial^2 \Psi}{\partial y^2} + \frac{\partial^2 \Psi}{\partial z^2} \tag{15}$$

The function

$$\Psi(x, y, z, t) = A \cos(\mathbf{k} \cdot \mathbf{r} - \omega t) \tag{16}$$

represents a plane wave propagating along the direction of **k**. At all points on a plane normal to **k**, the quantity $\mathbf{k} \cdot \mathbf{r}$ is a constant; thus the phase fronts are perpendicular to **k**. Further, if we substitute Eq. (16) in Eq. (15), we would obtain

$$\frac{\omega^2}{k^2} = v^2 \tag{17}$$

where,

$$k^2 = k_x^2 + k_y^2 + k_z^2 \tag{18}$$

Around the middle of the nineteenth century, Maxwell summed up all laws of electricity and magnetism in the form of four equations—which are now referred to as *Maxwell's equations*. In a charge free, homogeneous and isotropic dielectric, Maxwell's equations are given by

$$\nabla \cdot \mathbf{E} = 0 \tag{19}$$

$$\nabla \cdot \mathbf{H} = 0 \tag{20}$$

$$\nabla \times \mathbf{E} = -\mu \frac{\partial \mathbf{H}}{\partial t} \tag{21}$$

and

$$\nabla \times \mathbf{H} = \varepsilon \frac{\partial \mathbf{E}}{\partial t} \tag{22}$$

where **E** and **H** represent the electric field and the magnetic field respectively; ε and μ represent the dielectric permittivity and magnetic permeability of the dielectric. Using these equations, he showed that the Cartesian components of the electric and magnetic field satisfy the wave equation (see Problem 15.5):

$$\nabla^2 \Psi = \varepsilon \mu \frac{\partial^2 \Psi}{\partial t^2} \tag{23}$$

After deriving the wave equation, Maxwell could predict the existence of electromagnetic waves whose velocity will be given by:

$$v = \frac{1}{\sqrt{\varepsilon \mu}} \tag{24}$$

In free space,

$$\varepsilon = \varepsilon_0 = 8.854 \times 10^{-12}\ \mathrm{C^2 N^{-2} m^{-2}} \quad \text{and} \quad \mu = \mu_0 = 4\pi \times 10^{-7}\ \mathrm{Ns^2 C^{-2}} \tag{25}$$

and we obtain

$$v = c = \frac{1}{\sqrt{\varepsilon_0 \mu_0}} = 2.99794 \times 10^8\ \mathrm{ms^{-1}} \tag{26}$$

which is the speed of light in free space. Maxwell argued that since the (predicted) speed of electromagnetic waves was very close to the measured value of speed of light,

Light must be an electromagnetic wave

In a dielectric

$$v = \frac{c}{n} \tag{27}$$

where the refractive index (n) of a dielectric (characterised by dielectric permittivity ε and magnetic permeability μ) would be given by

$$n = \frac{c}{v} = \sqrt{\frac{\varepsilon \mu_0}{\varepsilon_0 \mu_0}} = \sqrt{\frac{\varepsilon}{\varepsilon_0}} = \sqrt{\kappa} \tag{28}$$

where we have assumed $\mu \approx \mu_0$ which is true for almost all dielectrics and

$$\kappa = \frac{\varepsilon}{\varepsilon_0} \tag{29}$$

is known as the *dielectric constant of the medium*.

PROBLEMS

15.1 The displacement associated with a wave is given by

(a) $y(x, t) = 0.1 \cos (0.2x - 2t)$

(b) $y(x, t) = 0.2 \sin (0.5x + 3t)$

(c) $y(x, t) = 0.5 \sin 2\pi (0.1x - t)$

where in each case x and y are measured in centimeters and t in seconds. Calculate the wavelength, amplitude, frequency and the velocity in each case.

15.2 Show that the functions

$$\Psi(x, t) = Ae^{i(kx + \omega t)},\ \Psi(x, t) = Ae^{i(\omega t + kx)} \quad \text{and} \quad \Psi(x, t) = Ae^{-\frac{(x - vt)^2}{\sigma^2}}$$

satisfy Eq. (1) and therefore each of the above functions would represent a wave.

15.3 Show that the functions

$$\Psi(x, t) = Ae^{-\frac{x^2}{\sigma^2}} \cos \omega t \quad \text{and} \quad \Psi(x, t) = Ae^{-\frac{x^2}{\sigma^2}} e^{-\frac{t^2}{\tau^2}}$$

do not satisfy Eq. (1) and hence do not represent waves.

15.4 Show that the plane wave solutions

$$\mathbf{E} = \mathbf{E_0} \exp [i(\mathbf{k} \cdot \mathbf{r} - \omega t)] \tag{30}$$

and

$$\mathbf{H} = \mathbf{H_0} \exp [i(\mathbf{k} \cdot \mathbf{r} - \omega t)] \tag{31}$$

where $\mathbf{E_0}$ and $\mathbf{H_0}$ are space and time independent vectors satisfy Maxwell's equations [Eqs (19)-(22)] and show that

$$\mathbf{k} \cdot \mathbf{E} = 0 \tag{32}$$

$$\mathbf{k} \cdot \mathbf{H} = 0 \tag{33}$$

$$\mathbf{H} = \frac{\mathbf{k} \times \mathbf{E}}{\omega \mu} \tag{34}$$

and

$$\mathbf{E} = \frac{\mathbf{H} \times \mathbf{k}}{\omega \varepsilon} \tag{35}$$

Thus **E**, **H** and **k** are at right angles to each other showing the transverse nature of the waves. Using the above equations show that

$$H_0 = \frac{k}{\omega \mu} E_0 \tag{36}$$

(b) Substitute Eq. (35) in Eq. (34) to obtain the following expression for the velocity of electromagnetic waves

$$v = \frac{\omega}{k} = \frac{1}{\sqrt{\varepsilon \mu}} \tag{37}$$

15.5 In continuation of the previous problem, use Maxwell's equations to derive the wave equation, determine the velocity of electromagnetic waves in a dielectric and hence derive an expression for the refractive index of the dielectric.

15.6 For an x-polarised plane electromagnetic wave propagating in the $+z$ direction, we may write the electric field as

$$\mathbf{E} = \hat{\mathbf{x}} E_0 \exp[i(kz - \omega t)] \tag{38}$$

where the actual electric field is given by

$$E_x = E_0 \cos(kz - \omega t),\ E_y = 0,\ E_z = 0 \tag{39}$$

Calculate the corresponding magnetic field.

15.7 (a) Define two variables

$$\xi = x - vt \quad \text{and} \quad \eta = x + vt \tag{40}$$

Using the above variables, show that Eq. (1) transforms to the following equation

$$\frac{\partial^2 \Psi}{\partial \xi \, \partial \eta} = 0 \tag{41}$$

(b) Integrate the above equation to obtain the general solution given by Eq. (2).

15.8 Write the three-dimensional wave equation [Eq. (13)] in spherical coordinates (r, θ, ϕ). Assume Ψ to be a function only of r and t, and obtain the general solution of the wave equation.

15.9 A Gaussian pulse is propagating in the $+x$-direction and at $t = t_0$ the displacement is given by

$$y(x, t = t_0) = a \exp\left[-\frac{(x-b)^2}{\sigma^2}\right]$$

Find $y(x, t)$.

15.10 A sonometer wire is stretched with a tension of 1 N. Calculate the velocity of transverse waves if $\rho = 0.2$ g/cm.

15.11 The displacement associated with a three-dimensional wave is given by

$$\psi(x, y, z, t) = a \cos\left[\frac{\sqrt{3}}{2} kx + \frac{1}{2} ky - \omega t\right]$$

Show that the wave propagates along a direction making an angle 30° with the x-axis.

15.12 Obtain the unit vector along the direction of propagation for a wave, the displacement of which is given by

$$\psi(x, y, z, t) = a \cos[2x + 3y + 4z - 5t]$$

where x, y and z are measured in centimeters and t in seconds. What will be the wavelength and the frequency of the wave?

SOLUTIONS

15.1 (a) $a = 0.1$ cm; $k = \frac{2\pi}{\lambda} = 0.2\ \text{cm}^{-1} \Rightarrow \lambda \approx 31.4$ cm

$\omega = 2\ \text{s}^{-1} \Rightarrow \nu \approx 0.32\ \text{s}^{-1}$; $v = \frac{2}{0.2} = 10$ cm/s.

Wave propagating in the $+x$ direction.

(b) $a = 0.2$ cm; $k = \frac{2\pi}{\lambda} = 0.5\ \text{cm}^{-1} \Rightarrow \lambda \approx 12.6$ cm

$\omega = 3\ \text{s}^{-1} \Rightarrow \nu \approx 0.48\ \text{s}^{-1}$; $v = \frac{3}{0.5} = 6$ cm/s.

Wave propagating in the $-x$ direction.

(c) $a = 0.5$ cm; $k = 0.2\pi = \frac{2\pi}{\lambda} \Rightarrow \lambda \approx 10$ cm

$\nu = 1\ \text{s}^{-1}$; $v = 10$ cm/s.

Wave propagating in the $+x$ direction

15.4 Maxwell's equations are given by Eqs (19)-(22). Now,

$$\nabla \cdot \mathbf{E} = \frac{\partial E_x}{\partial x} + \frac{\partial E_y}{\partial y} + \frac{\partial E_z}{\partial z}$$

Since, $E_x = E_{0x} \exp[i(\mathbf{k}\cdot\mathbf{r} - \omega t)] = E_{0x} \exp[i(k_x x + k_y y + k_z z - \omega t)]$

we get $\frac{\partial E_x}{\partial x} = i k_x E_{0x} \exp[i(k_x x + k_y y + k_z z - \omega t)]$

Thus, the Maxwell's equation $\nabla \cdot \mathbf{E} = 0$ would give us

$$i[k_x E_{0x} + k_y E_{0y} + k_z E_{0z}] \exp[i(\mathbf{k}\cdot\mathbf{r} - \omega t)] = 0$$

implying

$$\mathbf{k}\cdot\mathbf{E} = 0 \qquad (42)$$

Similarly, the equation $\nabla \cdot \mathbf{H} = 0$ would give us

$$\mathbf{k}\cdot\mathbf{H} = 0 \qquad (43)$$

The above two equations tell us that **E** and **H** are at right angles to **k**, thus the waves are transverse in nature. Now, using Eq. (29)

$$(\nabla \times \mathbf{E})_x = \left(\frac{\partial E_z}{\partial y} - \frac{\partial E_y}{\partial z}\right) = i[k_y E_{0z} - k_z E_{0y}] \exp[i(\mathbf{k}\cdot\mathbf{r} - \omega t)]$$

$$= i(\mathbf{k} \times \mathbf{E})_x$$

Thus, Eq. (21) gives us

$$i(\mathbf{k} \times \mathbf{E})_x = i\omega\mu H_x \quad \Rightarrow \quad H_x = \frac{(\mathbf{k} \times \mathbf{E})_x}{\omega\mu} \qquad (44)$$

Similarly we can write for the y and z components of Eq. (21) and obtain the vector equation

$$\mathbf{H} = \frac{\mathbf{k} \times \mathbf{E}}{\omega\mu} \qquad (45)$$

Similarly, Eq. (22) would give us

$$\mathbf{E} = \frac{\mathbf{H}\times\mathbf{k}}{\omega\varepsilon} \tag{46}$$

showing that **k**, **E** and **H** are at right angles to each other. From Eqs (45) one readily gets

$$H_0 = \frac{k}{\omega\mu}E_0 \tag{47}$$

Substituting for **H** from Eq. (34) in Eq. (35), we get

$$\begin{aligned}\mathbf{E} &= \frac{1}{\omega^2\varepsilon\mu}[(\mathbf{k}\times\mathbf{E})\times\mathbf{k}] \\ &= \frac{1}{\omega^2\varepsilon\mu}[(\mathbf{k}\cdot\mathbf{k})\mathbf{E}-(\mathbf{k}\cdot\mathbf{E})\times\mathbf{k}]\end{aligned} \tag{48}$$

where we have used the vector identity

$$(\mathbf{A}\times\mathbf{B})\times\mathbf{C} = (\mathbf{A}\cdot\mathbf{C})\mathbf{B}-(\mathbf{B}\cdot\mathbf{C})\mathbf{A} \tag{49}$$

Since $\mathbf{k}\cdot\mathbf{E} = 0$, we get

$$\mathbf{E} = \frac{k^2}{\omega^2\varepsilon\mu}\mathbf{E}$$

Thus,

$$k = \omega\sqrt{\varepsilon\mu} \tag{50}$$

and the speed of propagation of the electromagnetic wave is given by

$$v = \frac{\omega}{k} = \frac{1}{\sqrt{\varepsilon\mu}} \tag{51}$$

15.5 If we take the curl of Eq. (21), we would obtain

$$\text{curl curl }\mathbf{E} = -\mu\frac{\partial}{\partial t}\text{curl }\mathbf{H} = -\varepsilon\mu\frac{\partial^2\mathbf{E}}{\partial t^2} \tag{52}$$

where we have used Eq. (22). Now, the operator $\nabla^2\mathbf{E}$ is *defined* by the following equation;

$$\nabla^2\mathbf{E} \equiv \text{grad div }\mathbf{E} - \text{curl curl }\mathbf{E} \tag{53}$$

Using Cartesian coordinates, one can easily show that

$$(\nabla^2\mathbf{E})_x = \frac{\partial^2 E_x}{\partial x^2}+\frac{\partial^2 E_x}{\partial y^2}+\frac{\partial^2 E_x}{\partial z^2} = \text{div grad }E_x \tag{54}$$

i.e., a Cartesian component of $\nabla^2\mathbf{E}$ is the div grad of the Cartesian component[1]. Thus, using

$$\text{curl curl }\mathbf{E} = \text{grad div }\mathbf{E} - \nabla^2\mathbf{E}$$

1. However, $(\nabla^2\mathbf{E})_r \neq \text{grad div } E_r$

we obtain

$$\text{grad div } \mathbf{E} - \nabla^2 \mathbf{E} = -\varepsilon\mu_0 \frac{\partial^2 \mathbf{E}}{\partial t^2} \tag{55}$$

or,

$$\nabla^2 \mathbf{E} = \varepsilon\mu \frac{\partial^2 \mathbf{E}}{\partial t^2} \tag{56}$$

where we have used the equation div $\mathbf{E} = 0$. Equation (56) is known as the three-dimensional wave equation and each Cartesian component of **E** satisfies the scalar wave equation:

$$\nabla^2 \Psi = \varepsilon\mu \frac{\partial^2 \Psi}{\partial t^2} \tag{57}$$

In a similar manner, one can derive the wave equation satisfied by **H**

$$\nabla^2 \mathbf{H} = \varepsilon\mu \frac{\partial^2 \mathbf{H}}{\partial t^2} \tag{58}$$

The velocity of propagation (v) of the wave is simply given by

$$v = \frac{1}{\sqrt{\varepsilon\mu}} \tag{59}$$

15.6 The corresponding magnetic field is given by

$$H_x = 0,\ H_y = H_0 \cos(kz - \omega t),\ H_z = 0 \tag{60}$$

with

$$H_0 = \frac{k}{\omega\mu} E_0 \tag{61}$$

15.7 We introduce the new variables

$$\xi = x - vt \quad \text{and} \quad \eta = x + vt.$$

Thus,

$$\frac{\partial \xi}{\partial x} = 1 \quad \text{and} \quad \frac{\partial \eta}{\partial x} = 1$$

Now, in terms of the independent variables ξ and η

$$\frac{\partial \Psi}{\partial x} = \frac{\partial \Psi}{\partial \xi}\frac{\partial \xi}{\partial x} + \frac{\partial \Psi}{\partial \eta}\frac{\partial \eta}{\partial x} = \frac{\partial \Psi}{\partial \xi} + \frac{\partial \Psi}{\partial \eta}$$

Further,

$$\frac{\partial^2 \Psi}{\partial x^2} = \frac{\partial}{\partial \xi}\left(\frac{\partial \Psi}{\partial \xi} + \frac{\partial \Psi}{\partial \eta}\right)\frac{\partial \xi}{\partial x} + \frac{\partial}{\partial \eta}\left(\frac{\partial \Psi}{\partial \xi} + \frac{\partial \Psi}{\partial \eta}\right)\frac{\partial \eta}{\partial x}$$

$$= \frac{\partial^2 \Psi}{\partial \xi^2} + 2\frac{\partial^2 \Psi}{\partial \xi \partial \eta} + \frac{\partial^2 \Psi}{\partial \eta^2} \tag{62}$$

Similarly, since $\frac{\partial \xi}{\partial t} = -v$ and $\frac{\partial \eta}{\partial t} = +v$, we get

$$\frac{\partial \Psi}{\partial t} = \frac{\partial \Psi}{\partial \xi}\frac{\partial \xi}{\partial t} + \frac{\partial \Psi}{\partial \eta}\frac{\partial \eta}{\partial t} = -v\frac{\partial \Psi}{\partial \xi} + v\frac{\partial \Psi}{\partial \eta}$$

and $$\frac{\partial^2 \Psi}{\partial t^2} = -v\left[\frac{\partial}{\partial \xi}\left(\frac{\partial \Psi}{\partial \xi}\right)\frac{\partial \xi}{\partial t} + \frac{\partial}{\partial \eta}\left(\frac{\partial \Psi}{\partial \xi}\right)\frac{\partial \eta}{\partial t}\right]$$
$$+v\left[\frac{\partial}{\partial \xi}\left(\frac{\partial \Psi}{\partial \eta}\right)\frac{\partial \xi}{\partial t} + \frac{\partial}{\partial \eta}\left(\frac{\partial \Psi}{\partial \eta}\right)\frac{\partial \eta}{\partial t}\right]$$

or, $$\frac{\partial^2 \Psi}{\partial t^2} = v^2\left(\frac{\partial^2 \Psi}{\partial \xi^2} - 2\frac{\partial^2 \Psi}{\partial \xi \partial \eta} + \frac{\partial^2 \Psi}{\partial \eta^2}\right) \tag{63}$$

Substituting Eqs (62) and (63) in Eq. (1) we get

$$\frac{\partial}{\partial \eta}\left(\frac{\partial \Psi}{\partial \xi}\right) = 0$$

Thus, $\frac{\partial \Psi}{\partial \xi}$ has to be independent of η; however it can be an arbitrary function of ξ:

$$\frac{\partial \Psi}{\partial \xi} = F(\xi)$$

or, $$\Psi(\xi, \eta) = \int F(\xi)d\xi + \text{constant of integration}$$

The constant of integration can be an arbitrary function of η. Further, since the integral of an arbitrary function is again an arbitrary function, we obtain as the general solution of the wave equation

$$\Psi(\xi, \eta) = f(\xi) + g(\eta) = f(x - vt) + g(x + vt) \tag{64}$$

where f and g are arbitrary functions of their arguments.

15.8 In spherical coordinates

$$\nabla^2 \Psi = \frac{1}{r^2}\frac{\partial}{\partial r}\left(r^2 \frac{\partial \Psi}{\partial r}\right) + \frac{1}{r^2 \sin\theta}\frac{\partial}{\partial \theta}\left(\sin\theta \frac{\partial \Psi}{\partial \theta}\right) + \frac{1}{r^2 \sin^2\theta}\frac{\partial^2 \Psi}{\partial \phi^2} \tag{65}$$

Since Ψ is a function only of r and t, we have

$$\nabla^2 \Psi(r, t) = \frac{1}{r^2}\frac{\partial}{\partial r}\left(r^2 \frac{\partial \Psi}{\partial r}\right) \tag{66}$$

Thus, the wave equation becomes

$$\frac{1}{r^2}\frac{\partial}{\partial r}\left(r^2 \frac{\partial \Psi}{\partial r}\right) = \frac{1}{v^2}\frac{\partial^2 \Psi}{\partial t^2} \tag{67}$$

Making the transformation

$$\Psi(r, t) = \frac{u(r,t)}{r} \tag{68}$$

we obtain

$$r^2 \frac{\partial \Psi}{\partial r} = r^2 \frac{\partial}{\partial r}\left(\frac{u}{r}\right) = r\frac{\partial u}{\partial r} - u$$

Thus, $$\frac{1}{r^2}\frac{\partial}{\partial r}\left(r^2 \frac{\partial \Psi}{\partial t}\right) = \frac{1}{r}\frac{\partial^2 u}{\partial r^2}$$

and the wave equation [Eq. (67)] will take the form

$$\frac{\partial^2 u}{\partial r^2} = \frac{1}{v^2}\frac{\partial^2 u}{\partial t^2} \tag{69}$$

which is of the same form as Eq. (1). Thus, the general solution will be

$$\Psi(r, t) = \frac{f(r - vt)}{r} + \frac{g(r + vt)}{r} \tag{70}$$

The function $\frac{f(r - vt)}{r}$ will represent an outgoing spherical wave with amplitude decreasing as $\frac{1}{r}$ and therefore the intensity decreasing as $\frac{1}{r^2}$. Similarly the term $\frac{g(r + vt)}{r}$ will represent an incoming spherical wave.

15.9 $y(x, t) = a \exp\left[-\frac{(x - b - v(t - t_0))^2}{\sigma^2}\right]$

15.10 $T = 1$ N $= 10^5$ dynes, $\rho = 0.2$ g/cm, $v = \sqrt{\frac{T}{\rho}} = \sqrt{\frac{10^5}{0.2}} \approx 707$ cm/s

15.11 $k_x = \frac{\sqrt{3}}{2}k;\quad k_y = \frac{1}{2}k;\quad k_z = 0$

$$\tan\theta = \frac{k_y}{k_x} = \frac{1}{\sqrt{3}} \Rightarrow \theta = 30°$$

15.12 $k_x = 2$ cm^{-1}; $k_y = 3$ cm^{-1} and $k_z = 4$ cm^{-1}

$$k = \sqrt{29} = \frac{2\pi}{\lambda} \Rightarrow \lambda = \frac{2\pi}{\sqrt{29}} \approx 1.17 \text{ cm}$$

$$\omega = 5 \text{ s}^{-1} \Rightarrow \nu = \frac{5}{2\pi} \approx 0.796 \text{ s}^{-1}$$

Unit vector along the direction of propagation will be given by

$$\hat{\mathbf{k}} = \frac{k_x}{k}\hat{\mathbf{x}} + \frac{k_y}{k}\hat{\mathbf{y}} + \frac{k_z}{k}\hat{\mathbf{z}} = \frac{2}{\sqrt{29}}\hat{\mathbf{x}} + \frac{3}{\sqrt{29}}\hat{\mathbf{y}} + \frac{4}{\sqrt{29}}\hat{\mathbf{z}}$$

Group Velocity and Pulse Dispersion 16

A Quick Review

Consider a plane wave propagating along the $+z$ direction:

$$\Psi(z, t) = Ae^{i(\omega t - kz)} \tag{1}$$

If the wave is propagating in a medium characterised by the refractive index variation $n(\omega)$, then

$$k(\omega) = \frac{\omega}{c}n(\omega) \tag{2}$$

The phase velocity of the wave is given by

$$v_p = \frac{\omega}{k} \tag{3}$$

A temporal pulse travels with the group velocity given by

$$v_g = \frac{1}{dk/d\omega} \tag{4}$$

Thus,

$$\frac{1}{v_g} = \frac{dk}{d\omega} = \frac{1}{c}\left[n(\omega) + \omega\frac{dn}{d\omega}\right] \tag{5}$$

In free space $n(\omega) = 1$ at all frequencies; hence

$$v_g = v_p = c \tag{6}$$

Since,

$$\omega = \frac{2\pi c}{\lambda_0} \tag{7}$$

we get

$$\frac{1}{v_g} = \frac{1}{c}\left[n(\lambda_0) - \lambda_0\frac{dn}{d\lambda_0}\right] \tag{8}$$

and the time taken by a pulse to traverse a length L of the dispersive medium is given by

$$\tau = \frac{L}{v_g} = \frac{L}{c}\left[n(\lambda_0) - \lambda_0\frac{dn}{d\lambda_0}\right] \tag{9}$$

For a source having a spectral width of $\Delta\lambda_0$, the temporal broadening of a pulse will therefore be given by

$$\Delta\tau_m = \frac{d\tau}{d\lambda_0}\Delta\lambda_0 = -\frac{L\Delta\lambda_0}{\lambda_0 c}\left[\lambda_0^2\frac{d^2n}{d\lambda_0^2}\right] \tag{10}$$

The quantity $\Delta\tau_m$ is usually referred as material dispersion because it is due to the material properties of the medium–hence the subscript m. Indeed, after propagating through a length L of the dispersive medium, a pulse of temporal width τ_0 will get broadened to τ_f where

$$\tau_f^2 = \tau_0^2 + (\Delta\tau_m)^2 \tag{11}$$

In Eq. (10), we assume $L = 1$ km (= 1000 m), $\Delta\lambda_0 = 1$ nm (= 10^{-9} m) to obtain the following expression for the material dispersion coefficient (which is measured in ps/km-nm):

$$D_m = \frac{\Delta\tau_m}{L\Delta\lambda_0} = -\frac{10^4}{3\lambda_0}\left[\lambda_0^2 \frac{d^2 n}{d\lambda_0^2}\right] \text{ps/km.nm (Material Dispersion Coefficient)} \tag{12}$$

where we have used $c \approx 3 \times 10^8$ m/s $= 3 \times 10^{-7}$ km/ps and λ_0 [in Eq. (12)] is measured in μm and the quantity inside the square brackets is dimensionless. The quantity D_m is usually referred as the material dispersion coefficient (because it is due to the material properties of the medium) and hence the subscript m on D; it is tabulated (for pure silica) in Table 18.1.

For pure silica, the refractive index variation can be assumed to be given by the following convenient approximate empirical formula (in the wavelength domain 0.5 μm < λ_0 < 1.6 μm)

$$n(\lambda_0) \approx C_0 - a\lambda_0^2 + \frac{a}{\lambda_0^2} \tag{13}$$

where $C_0 \approx 1.451$, $a \approx 0.003$ and λ_0 is measured in μm. [A more accurate expression for $n(\lambda_0)$ is given in Problem 16.5].

PROBLEMS

16.1 Using the empirical formula given by Eq. (13) calculate the phase and group velocities in silica at λ_0 = 0.7 μm, 0.8 μm, 1.0 μm, 1.2 μm and 1.4 μm. Compare with the (more accurate) values given in Table 18.1.

16.2 Using the empirical formula given by Eq. (13)

(a) Calculate the zero dispersion wavelength.

(b) Calculate the material dispersion at 800 nm in ps/km.nm.

[1.32 μm; –101 ps/km.nm]

16.3 Let,

$$n(\lambda_0) = n_0 + A\lambda_0 \tag{14}$$

where λ_0 is the free space wavelength. Derive expressions for phase and group velocities.

16.4 In 1836, Cauchy gave the following approximate formula to describe the wavelength dependence of refractive index in glass in the visible region of the spectrum

$$n(\lambda) = A + \frac{B}{\lambda_0^2}$$

Now, $n(\lambda_1) = 1.50883$; $n(\lambda_2) = 1.51690$ for borosilicate glass
$n(\lambda_1) = 1.45640$; $n(\lambda_2) = 1.46318$ for vitreous quartz
where $\lambda_1 = 0.6563$ μm and $\lambda_2 = 0.4861$ μm.

(a) Calculate the values of A and B.
(b) Using the Cauchy formula calculate the refractive index at 0.5890 μm and 0.3988 μm and compare with the corresponding experimental values:
   (i) (1.51124 and 1.52546) for borosilicate glass and
   (ii) (1.45845 and 1.47030) for vitreous quartz.

16.5 The refractive index variation for pure silica in the wavelength region 0.5 μm < λ_0 < 1.6 μm is accurately described by the following empirical formula [Ref. Pa 1]:

$$n(\lambda_0) = C_0 + C_1\lambda_0^2 + C_2\lambda_0^4 + \frac{C_3}{(\lambda_0^2 - l)} + \frac{C_4}{(\lambda_0^2 - l)^2} + \frac{C_5}{(\lambda_0^2 - l)^3} \tag{15}$$

where $C_0 = 1.4508554$, $C_1 = -0.0031268$, $C_2 = -0.0000381$, $C_3 = 0.0030270$, $C_4 = -0.0000779$, $C_5 = 0.0000018$, $l = 0.035$ and λ_0 is measured in μm. Calculate and plot $n(\lambda_0)$ and $\frac{d^2n}{d\lambda_0^2}$ in the wavelength domain 0.5 μm < λ_0 < 1.6 μm.

16.6 (a) For a Gaussian pulse given by

$$E(z=0, t) = E_0 e^{-\frac{t^2}{\tau_0^2}} e^{i\omega_0 t} \tag{16}$$

show that the spectral width is approximately given by

$$\Delta\omega \approx \frac{2}{\tau_0} \tag{17}$$

(b) Assume $\lambda_0 = 8000$ Å. Calculate $\frac{\Delta\omega}{\omega_0}$ for $\tau_0 = 1$ ns and for $\tau_0 = 1$ ps.

16.7 The time evolution of a Gaussian pulse in a dispersive medium is given by (see Ref. Gh1 and Gh2):

$$E(z, t) = \frac{E_0}{\sqrt{1+ip}} e^{i(\omega_0 t - k_0 z)} \exp\left[-\frac{\left(t - \frac{z}{v_g}\right)^2}{\tau_0^2(1+ip)}\right] \tag{18}$$

where, $$p \equiv \frac{2\gamma z}{\tau_0^2} \quad \text{and} \quad \gamma = \frac{d^2k}{d\omega^2} \tag{19}$$

(a) Show that the pulse broadening is given by

$$\Delta\tau = \frac{2z}{\tau_0}|\gamma| \tag{20}$$

(b) Using Eq. (2), show that

$$\gamma = \frac{\lambda_0}{2\pi c^2}\left[\lambda_0^2 \frac{d^2n}{d\lambda_0^2}\right] \tag{21}$$

(c) Using the above equation, calculate $\Delta\tau$ and show that the results are consistent with Eq. (10).

16.8 (a) For pure silica, at $\lambda_0 = 1.55$ μm, $\dfrac{d^2n}{d\lambda_0^2} \approx -0.004165$ μm^{-2}. Calculate (with proper units the value of γ.

(b) Calculate the value of $\Delta\tau$ for a 100 ps pulse propagating through a 2 km long fiber.

16.9 (a) For the propagating Gaussian pulse given by Eq. (18) show that the frequency chirp is given by

$$\Delta\omega = \frac{2p}{\tau_0^2(1+p^2)}\left(t - \frac{z}{v_g}\right) \tag{22}$$

where p is defined in Eq. (19).

(b) Assume a 100 ps ($= \tau_0$) pulse at $\lambda_0 = 1$ μm. Calculate the frequency chirp $\dfrac{\Delta\omega}{\omega_0}$ at $t - z/v_g = -100$ ps, -50 ps, $+50$ ps and $+100$ ps. Assume $z = 1$ km and other values from Table 18.1.

16.10 Repeat the previous problem for $\lambda_0 = 1.5$ μm; the values of τ_0 and z remain the same. Discuss the qualitative difference in the results obtained in the previous problem.

16.11 The frequency spectrum of $E(0, t)$ is given by the function $A(\omega)$. Show that the frequency spectrum of $E(z, t)$ is simply

$$A(\omega)\, e^{-ik(\omega)z}$$

implying that no new frequencies are generated–different frequencies superpose with different phases at different values of z.

SOLUTIONS

16.1 $n(\lambda_0) = C_0 - a\lambda_0^2 + \dfrac{a}{\lambda_0^2} = C_0 - a\left[\lambda_0^2 - \dfrac{1}{\lambda_0^2}\right]$; $\dfrac{dn}{d\lambda_0} = -2a\lambda_0 - \dfrac{2a}{\lambda_0^3}$

$$n_g = \frac{c}{v_g} = C_0 - a\lambda_0^2 + \frac{a}{\lambda_0^2} + 2a\lambda_0^2 + \frac{2a}{\lambda_0^2} = C_0 + a\left[\lambda_0^2 + \frac{3}{\lambda_0^2}\right]$$

Now, $C_0 \approx 1.451$, $a = 0.003$ and λ_0 should be in μm. Thus at $\lambda_0 = 0.7$ μm, 0.8 μm, 1.0 μm, 1.2 μm and 1.4 μm, we get

$$\lambda_0^2 - \frac{1}{\lambda_0^2} \approx -1.551, -0.923, 0, 0.746, 1.450$$

Thus, $n(\lambda_0) \approx 1.456, 1.454, 1.451, 1.449, 1.455$

Similarly, $n_g(\lambda_0) \approx 1.4708, 1.4670, 1.4630, 1.4616, 1.4615$

Actually the group index n_g attains a minimum around $\lambda_0 \approx 1.32$ μm. The phase and group velocities are c/n and c/n_g respectively.

16.2 (a) $n(\lambda_0) \approx 1.451 - 0.003\left(\lambda_0^2 - \frac{1}{\lambda_0^2}\right)$

$$\frac{dn}{d\lambda_0} \approx -0.003\left(2\lambda_0 + \frac{2}{\lambda_0^3}\right)(\mu\text{m})^{-1}$$

$$\frac{d^2n}{d\lambda_0^2} \approx -0.006\left(1 - \frac{3}{\lambda_0^4}\right)(\mu\text{m})^{-2}$$

Thus, $\frac{d^2n}{d\lambda_0^2}$ (and hence material dispersion) vanishes when

$$\lambda_0 \approx 3^{1/4}\ \mu\text{m} \approx 1.32\ \mu\text{m}.$$

(b) At $\lambda_0 = 800$ nm $= 0.8$ μm

$$\lambda_0^2 \frac{d^2n}{d\lambda_0} = 0.64 \times \left[-0.006\left(1 - \frac{3}{0.8^4}\right)\right] \approx +0.0243$$

Thus, using Eq. (19)

$$D_m \approx -\frac{10^4}{3 \times 0.8}(+0.0243) \approx -101 \text{ ps/km.nm}$$

which may be compared with the more accurate value of -106.6 ps/km.nm (see Table 18.1).

16.3 $\frac{dn}{d\lambda_0} = A$

Thus,

$$\frac{1}{v_g} = \frac{1}{c}\left[n(\lambda_0) - \lambda_0\frac{dn}{d\lambda_0}\right] = \frac{1}{c}[n_0 + A\lambda_0 - \lambda_0 A] = \frac{n_0}{c} \Rightarrow v_g = \frac{c}{n_0}$$

$$v_p = \frac{c}{n(\lambda_0)} = \frac{c}{n_0 + A\lambda_0}$$

16.4 $n(\lambda_1) = A + \frac{B}{\lambda_1^2}$ and $n(\lambda_2) = A + \frac{B}{\lambda_2^2}$

Thus, $B\left(\frac{1}{\lambda_2^2} - \frac{1}{\lambda_1^2}\right) = n(\lambda_2) - n(\lambda_1) \Rightarrow B = \frac{\lambda_1^2\lambda_2^2}{\lambda_1^2 - \lambda_2^2}[n(\lambda_2) - n(\lambda_1)]$

For $\lambda_1 = 0.6563$ μm and $\lambda_2 = 0.4861$ μm $\frac{\lambda_1^2\lambda_2^2}{\lambda_1^2 - \lambda_2^2} \approx 5.23 \times 10^{-13}\ \text{m}^2$

Thus, for borosilicate glass

$$B = 5.23 \times 10^{-13}[1.51690 - 1.50883] \approx 4.22 \times 10^{-15}\ \text{m}^2 \Rightarrow A = 1.499$$

Thus, at $\lambda = 0.5890$ μm, $n = 1.51120$ and at $\lambda = 0.3988$ μm, $n = 1.52557$
Similarly, for vitreous quartz

$$B = 5.23 \times 10^{-13}[1.46318 - 1.45640] \approx 3.546 \times 10^{-15}\ \text{m}^2 \Rightarrow A = 1.44817$$

Thus, at $\lambda = 0.5890$ μm, $n = 1.45839$ and at $\lambda = 0.3988$ μm, $n = 1.47047$

16.5 We define

$$y = \lambda_0^2 \quad \text{and} \quad z = \lambda_0^2 - l$$

Then,

$$n(\lambda_0) = C_0 + C_1 y + C_2 y^2 + \frac{C_3}{z} + \frac{C_4}{z^2} + \frac{C_5}{z^3}$$

$$\frac{dn}{d\lambda_0} = \left[C_1 + 2C_2 y - \frac{C_3}{z^2} - \frac{2C_4}{z^3} - \frac{3C_5}{z^4}\right] 2\lambda_0$$

and

$$\frac{d^2 n}{d\lambda_0^2} = \left[2C_2 + \frac{2C_3}{z^3} + \frac{6C_4}{z^4} + \frac{12C_5}{z^5}\right] 4\lambda_0^2 + 2\left[C_1 + 2C_2 y - \frac{C_3}{z^2} - \frac{2C_4}{z^3} - \frac{3C_5}{z^4}\right]$$

A GNUPLOT program for calculating $n(\lambda_0)$ [denoted as $n(x)$], $\frac{dn}{d\lambda_0}$ [denoted as $np(x)$], and $\frac{d^2 n}{d\lambda_0^2}$ [denoted as $npp(x)$] is given below. The respective plots are shown in the diagrams.

GNUPLOT Program

```
#Variation of the group velocity and npp for pure
silica
set multiplot
set nokey
#set yrange [1.440:1.458]
#set yrange [-0.02:0.12]
#set yrange [-0.025:-0.010]
set xrange [0.5:1.65]
set xtics
set ytics
c=2.99792
c0=1.4508554
c1=-0.0031268
c2=-0.0000381
c3=0.0030270
c4=-0.0000779
c5=0.0000018
```

```
el=0.035
y(x)=1/(x*x - el)
n(x)=c0+ c1*x*x +c2*x*x*x*x +c3*y(x) +c4*y(x)*y(x)+c5*
y(x)*y(x)*y(x)
np(x)=2*c1*x+4*c2*x*x*x-2*x*c3*y(x)*y(x)
-4*x*c4*y(x)*y(x)*y(x)-6*x*c5*y(x)*y(x)*y(x)*y(x)

npp(x)=2*c1+12*c2*x*x-2*c3*y(x)*y(x)*(1
-4*x*x*y(x))-4*c4*y(x)*y(x)*y(x)*(1-6*x*x*y(x))
-6*c5*y(x)*y(x)*y(x)*y(x)*(1-8*x*x*y(x))

ng(x)=n(x) - x*npp(x)
vg(x)=c/ng(x)
nppn=npp(.85)
f0(x)=0.
f1(x)=x*x*npp(x)
f2(x)=-10000.0*f1(x)/(c*x)
#plot n(x),f0(x) w l 1
#plot npp(x),f0(x) w l 1
#plot np(x)w l 1
```

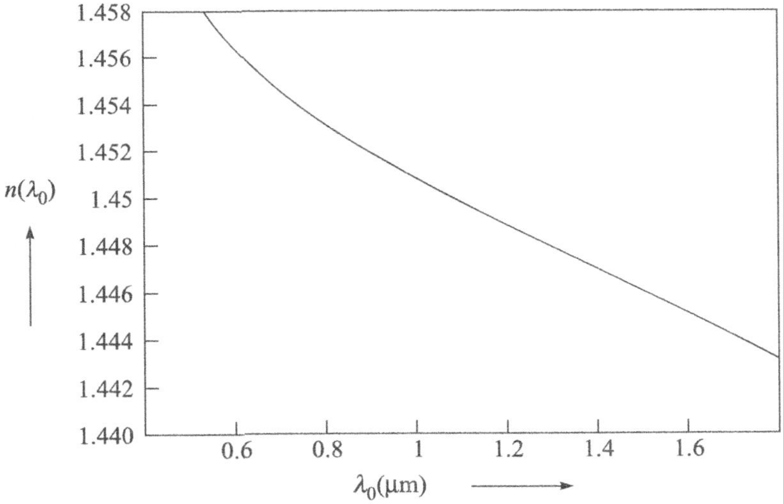

Fig. 16.1 Variation of $n(\lambda_0)$ with λ_0 for pure silica with $n(\lambda_0)$ given by Eq. (15)

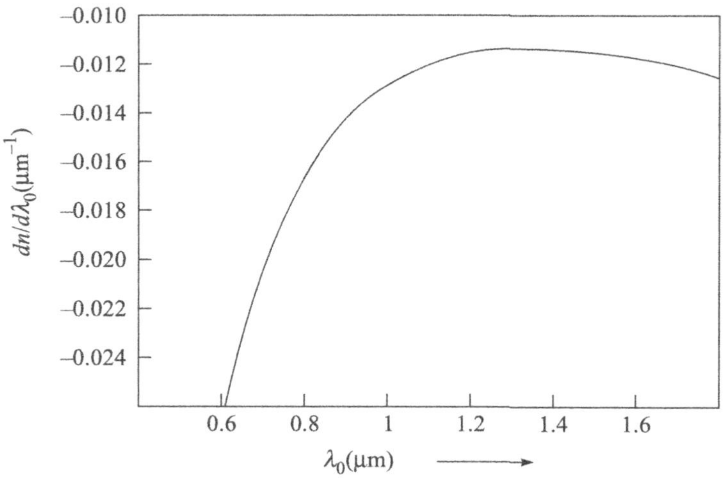

Fig. 16.2 Variation of $\frac{dn}{d\lambda_0}$ with λ_0 for pure silica with $n(\lambda_0)$ given by Eq. (15)

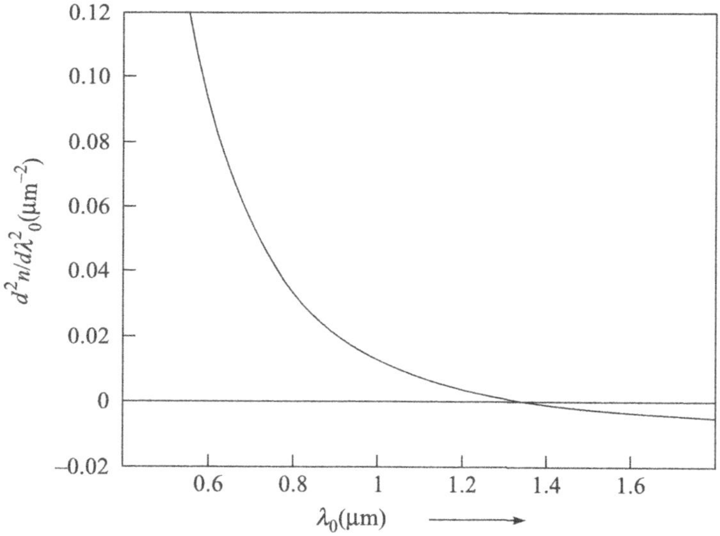

Fig. 16.3 Variation of $\frac{d^2n}{d\lambda_0^2}$ with λ_0 for pure silica with $n(\lambda_0)$ given by Eq. (15)

16.6 (a) Consider a Gaussian pulse for which we may write

$$E(z=0,t) = E_0\, e^{-\frac{t^2}{\tau_0^2}}\, e^{+i\omega_0 t} \tag{23}$$

A wave packet can always be expressed as a superposition of plane waves of different frequencies:

$$E(z,t) = \int_{-\infty}^{+\infty} A(\omega)\, e^{i[\omega t - kz]} d\omega \tag{24}$$

Obviously,

$$E(z=0,t) = \int_{-\infty}^{+\infty} A(\omega)e^{+i\omega t}d\omega \tag{25}$$

Thus,

$$A(\omega) = \frac{1}{2\pi}\int_{-\infty}^{+\infty} E(z=0,t)e^{-i\omega t}dt \tag{26}$$

Substituting from Eq. (23) we get

$$A(\omega) = \frac{E_0}{2\pi}\int_{-\infty}^{+\infty} e^{-\frac{t^2}{\tau_0^2}}\, e^{-i(\omega-\omega_0)t}\, dt$$

$$= \frac{E_0\tau_0}{2\sqrt{\pi}}\exp\left[-\frac{1}{4}(\omega-\omega_0)^2\,\tau_0^2\right] \tag{27}$$

where we have used

$$\int_{-\infty}^{+\infty} e^{-\alpha x^2+\beta x}dx = \sqrt{\frac{\pi}{\alpha}}\; e^{\beta^2/4\alpha} \tag{28}$$

In general, $A(\omega)$ can be complex and as such one defines the power spectral density

$$S(\omega) = |A(\omega)|^2 \tag{29}$$

For the Gaussian pulse,

$$S(\omega) = \frac{E_0^2\tau_0^2}{4\pi}\exp\left[-\frac{1}{2}(\omega-\omega_0)^2\,\tau_0^2\right] \tag{30}$$

If the FWHM (Full Width at Half Maximum) is $\Delta\omega$, then

$$\exp\left[-\frac{1}{4}\left(\frac{\Delta\omega}{2}\right)^2\tau_0^2\right] = \frac{1}{2}$$

which would give

$$\Delta\omega\;\tau_0 \approx 2.4$$

Thus, the spectral width is approximately given by

$$\Delta\omega \approx \frac{2}{\tau_0} \tag{31}$$

(b) $\lambda_0 = 8\times10^{-7}\,\text{m} \Rightarrow \omega = \dfrac{2\pi c}{\lambda_0} = \dfrac{2\pi\times3\times10^{8}}{8\times10^{-7}} \approx 2.36\times10^{15}\,\text{s}^{-1}$

$\Delta\omega = \dfrac{2}{\tau_0} = 2\times10^{9}\,\text{s}^{-1}$ and $2\times10^{12}\,\text{s}^{-1}$ for $\tau_0 = 1$ ns and 1 ps respectively.

Thus,

$\dfrac{\Delta\omega}{\omega} \approx 10^{-6}$ and 10^{-3} which represent the spectral purity of the pulse.

16.7 The intensity distribution corresponding to Eq. (18) would be given by

$$I(z,t) = \frac{I_0}{\tau(z)/\tau_0} \exp\left[-\frac{\left(t - \dfrac{z}{v_g}\right)^2}{\tau^2(z)}\right] \tag{32}$$

where, $$\tau^2(z) \equiv \tau_0^2(1+p^2) \tag{33}$$

Thus, the pulse broadening will be given by

$$\Delta\tau = \sqrt{\tau^2(z) - \tau_0^2}$$
$$= |p|\tau_0 = \frac{2|\gamma|z}{\tau_0} \tag{34}$$

(b) Now, $$\gamma = \frac{d^2k}{d\omega^2} = \frac{d}{d\omega}\left[\frac{1}{c}\left(n - \lambda_0 \frac{dn}{d\lambda_0}\right)\right]$$
$$= \frac{1}{c}\frac{d}{d\lambda_0}\left[n(\lambda_0) - \lambda_0 \frac{dn}{d\lambda_0}\right]\frac{d\lambda_0}{d\omega}$$

or, $$\gamma = \frac{\lambda_0}{2\pi c^2}\left[\lambda_0^2 \frac{d^2n}{d\lambda_0^2}\right] \tag{35}$$

where the quantity inside the square brackets is dimensionless. Further, since the spectral width of the Gaussian pulse is given by [see Eq. (31)]

$$\Delta\omega \approx \frac{2}{\tau_0}$$

we may write

$$\frac{1}{\tau_0} \approx \frac{1}{2}\Delta\omega \approx \frac{1}{2}\frac{2\pi c}{\lambda_0^2}|\Delta\lambda_0| \tag{36}$$

Substituting for τ_0 from the above equation and for γ [from Eq. (35)] in Eq. (34) we get

$$\Delta\tau = \frac{z}{\lambda_0 c}\left|\lambda_0^2 \frac{d^2n}{d\lambda_0^2}\right|\Delta\lambda_0 \tag{37}$$

which is the same as Eq. (10).

16.8 At $\lambda_0 = 1.55\ \mu\text{m}$, $\dfrac{d^2n}{d\lambda_0^2} \approx -0.004165\ \mu\text{m}^{-2}$ [see Table 18.1]

Thus,

$$\gamma = \frac{\lambda_0}{2\pi c^2}\left[\lambda_0^2 \frac{d^2n}{d\lambda_0^2}\right] \approx -\frac{1.55 \times 10^{-6}}{2\pi \times 9 \times 10^{16}}[1.55 \times 1.55 \times 0.004165]$$

$$\approx -2.743 \times 10^{-26}\ \text{m}^{-1}\,\text{s}^2$$

For a 100 ps pulse propagating through a 2 km long fiber

$$\Delta\tau = \frac{2|\gamma|z}{\tau_0} \approx \frac{2 \times 2.743 \times 10^{-26} \times 2 \times 10^3}{(10^{-10})} \approx 1.1 \text{ ps}$$

16.9 If we carry out simple manipulations, Eq. (18) can be written in the form:

$$E(z, t) = \frac{E_0}{[\tau(z)/\tau_0]^{1/2}} \exp\left[-\frac{\left(t - \frac{z}{v_g}\right)^2}{\tau^2(z)}\right] \exp[i(\Phi(z, t) - k_0 z)] \quad (38)$$

where the phase term is given by

$$\Phi(z, t) = \omega_0 t + \kappa\left(t - \frac{z}{v_g}\right)^2 - \frac{1}{2}\tan^{-1} p \quad (39)$$

and

$$\kappa(z) = \frac{p}{\tau_0^2(1 + p^2)} \quad (40)$$

Equation (39) represents the phase term and the instantaneous frequency is given by

$$\omega(t) = \frac{\partial\Phi}{\partial t} = \omega_0 + 2\kappa\left(t - \frac{z}{v_g}\right) \quad (41)$$

showing that $\omega(t)$ changes within the pulse. The frequency chirp is therefore given by

$$\Delta\omega = \omega(t) - \omega_0 = 2\kappa\left(t - \frac{z}{v_g}\right) \quad (42)$$

(b) $\lambda_0 = 1\ \mu\text{m} \Rightarrow \omega_0 = \dfrac{2\pi c}{\lambda_0} \approx 1.885 \times 10^{15}$ Hz; $\tau_0 = 100$ ps $= 10^{-10}$ s.

At $\lambda_0 = 1\ \mu\text{m}$, $\dfrac{d^2 n}{d\lambda_0^2} \approx +0.0120\ (\mu\text{m})^{-2}$ [see Table 18.1]

$$\text{Thus, } \gamma = \frac{d^2 k}{d\omega^2} = \frac{\lambda_0}{2\pi c^2}\left[\lambda_0^2 \frac{d^2 n}{d\lambda_0^2}\right] = \frac{10^{-6}}{2\pi \times (3 \times 10^8)^2}[1 \times 0.0120]$$

$$\approx 2.12 \times 10^{-26}\ \text{m}^{-1}\,\text{s}^2$$

$$p = \frac{2\gamma z}{\tau_0^2} \approx \frac{2 \times 2.12 \times 10^{-26} \times 10^3}{(10^{-10})^2} \approx 4.24 \times 10^{-3}$$

$$\frac{\Delta\omega}{\omega_0} = \frac{2p}{\omega_0 \tau_0^2 (1 + p^2)}\left(t - \frac{z}{v_g}\right) \approx 4.5 \times 10^{-10}\left(t - \frac{z}{v_g}\right)$$

where $\left(t - \dfrac{z}{v_g}\right)$ is measured in pico seconds. Thus, $\dfrac{\Delta\omega}{\omega_0} \approx -4.5 \times 10^{-8}, -2.25$

$\times 10^{-8}$, $+2.25 \times 10^{-8}$ and $+4.5 \times 10^{-8}$ at $\left(t - \dfrac{z}{v_g}\right) = -100$ ps, -50 ps, $+50$ ps and $+100$ ps respectively.

16.10 $\lambda_0 = 1.5\ \mu\text{m} \Rightarrow \omega_0 = \dfrac{2\pi c}{\lambda_0} \approx 1.257 \times 10^{15}$ Hz

$$\tau_0 = 100 \text{ ps} = 10^{-10} \text{ ps}, z = 1 \text{ km} = 10^3 \text{ m}$$

$$\text{At } \lambda_0 = 1.5\ \mu\text{m}, \frac{d^2n}{d\lambda_0^2} \approx -0.00365\ (\mu\text{m})^{-2}$$

$$\gamma = \frac{1.5 \times 10^{-6}}{2\pi \times (3 \times 10^8)^2}[-1.5 \times 1.5 \times 0.00365] \approx -2.18 \times 10^{-26} \text{ m}^{-1} \text{ s}^2$$

$$p = \frac{2\gamma z}{\tau_0^2} \approx -\frac{2 \times 2.18 \times 10^{-26} \times 10^3}{(10^{-10})^2} \approx -4.36 \times 10^{-3}$$

$$\frac{\Delta\omega}{\omega_0} = \frac{2p}{\omega_0 \tau_0^2 (1 + p^2)}\left(t - \frac{z}{v_g}\right) \approx -6.94 \times 10^{-10}\left(t - \frac{z}{v_g}\right)$$

where $\left(t - \dfrac{z}{v_g}\right)$ is measured in pico seconds. Thus, $\dfrac{\Delta\omega}{\omega_0} \approx +6.94 \times 10^{-8}$, $+3.47 \times 10^{-8}$, -3.47×10^{-8} and $+6.94 \times 10^{-8}$ at $\left(t - \dfrac{z}{v_g}\right) = -100$ ps, -50 ps, $+50$ ps and $+100$ ps respectively.

The qualitative difference in the results obtained in the previous and in the present problem is the fact that at $\lambda = 1\ \mu$m we have negative dispersion and the front end is red shifted ($\Delta\omega$ is negative) and the trailing end is blue shifted. The converse is true at $\lambda = 1.5\ \mu$m where we have positive dispersion.

16.11 We can write Eq. (24) as

$$E(z, t) = \int_{-\infty}^{+\infty} G(\omega, z) e^{i\omega t} d\omega$$

where $G(\omega, z) \equiv A(\omega)e^{-ikz}$. The inverse Fourier transform is given by

$$G(\omega, z) = \frac{1}{2\pi}\int_{-\infty}^{+\infty} E(z,t) e^{-i\omega t} dt$$

or, $$A(\omega)e^{-ikz} = \frac{1}{2\pi}\int_{-\infty}^{+\infty} E(z,t) e^{-i\omega t} d\omega$$

showing that the frequency spectrum of $E(z, t)$ is simply $A(\omega)e^{-ikz}$ implying that no new frequencies are generated.

Lasers

17

A Quick Review

17.1 EINSTEIN COEFFICIENTS

The quantities A_{21}, B_{12} and B_{21} are known as Einstein coefficients and are determined by the atomic system. The coefficient A_{21} is given by

$$A_{21} = \frac{1}{t_{sp}} \tag{1}$$

where t_{sp} represents the spontaneous emission lifetime of the upper level (see Fig. 17.1). Further,

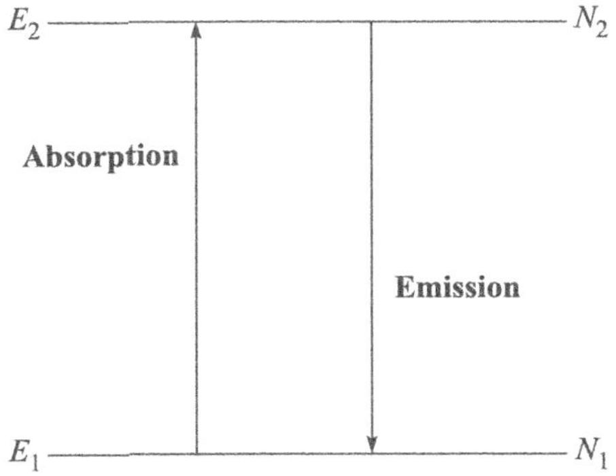

Fig. 17.1 E_1 and E_2 represent the energy levels of an atom. N_1 and N_2 represent the number of atoms (per unit volume) in the energy levels E_1 and E_2 respectively.

$$B_{12} = B_{21} = B \tag{2}$$

and

$$\frac{A_{21}}{B_{21}} = \frac{\hbar\omega^3 n_0^3}{\pi^2 c^3} \tag{3}$$

where n_0 represents the refractive index of the medium. At thermal equilibrium, the ratio of the number of spontaneous to stimulated emissions is given by

$$\frac{A_{21}}{B_{21}u(\omega)} = \exp\left(\frac{\hbar\omega}{k_B T}\right) - 1 \tag{4}$$

where $u(\omega)$ represents the radiation energy density and is given by Planck's law:

$$u(\omega) = \frac{\hbar\omega^3 n_0^3}{\pi^2 c^3} \frac{1}{\exp\left(\frac{\hbar\omega}{k_B T}\right) - 1} \tag{5}$$

17.2 LINESHAPE FUNCTIONS

17.2.1 Natural Broadening

The normalised lineshape function (corresponding to natural broadening) is given by

$$g(\omega) = \frac{2t_{sp}}{\pi} \frac{1}{1 + 4(\omega - \omega_0)^2 t_{sp}^2} \tag{6}$$

where $t_{sp} = 1/A_{21}$ is the spontaneous emission lifetime. The above functional form is referred to as a Lorentzian and the full width at half maximum (FWHM) of the Lorentzian is

$$\Delta\omega_N = \frac{1}{t_{sp}} \tag{7}$$

The normalisation condition satisfied by $g(\omega)$ is

$$\int g(\omega) d\omega = 1 \tag{8}$$

17.2.2 Collisonal Broadening

The normalised lineshape function corresponding to collisional broadening is given by

$$g(\omega) = \frac{\tau_c}{\pi} \frac{1}{1 + (\omega - \omega_0)^2 \tau_c^2} \tag{9}$$

where τ_c represents the mean collision time; the FWHM will be

$$\Delta\omega_c = \frac{2}{\tau_c} \tag{10}$$

Thus, a mean collision time of $\sim 10^{-6}$ s corresponds to a $\Delta\nu$ of about 0.3 MHz. The mean time between collisions depends on the mean free path and the average speed of the atoms in the gas which in turn would depend on the pressure and temperature of the gas as well as the mass of the atom. An approximate expression for the average collision time is

$$\tau_c \approx \frac{1}{8\pi}\left(\frac{2}{3}\right)^{1/2} \frac{(M k_B T)^{1/2}}{p a^2} \tag{11}$$

where M is the atomic mass, a is the radius of the atom (assumed to be a hard sphere) and p is the pressure of the gas.

17.2.3 Doppler Broadening

The normalised lineshape function corresponding to Doppler broadening is given by

$$g(\omega) = \frac{c}{\omega_0}\left(\frac{M}{2\pi k_B T}\right)^{\frac{1}{2}} \exp\left[-\frac{Mc^2}{2k_B T}\frac{(\omega-\omega_0)^2}{\omega_0^2}\right] \tag{12}$$

which corresponds to a Gaussian distribution. The lineshape function is peaked at ω_0, and the FWHM is given by

$$\Delta\omega_D = 2\omega_0\left(\frac{2k_B T}{Mc^2}\ln 2\right)^{\frac{1}{2}} \tag{13}$$

17.3 THE THRESHOLD CONDITION

Let N_1 and N_2 be the number of atoms per unit volume present in the energy levels E_1 and E_2 respectively (see Fig. 17.1). The atom in the lower energy level E_1 can absorb the incident radiation at a frequency $\omega = (E_2 - E_1)/\hbar$ and be excited to E_2. The threshold condition for the onset of laser action is approximately given by

$$(N_2 - N_1) \geq \frac{4\nu^2 n_0^3}{c^3}\frac{t_{sp}}{t_c}\frac{1}{g(\omega)} \tag{14}$$

where, $g(\omega)$ represents the lineshape function,

n_0 represents the refractive index of the medium enclosed by the cavity,

$t_{sp} = 1/A_{21}$ is the spontaneous lifetime associated with the transition $2 \to 1$,

t_c is the passive cavity lifetime (which is the time in which energy in the cavity reduces by a factor $1/e$) and is given by

$$\frac{1}{t_c} = \frac{c}{2dn_0}(2\alpha_1 d - \ln R_1 R_2) \tag{15}$$

with R_1 and R_2 represent the reflectivities of the mirrors forming the cavity and α_1 represents the average loss per unit length due to all loss mechanisms (other than the finite reflectivity) such as scattering loss, diffraction loss due to finite mirror sizes etc.

17.3.1 Absorption and Emission Cross Sections

The absorption cross section σ_a is related to the lineshape function and the spontaneous lifetime through the following equation:

$$\sigma_a = \frac{\pi^2 c^2}{\omega^2 n_0^2 t_{sp}} g(\omega) \tag{16}$$

The emission cross section σ_e is equal to the absorption cross section σ_a

17.3.2 Gain Co-efficient

In the presence of population inversion the gain co-efficient γ is given by

$$\gamma = \frac{\pi^2 c^2}{\omega^2 n_0^2 t_{sp}} g(\omega)(N_2 - N_1) \tag{17}$$

where N_1 and N_2 are the population densities in the lower and upper level.

17.4 MODES IN A CAVITY

The discrete frequencies of oscillation of the modes inside a cavity consisting of two mirrors separated by a distance d are given by

$$\nu = \nu_m = m\frac{c}{2n_0 d} \tag{18}$$

where m is an integer and n_0 represents the refractive index of the medium enclosed by the cavity (see Fig. 17.2)

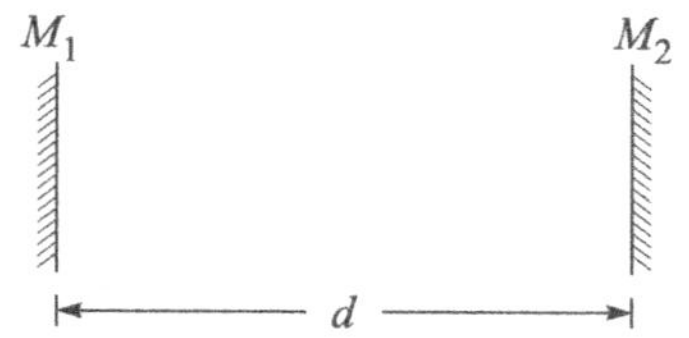

Fig. 17.2 An optical resonator consisting of two plane mirrors separated by a distance *d*.

17.4.1 Stability Condition

The stability condition for a general spherical mirror resonator consisting of two mirrors of radii of curvatures R_1 and R_2 and separated by a distance d (see Fig. 17.3) is

$$0 \le \left(1 - \frac{d}{R_1}\right)\left(1 - \frac{d}{R_2}\right) \le 1 \tag{19}$$

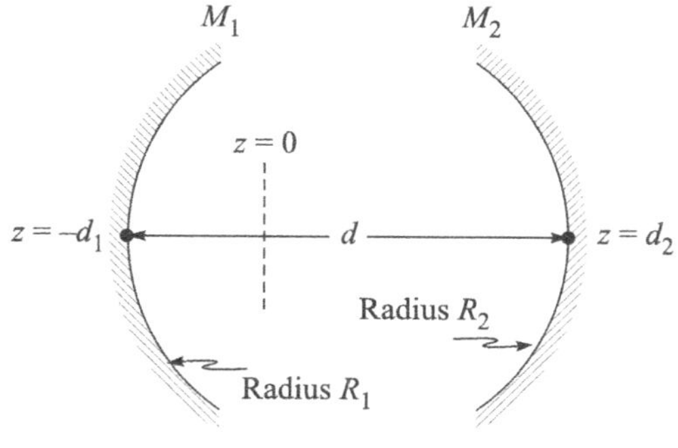

Fig. 17.3 A resonator consisting of two spherical mirrors.

The spot sizes of the Gaussian oscillating mode at the two mirrors is given by

$$w^2(z_1) = \frac{\lambda d}{\pi}\sqrt{\frac{g_2}{g_1(1 - g_1 g_2)}} \tag{20}$$

$$w^2(z_2) = \frac{\lambda d}{\pi}\sqrt{\frac{g_1}{g_2(1-g_1g_2)}} \tag{21}$$

where $w(z_1)$ and $w(z_2)$ represent the spot sizes at the two mirrors and,

$$g_1 = 1 - \frac{d}{R_1}; \quad g_2 = 1 - \frac{d}{R_2} \tag{22}$$

17.4.2 Quality Factor

The quality factor Q of a laser cavity is given by

$$Q = \frac{4\pi\nu_0 n_0 d}{c}\,\frac{1}{2\alpha_l d - \ln R_1R_2} \tag{23}$$

where R_1 and R_2 are the reflectivities of the two mirrors of the cavity, d is the separation between the mirrors and α_1 is the loss co-efficient of the cavity due to mechanisms other than the finite reflectivity of the mirror.

17.4.3 Mode Locking

The time variation of the intensity at the output of a mode locked laser is given by

$$I = I_0\left\{\frac{\sin[\pi(N+1)\delta\nu t]}{\sin[\pi\delta\nu t]}\right\}^2 \tag{24}$$

where $\delta\nu$ is the intermode spacing, and N represents the number of oscillating longitudinal modes of the resonator.

PROBLEMS

17.1 Consider a cavity consisting of two plane mirrors separated by a distance 60 cm in air (see Fig. 17.2). Calculate the mode number corresponding to the wavelength $\lambda = 6000$ Å. Also calculate the frequency spacing between the two longitudinal modes.

17.2 A laser cavity consists of two mirrors separated by a distance 10 cm in air. The laser beam has a central frequency of $\nu = \nu_0 = 6\times10^{14}$ Hz and two frequencies on either side of the central frequency. Calculate the frequency spacing between the longitudinal modes and the corresponding mode numbers.

17.3 Consider two concave mirrors M_1 and M_2 at $z = z_1 = -d_1$ and at $z = z_2 = +d_2$ respectively (see Fig. 17.3). We are assuming the origin somewhere between the mirrors so that both d_1 and d_2 are positive quantities. Thus, the distance between the two mirrors is given by $d = d_1 + d_2$. We assume a Gaussian beam propagating along the z-direction whose amplitude distribution on the plane $z = 0$ is given by

$$u(x, y) = a\exp\left[-\frac{x^2+y^2}{w_0^2}\right] \tag{25}$$

implying that the phase front is plane at $z = 0$; the parameter w_0 is the spot size and also called the *beam waist.* As the Gaussian beam propagates along the z-direction, the spot size and the radius of curvature of the wavefront change and are given by:

$$w(z) = w_0 \sqrt{1 + \frac{z^2}{\alpha}} \quad \text{and} \quad R(z) = z + \frac{\alpha}{z} \tag{26}$$

where $\alpha = \dfrac{\pi^2 w_0^4}{\lambda^2}$. For the Gaussian beam to resonate between the two mirrors, show that we must have

$$w_0^2 = \frac{\lambda d}{\pi (g_1 + g_2 - 2g_1 g_2)} \sqrt{g_1 g_2 (1 - g_1 g_2)} \tag{27}$$

where,

$$g_1 = 1 - \frac{d}{R_1} \quad \text{and} \quad g_2 = 1 - \frac{d}{R_2} \tag{28}$$

17.4 In continuation of the previous problem, show that for w_0 to be real we must have

$$0 \le \left(1 - \frac{d}{R_1}\right)\left(1 - \frac{d}{R_2}\right) \le 1 \tag{29}$$

Plot the stability diagram.

17.5 (a) Consider a simple resonator configuration consisting of a plane mirror and a spherical mirror separated by a distance d (see Fig. 17.4). Show that

$$w_0^2 = \frac{\lambda d}{\pi} \sqrt{\left(\frac{R}{d} - 1\right)} \tag{30}$$

(For $R < d$, w_0 will become imaginary and resonator will become unstable).

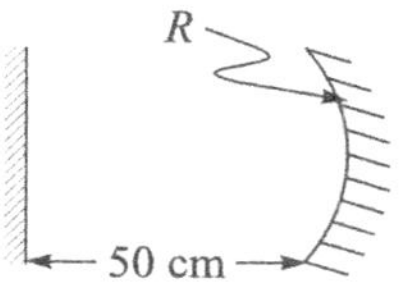

Fig. 17.4 A simple resonator consisting of a plane mirror and a concave mirror.

(b) For a typical He-Ne laser (λ = 0.6328 μm) we may have $d \approx 50$ cm, $R \approx 100$ cm. show that the resonator configuration is well within the shaded region of the stability diagram. Calculate the spot size w_0.

(c) If we increase R to 200 cm, what will be the spot size w_0?

17.6 Consider a simple resonator configuration consisting of two spherical mirrors separated by a distance $d = 1.5$ m with $R_1 = 1.0$ m and $R_2 = 0.75$ m. Show that the resonator configuration is very much stable. For $\lambda = 1$ μm, calculate the spot size w_0.

17.7 Calculate the values of g_1, g_2 and w_0 for

(a) A symmetric concentric resonator with $R_1 = R_2 = \dfrac{d}{2}$ so that the center of curvature of both mirrors are at the center.

(b) A symmetric confocal resonator with $R_1 = R_2 = d$ so that the center of curvature of both mirrors are at the pole of the other mirror.

(c) A resonator with plane parallel mirrors.

17.8 Determine the MKS units of $u(\omega)$, A and B.
[**Ans.** Js m^{-3}; Jm^{-3}; s^{-1}; m^3 J^{-1}s^{-2}].

17.9 For an optical source at thermal equilibrium, $T \sim 10^3$ °K with $\omega \approx 3.8 \times 10^{15}$ Hz (corresponding to $\lambda \approx 5000$ Å), calculate the ratio of the number of spontaneous to stimulated emissions.

17.10 For the $2P \to 1S$ transition in the hydrogen atom calculate the frequency of the transition ω. The lifetime of the $2P$ state for spontaneous emission is given by:

$$t_{sp} \approx 1.6 \times 10^{-9}\text{ s}$$

Calculate the Einstein A and B co-efficients. Assume $n_0 \approx 1$.
[**Ans.** $\omega \approx 1.5 \times 10^{16}$ Hz, $B_{21} \approx 4.2 \times 10^{20}$ m^3 J^{-1}s^{-2}]

17.11 Consider the D_1 line of Na ($\lambda \approx 5890$ Å)

(a) The spontaneous emission lifetime $t_{sp} \approx 16$ ns. Calculate the natural line width $\Delta\nu_N$ and $\Delta\lambda_N$.

(b) Assume $T = 500$ °K. Calculate $\Delta\nu_D$ and $\Delta\lambda_D$.
[$k_B \approx 1.38 \times 10^{-23}$ J/°K; $M_{Na} \approx 23\ M_H$; $M_H \approx 1.67 \times 10^{-27}$ kg].
[**Ans.** $\Delta\lambda_N \approx 10^{-4}$ Å; $\Delta\lambda_D \approx 0.02$ Å]

17.12 In a He-Ne laser the pressure of gas is typically 0.5 Torr; Torr is a unit of pressure and 1 Torr = 1 mm of Hg = 133 Nm^{-2}. Assuming the atomic mass $M = 20 \times 1.67 \times 10^{-27}$ kg, the radius of the atom $a \sim 0.1$ nm, $T = 300$ K calculate the mean collision time τ_c.

17.13 In a CO_2 laser ($\lambda_0 \approx 10.6$ μm), the laser transition occurs between the vibrational states of the CO_2 molecule. At $T \approx 300°$ K, calculate the Doppler linewidth $\Delta\nu_D$ and $\Delta\lambda_D$ [$M_{CO_2} \approx 44\ M_H$].
[**Ans.** $\Delta\nu_D \approx 53$ MHz; $\Delta\lambda_D \approx 0.2$ Å]

17.14 (a) Consider a He-Ne laser with cavity life time $t_c \approx 5 \times 10^{-8}$ s. If $R_1 = 1.0$ and $R_2 = 0.98$, calculate the cavity length d; assume $n_0 \approx 1$ and $\alpha_1 \approx 0$.

(b) Calculate the passive cavity line width $\Delta\nu_p$ and compare with the longitudinal mode spacing $\delta\nu$.
[**Ans.** (a) $d \approx 15$ cm (b) $\Delta\nu_p \approx 3.2$ MHz; $\delta\nu \approx 1$ GHz]

17.15 In a typical He-Ne laser ($\lambda = 6328$ Å) we have $d \approx 20$ cm, $R_1 \approx R_2 \approx 0.98$, $\alpha_1 \approx 0$, $t_{sp} \approx 10^{-7}$ s, $\Delta\nu_D \approx 1.3 \times 10^9$ Hz and $n_0 = 1$. Calculate t_c and $(N_2 - N_1)_{th}$.
[**Ans.** 33 ns; 8.8×10^8 cm^{-3}]

17.16 Consider a He-Ne laser ($\lambda_0 = 0.6328$ μm) with $d = 30$ cm, $n_0 \approx 1$, $R_1 \approx 1$, $R_2 \approx 0.99$. Calculate the passive cavity linewidth $\Delta\nu_p$ and the passive cavity life time t_c. You may assume $\alpha_c \approx 0$. [**Ans.** 0.8 MHz, 0.2 μs]

17.17 (a) For the He-Ne laser described in the previous problem, if the power level is 0.5 mW, calculate the ultimate linewidth $(\delta\nu)_{sp}$.

(b) Discuss the stability of the mirror position Δd to obtain the ultimate linewidth.

17.18 Limiting aperture are used to suppress higher order transverse mode oscillation. Consider a Gaussian beam of waist size $w_0 = 0.5$ mm and a total power of 1 mW. An aperture of radius 1 mm is introduced at the position of the waist. Calculate the power which goes through the aperture.

17.19 A 100 cm long laser having an oscillating bandwidth of 1500 MHz is mode locked. What would be (a) the approximate pulse width of the mode locked pulses, (b) the pulse repetition rate, and (c) the length of the mode locked pulse in free space.

17.20 Consider a He-Ne laser with Doppler broadened linewidth of 1500 MHz. What should be the length of the resonator cavity so that only a single longitudinal mode would oscillate? Assume that there is an oscillating mode at the line center.

17.21 An atomic transition has a time width of $\Delta\nu = 10^8$ Hz. Estimate the approximate value of $g(\omega)$ at the center of line.

17.22 There are 10^{11} photons in the cavity of an Ar-ion laser oscillating in steady state at the wavelength of 514 nm. If the laser resonator is formed by two plane mirrors of reflectivities 100% and 90%, separated by a distance of 50 cm, calculate the output power and the energy inside the cavity. Neglect internal losses of the cavity. [**Ans.** $P_o = 1.22$ W, $E = 38$ nJ]

17.23 Consider a laser with plane mirrors having reflectivities of 0.9 each and of length 50 cm filled with the gain medium (see Fig. 17.2). Neglecting scattering and other cavity losses, estimate the threshold gain coefficient (in m^{-1}) required to start laser oscillation.

17.24 In a typical He-Ne laser the threshold population inversion density is $10^9\,cm^{-3}$. What is the population inversion density when the laser is oscillating in steady state with an output of 1 mW?

17.25 Figure 17.5 shown the output from a mode locked laser as a function of time.

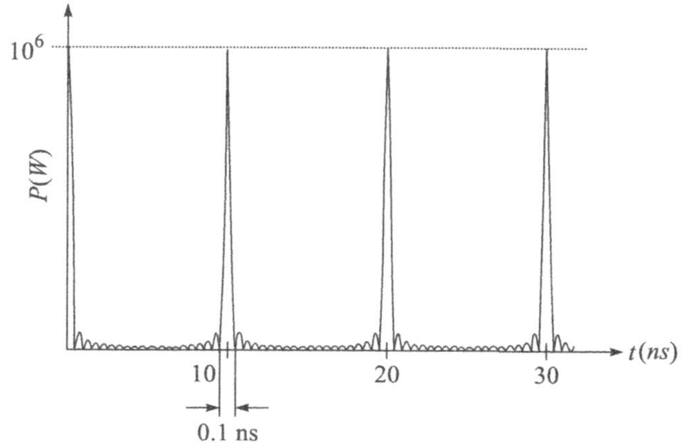

Fig. 17.5

(a) What is the length of the laser resonator (assuming an internal refractive index of unity)?

(b) What is the approximate number of oscillating modes?

(c) What would be the average output power when the same laser is operated without mode locking?

(d) What should be the frequency of a loss modulator placed inside the cavity for mode locking?

17.26 The gain co-efficient (in m^{-1}) of a laser medium with a center wavelength of 500 nm depends on frequency through the following equation:

$$\gamma(\nu) = \gamma(\nu_0)\exp\left[-4\left\{\frac{\nu - \nu_0}{\Delta\nu}\right\}^2\right]$$

where ν_0 is the center frequency, $\gamma(\nu_0) = 1\ \text{m}^{-1}$ and $\Delta\nu = 3$ GHz. The length of the laser cavity is 1 m and the mirror reflectivities are 99% each. Obtain the number of longitudinal modes that will oscillate in the laser. Neglect all other losses in the cavity.

17.27 Consider a three level laser system with lasing between levels E_2 and E_1. The level E_2 has a lift time of 1 μs. Assuming the transition E_3 -> E_2 to be very rapid, estimate the number of atoms that needs to be pumped per unit time per unit volume from level E_1 to reach threshold for achieving population inversion. Given that the total population density of the atoms is $10^{19}\ \text{cm}^{-3}$.

17.28 Consider an atomic system as shown below:

3 —— $E_3 = 3$ eV
2 —— $E_2 = 1$ eV
1 —— $E_1 = 0$ eV

The A co-efficient of the various transitions are given by

$$A_{32} = 7\times 10^7\ \text{s}^{-1}, A_{31} = 10^7\text{s}^{-1}, A_{21} = 10^8\ \text{s}^{-1}$$

(a) Show that this system cannot be used for continuous wave laser oscillation between levels 2 and 1.

(b) What is the spontaneous lifetime of level 3?

(c) If the steady state population of level 3 is 10^{15} atoms/cm^3, what is the power emitted spontaneously in the $3 \rightarrow 2$ transition?

SOLUTIONS

17.1 $\nu \approx \dfrac{3\times 10^8\ \text{m/s}}{6\times 10^{-7}\ \text{m}} = 5\times 10^{14}$ Hz

Thus, $m \approx \dfrac{5\times 10^{14}\times 2\times 60}{3\times 10^{10}} = 2\times 10^6$ and the frequency spacing between two adjacent modes will be given by

$$\delta\nu = \frac{c}{2d} \approx \frac{3\times 10^8\ \text{m/s}}{2\times 0.6\ \text{m}} = 2.5\times 10^8\ \text{Hz} = 250\ \text{MHz}.$$

17.2 The spacing of two adjacent modes will be

$$\delta\nu = \frac{c}{2d} = 1500 \text{ MHz}$$

Thus, the output beam will have frequencies

$$\nu_0 - 2\delta\nu,\ \nu_0 - \delta\nu,\ \nu_0,\ \nu_0 + \delta\nu \quad \text{and} \quad \nu_0 + 2\delta\nu$$

corresponding to m = 399998, 399999, 400000, 400001 and 400002, respectively. Figure 17.6 shows typical longitudinal modes of a laser.

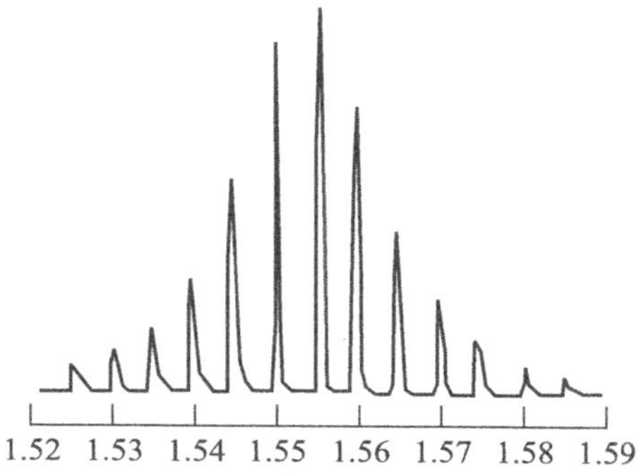

Fig. 17.6 The output of a typical multi-longitudinal mode (MLM) laser [Adapted from Ref. Lil]..

17.3 The poles of the mirrors M_1 and M_2 are at $z = z_1 = -d_1$ and at $z = z_2 = +d_2$ respectively (see Fig. 17.3). Now, as the Gaussian beam propagates along the z-direction, the spot size and the radius of curvature of the wavefront change and are given by:

$$w(z) = w_0\sqrt{1 + \frac{z^2}{\alpha}} \quad \text{and} \quad R(z) = z + \frac{\alpha}{z}$$

where $\alpha = \dfrac{\pi^2 w_0^4}{\lambda^2}$. For the Gaussian beam to resonate between the two mirrors, the radii of the phase front (at the mirrors) should be equal to the radii of curvatures of the mirrors:

$$-R_1 = -d_1 - \frac{\alpha}{d_1} \quad \text{and} \quad R_2 = d_2 + \frac{\alpha}{d_2}$$

We have used the sign convention such that for the type of mirrors shown in Fig. 17.3, both R_1 and R_2 are positive. Thus,

$$\alpha = d_1(R_1 - d_1) = d_2(R_2 - d_2)$$

If we use the relation $d_2 = d - d_1$, we would readily get

$$d_1 = \frac{(R_2 - d)d}{R_1 + R_2 - 2d} \quad \text{and} \quad d_2 = \frac{(R_1 - d)d}{R_1 + R_2 - 2d}$$

We define

$$g_1 = 1 - \frac{d}{R_1} \quad \text{and} \quad g_2 = 1 - \frac{d}{R_2}$$

From the above equations we may write $R_1 = \frac{d}{1-g_1}$ and $R_2 = \frac{d}{1-g_2}$ and we would obtain

$$d_1 = \frac{g_2(1-g_1)d}{g_1+g_2-2g_1g_2} \quad \text{and} \quad d_2 = \frac{g_1(1-g_2)d}{g_1+g_2-2g_1g_2}$$

Thus,

$$\alpha = d_1(R_1 - d_1) = \frac{g_1g_2d^2(1-g_1g_2)}{(g_1+g_2-2g_1g_2)^2}$$

Since $\alpha = \frac{\pi^2 w_0^4}{\lambda^2}$, we get for the spot size at the waist

$$w_0^2 = \frac{\lambda d}{\pi(g_1+g_2-2g_1g_2)}\sqrt{g_1g_2(1-g_1g_2)}$$

17.4 For w_0 to be real we must have $0 \le g_1g_2 \le 1$, or

$$0 \le \left(1-\frac{d}{R_1}\right)\left(1-\frac{d}{R_2}\right) \le 1$$

The above equation represents the stability condition for a resonator consisting of two spherical mirrors. Figure 17.7 shows the stability diagram and the shaded region correspond to stable resonator configurations.

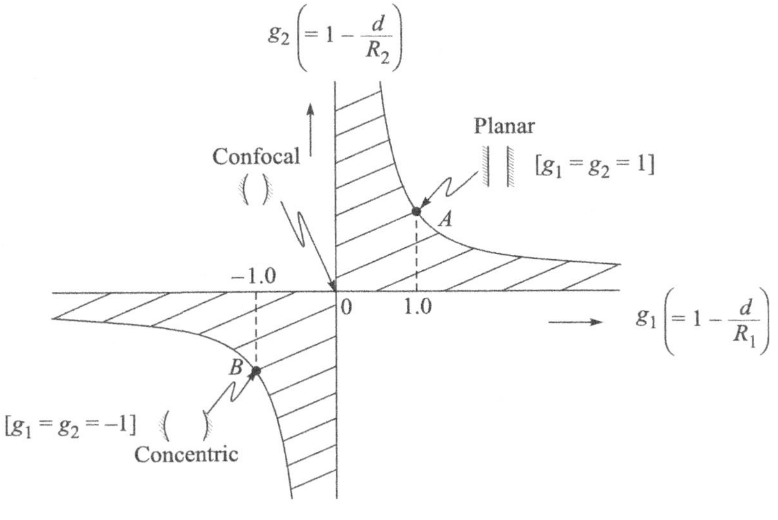

Fig. 17.7 The stability diagram for optical resonators. The shaded region corresponds to stable configurations.

17.5 (a) For a resonator configuration consisting of a plane mirror and a spherical mirror separated by a distance d, we will have $R_1 = \infty$ and $R_2 = R$ giving

$g_1 = 1$ and $g_2 = 1 - \dfrac{d}{R}$. Simple manipulations of Eq. (27) would give

$$w_0^2 = \frac{\lambda d}{\pi}\sqrt{\left(\frac{R}{d} - 1\right)}$$

(b) When $d \approx 50$ cm, $R \approx 100$ cm, we get $g_1 = 1$, $g_2 = 0.5$ and the resonator configuration is well within the shaded region of Fig. 17.7 and is very much stable. Further, $g_1 g_2 = 0.5$ and $w_0 \approx 0.32$ mm.

(c) If we increase R to 200 cm, we will get $w_0 \approx 0.38$ mm.

17.6 For $d = 1.5$ m, $R_1 = 1.0$ m and $R_2 = 0.75$ m, we get $g_1 = -0.5$, $g_2 = -1.0$ and $g_1 g_2 = 0.5$. Thus, the values of g_1 and g_2 are such that the resonator configuration is well within the shaded region of Fig. 17.7 and is very much stable. For $\lambda = 1$ µm, we get $w_0 \approx 0.31$ mm.

17.7 When $g_1 = g_2 = g$, Eq. (27) simplifies to

$$w_0^2 = \frac{\lambda d}{\pi}\sqrt{\frac{1+g}{1-g}}$$

(a) For the symmetric concentric resonator $R_1 = R_2 = \dfrac{d}{2}$ and $g_1 = g_2 = -1$. Thus, $g_1 g_2 = 1$ and w_0 becomes zero.

(b) For the symmetric confocal resonator $R_1 = R_2 = d$ and $g_1 = g_2 = 0$. Thus, $g_1 g_2 = 0$ and

$$w_0 = \sqrt{\frac{\lambda d}{\pi}}$$

(c) For plane parallel mirrors $R_1 = R_2 = \infty$, $g_1 = g_2 = 1$ and w_0 becomes infinity.

All three configurations discussed above (concentric, confocal and planar) lie on the boundary of the stability diagram so that a small variation of the parameters can make the system unstable and will have very large loss.

17.8 (a) $u(\omega)d\omega$ = Radiation energy per unit volume in the frequency interval ω and $\omega + d\omega$

Thus, unit of $u(\omega)d\omega = \text{J m}^{-3} \Rightarrow [u(\omega)] = \text{J s m}^{-3}$

(b) $[A] = \text{s}^{-1}$

(c) Number of stimulated emissions per unit time per unit volume $= N_2 B_{21} u(\omega)$

$\Rightarrow [N_2 B_{21} u(\omega)] = \text{s}^{-1}\text{m}^{-3}$. Since $[N_2] = \text{m}^{-3}$; $[u(\omega)] = \text{J s m}^{-3}$, we have

$$\text{m}^{-3}\,\text{J s m}^{-3}\,[B_{21}] = \text{s}^{-1}\text{m}^{-3} \quad \Rightarrow \quad [B_{21}] = \text{m}^3\,\text{J}^{-1}\text{s}^{-2}$$

17.9 At thermal equilibrium, the ratio of the number of spontaneous to stimulated emissions is given by;

$$R = \frac{A_{21}N_2}{B_{21}N_2 u(\omega)} = e^{\hbar\omega/k_B T} - 1$$

Thus, at thermal equilibrium at a temperature T, for frequencies, $\omega >> k_B T/\hbar$, the number of spontaneous emissions far exceeds the number of stimulated emissions. For an optical source at $T = 1000$ K

$$\frac{k_B T}{\hbar} = \frac{1.38 \times 10^{-23}(\text{J/K}) \times 10^3(\text{K})}{1.054 \times 10^{-34}(\text{Js})} \approx 1.3 \times 10^{14}\,\text{s}^{-1}$$

Thus for $\omega >> 1.3 \times 10^{14}\,\text{s}^{-1}$, the radiation would be mostly due to spontaneous emission. For $\lambda \cong 500$ nm (which corresponds to the yellow region of the spectrum) $\omega \approx 3.8 \times 10^{15}\,\text{s}^{-1}$ and

$$R \approx e^{29.2} \approx 5.0 \times 10^{12}$$

Thus at optical frequencies, the emission from a hot body is predominantly due to spontaneous transitions and hence the light from usual light sources is incoherent.

17.10 The energy levels of the hydrogen atom are given by

$$E_n = -\frac{\mu Z^2}{2n^2\hbar^2}\left(\frac{q^2}{4\pi\varepsilon_0}\right)^2$$

where μ is the reduced mass, $q (\approx 1.6 \times 10^{-19}\,\text{C})$ represents the charge of electron, $\varepsilon_0 (\approx 8.854 \times 10^{-12}$ MKS units) is the dielectric permittivity of free space and $\hbar = \dfrac{h}{2\pi}$

where $h \approx 6.626 \times 10^{-34}$ Js is the *Planck's constant.* For the $n = n_1$ to $n = n_2$ transition the frequency of the emitted line will be given by

$$h\nu = E_{n_1} - E_{n_2} = -\frac{\mu Z^2}{2\hbar^2}\left(\frac{q^2}{4\pi\varepsilon_0}\right)^2 \cdot \left(\frac{1}{n_2^2} - \frac{1}{n_1^2}\right)$$

Thus,
$$\frac{1}{\lambda} = \frac{\nu}{c} = \frac{1}{hc}(E_{n_1} - E_{n_2}) = RZ^2\left(\frac{1}{n_2^2} - \frac{1}{n_1^2}\right)$$

where $R = \dfrac{\mu}{4\pi\hbar^3 c}\left(\dfrac{q^2}{4\pi\varepsilon_0}\right)^2 \approx 1.0973 \times 10^7\,\text{m}^{-1}$ is known as the *Rydberg constant.*

For the $2P \to 1S$ transition in the hydrogen atom $n_1 = 2$, $n_2 = 1$ and $Z = 1$ and we get

$\lambda \approx 1.215 \times 10^{-7}$ m and $\omega = 2\pi\nu = \dfrac{2\pi c}{\lambda} \approx 1.55 \times 10^{16}$ Hz. ($\hbar\omega \approx 10.2$ eV)

The lifetime of the $2P$ state for spontaneous emission is given by:

$$t_{sp} = \frac{1}{A_{21}} \approx 1.6 \times 10^{-9}\,\text{s}$$

Thus,
$$A_{21} \approx 6 \times 10^8\,\text{s}^{-1}$$

and $$B_{21} = \frac{\pi^2 c^3}{\hbar \omega^3 n_0^3} A_{21} \approx 4.1 \times 10^{20}\, \mathrm{m^3\, J^{-1}\, s^{-2}}$$

where we have assumed $n_0 \approx 1$.

17.11 The spontaneous lifetime of the sodium level leading to a D_1 line ($\lambda = 589.1$ nm) is 16 ns. Thus, the natural line width (FWHM) will be

$$\Delta\omega_N = \frac{1}{t_{sp}} = \frac{1}{16 \times 10^{-9}\,\mathrm{s}} = 6.25 \times 10^7\,\mathrm{s^{-1}} \Rightarrow \Delta\nu_N = 10^7\,\mathrm{s^{-1}} = 10\ \mathrm{MHz}.$$

Now, $$\nu = \frac{c}{\lambda} \quad \Rightarrow \quad \Delta\nu = \frac{c}{\lambda^2}\Delta\lambda \qquad \text{(ignoring the sign)}$$

Thus, $$\Delta\lambda_N \approx \frac{\lambda^2}{c}\Delta\nu \approx \frac{(5.89 \times 10^{-5})^2}{3 \times 10^{10}} \times 10^7\,\mathrm{cm} \approx 10^{-12}\,\mathrm{cm} = 10^{-4}\,\text{Å}$$

$$\Delta\omega_D = \frac{4\pi}{\lambda_0}\left[\frac{2k_B T}{M}\ln 2\right]^{1/2}$$

$$= \frac{4\pi}{5.89 \times 10^{-7}}\left[\frac{2 \times 1.38 \times 10^{-23} \times 500 \times \ln 2}{23 \times 1.67 \times 10^{-27}}\right]^{1/2}$$

Thus, $$\Delta\omega_D \approx 10^{10}\,\mathrm{s^{-1}} \quad \Rightarrow \quad \Delta\nu_D = \frac{\Delta\omega_D}{2\pi} \approx 1.6 \times 10^9\,\mathrm{s^{-1}}$$

and $$\Delta\lambda_D \approx \frac{\lambda^2}{c}\Delta\nu_D \approx 2 \times 10^{-12}\,\mathrm{m} = 0.02\,\text{Å}$$

17.12 $$\tau_c = \frac{1}{8\pi}\left(\frac{2}{3}\right)^{1/2}\frac{(Mk_BT)^{1/2}}{pa^2}$$

$p = 0.5$ Torr $= 0.5 \times 133\ \mathrm{N\,m^{-2}}$; $M = 20 \times 1.67 \times 10^{-27}$ kg, $a = 0.1$ nm, $T = 300$ K.

Thus,

$$\tau_c = \frac{1}{8\pi}\left(\frac{2}{3}\right)^{1/2}\frac{\sqrt{20 \times 1.67 \times 10^{-27}(\mathrm{kg}) \times 1.38 \times 10^{-23}(\mathrm{JK^{-1}}) \times 300(\mathrm{K})}}{0.5 \times 133(\mathrm{Nm^{-2}}) \times [0.1 \times 10^{-9}(\mathrm{m})]^2}$$

$$\approx 10^{-6}\,\mathrm{s} = 1000\ \mathrm{ns}.$$

17.13 $\lambda_0 = 1.06 \times 10^{-5}$ m

$$\Delta\nu_D = \frac{2}{\lambda_0}\left[\frac{2k_BT}{M}\ln 2\right]^{1/2} = \frac{2}{1.06 \times 10^{-5}}\left[\frac{2 \times 1.38\,10^{-23} \times 300}{44 \times 1.67 \times 10^{-27}} \times 0.693\right]^{1/2}$$

$$\approx 53 \times 10^6\ \mathrm{Hz} = 53\ \mathrm{MHz}$$

$$\Delta\lambda_D \approx \frac{\lambda^2}{c}\Delta\nu_D \approx 0.2 \times 10^{-10}\,\mathrm{m} = 0.2\,\text{Å}$$

17.14 (a) Since $\alpha_1 \approx 0$, $\dfrac{1}{t_c} = \dfrac{c/n_0}{2d}[-\ln R_1 R_2]$

$$\Rightarrow \qquad d = \frac{3\times 10^{10} \times 5 \times 10^{-8}}{2}[-\ln(0.98)] \approx 15\,\text{cm}$$

(b) $\Delta\omega_p \sim \dfrac{1}{t_c} \approx 2\times 10^7\,\text{s}^{-1} \Rightarrow \Delta\nu_p = \dfrac{\Delta\omega_p}{2\pi} \approx 3.2 \times 10^6\,\text{s}^{-1}$. The longitudinal mode spacing is giving by:

$$\delta\nu = \frac{c}{2d} = \frac{3\times 10^{10}}{30} = 10^9\,\text{Hz} = 1\,\text{GHz}$$

17.15 $\dfrac{1}{t_c} = \dfrac{c/n_0}{2d}[-\ln R_1 R_2]$

$$= \frac{3\times 10^{10}}{2\times 20}[-\ln(0.98\times 0.98)] \approx 33\,\text{ns}$$

$$(N_2 - N_1)_{\text{th}} = \frac{\omega^2 n_0^3 t_{sp}}{\pi^2 c^3 t_c g(\omega_0)} = \frac{4\pi^2 c^2}{\lambda^2}\cdot\frac{n_0^3 t_{sp}}{\pi^2 c^3 t_c g(\omega_0)} = \frac{4t_{sp}}{\lambda^2 c t_c g(\omega_0)}$$

$$g(\omega_0) = \frac{2}{2\pi\Delta\nu}\left(\frac{\ln 2}{\pi}\right)^{1/2} \approx 1.15\times 10^{-10}\,\text{s}$$

Thus, $(N_2 - N_1)_{\text{th}} = \dfrac{4\times 10^{-7}}{(6.328\times 10^{-5})\times 3\times 10^{10}\times 33\times 10^{-9}\times 1.15\times 10^{-10}}$

$$\approx 8.8\times 10^8\,\text{cm}^{-3}$$

17.16 $\dfrac{1}{t_c} = \dfrac{c/n_0}{2d}[2\alpha_c d - \ln R_1 R_2] \approx \dfrac{3\times 10^{10}}{2\times 30}[-\ln(0.99)] \Rightarrow t_c \approx 2\times 10^{-7}\,\text{s} = 0.2\,\mu\text{s}$.

$$\Delta\omega_p \sim \frac{1}{t_c} \quad \Rightarrow \quad \Delta\nu_p \sim \frac{1}{2\pi t_c} \approx 0.8\,\text{MHz}.$$

17.17 (a) $(\delta\nu)_{sp} = \dfrac{2\pi(\Delta\nu_p)^2 h\nu_0}{P} = \dfrac{2\pi(0.8\times 10^6)^2\times 6.626\times 10^{-34}\times 3\times 10^8}{0.5\times 10^{-3}\times 0.6328\times 10^{-6}}$

$$\approx 2.5\times 10^{-3}\,\text{Hz}$$

(b) Change in frequency $\Delta\nu$ caused by a change in length Δd is given by $\dfrac{\Delta\nu}{\nu} = \dfrac{\Delta d}{d}$

Thus, for $\Delta\nu$ to be $\approx (\delta\nu)_{sp}$ we must have

$$\frac{\Delta d}{d} \approx \frac{(\delta\nu)_{sp}}{\nu} \Rightarrow \Delta d \approx \frac{(\delta\nu)_{sp}}{c/\lambda_0}\times d \approx \frac{2.5\times 10^{-3}}{3\times 10^8/0.6328\times 10^{-6}}\times 30$$

$$\approx 1.6\times 10^{-16}\,\text{m}$$

which is less than nuclear dimensions.

17.18 We assume a Gaussian intensity distribution of the form

$$I = I_0 e^{-r^2/w_0^2}$$

The power transmitted through an aperture of radius a centered on the Gaussian beam is given by

$$P = \int_0^a r dr \int_0^{2\pi} d\theta I(r) = 2\pi \int_0^a I(r) r dr$$

Substituting and integrating we obtain

$$P = P_0 (1 - e^{-a^2/w_0^2})$$

where P_0 is the total power in the beam. Using the values given in the problem we obtain $P =$ mW.

17.19 The mode spacing of the laser cavity is

$$\Delta\nu = \frac{c}{2L} = 300 \text{ MHz}$$

In a gain bandwidth of 1500 MHz, the number of oscillating modes would be 5.

(a) The approximate pulse width is the inverse of the oscillating bandwidth and is 0.67 ns.

(b) The pulse repetition rate is the inverse of the time taken for one roundtrip through the cavity. For a 1 m long cavity this corresponds to 3×10^8 pulses per second.

(c) The length of the mode locked pulse is the product of the speed of light and the pulse duration. Hence for this case it is 200 km.

17.20 The intermode spacing should be larger than half of the gain bandwidth since there is a mode at the center of the line. Hence

$$\frac{c}{2L} > 1.5 \times 10^8$$

which gives $L < 20$ cm

17.21 Since $g(\omega)$ is normalised the approximate value of $g(\omega)$ at the line center would be the inverse of the transition bandwidth. The given bandwidth is $\Delta\omega = 2\pi \times 10^8 \text{ s}^{-1}$. Hence,

$$g(\omega_0) \approx \frac{1}{2\pi} \times 10^8 \text{ s.}$$

17.22 The output power is given by

$$P_{\text{out}} = \frac{nh\nu}{\tau_c}$$

where τ_c is the cavity lifetime, n is the number of photons present inside the cavity and ν is the frequency. The cavity lifetime is given by

$$\frac{1}{\tau_c} = \frac{c}{2L}(-2\alpha L - \ln R_1 R_2)$$

Substituting the given values we obtain $P_{\text{out}} = 1.22$ W

Energy inside the cavity is $nh\nu = 38.6$ nJ

17.23 The threshold gain co-efficient is the value of gain when the gain compensates the losses. Hence, for the given cavity, it satisfied the following equation:

$$e^{\gamma L} \cdot R \cdot e^{\gamma L} \cdot R = 1$$

where γ is the gain co-efficient and R the reflectivity of each mirror. Hence, we obtain

$$\gamma = -\frac{1}{2} \ln R = -\frac{1}{0.5} \ln 0.9 \text{ m}^{-1}$$

17.24 The population inversion density when the laser is oscillating in steady state is equal to the population inversion density at threshold. Hence, in this case, the population inversion density is 10^9 cm^{-3}.

17.25 (a) The repetition frequency is given by Eq. $\delta\nu = c/2L$. From Fig. 17.5 we see that $1/\delta\nu = 10$ ns. Hence, the length of the cavity is 1.5 m.

(b) The time between the peak of the pulse and the first zero is given by the inverse of the bandwidth of the pulse. In this case it is 0.05 ns. The time between the pulses is the inverse of the intermode spacing which from the figure is 10 ns. The number of oscillating modes is given by the ratio of the oscillating bandwidth to the mode spacing. Hence, the number of oscillating modes is 200.

(c) The peak power under mode locking is N times the average power without mode locking where N is the number of modes. Hence, in this case the average power when the laser is operated without mode locking would be $10^6/200 = 5$ kW.

(d) The frequency of the mode locker should be equal to the mode spacing. Hence, in this case it should be 100 MHz.

17.26 We need to first calculate the bandwidth over which the gain exceeds the loss of the cavity. For this we first need to evaluate the loss co-efficient. Since we can neglect the internal losses of the cavity, if we represent the loss co-efficient as α we have

$$\alpha = -\frac{1}{2L} \ln R_1 R_2$$

Substituting the values we obtain $\alpha \sim 0.01 \text{ m}^{-1}$.

The gain bandwidth will be the frequency range where the gain exceeds the loss. Hence, if the gain becomes equal to loss at $\nu = \nu_0 \pm \delta\nu$ then we have

$$0.01 = \gamma(\nu_0) \exp\left[-\frac{(\delta\nu)^2}{4(\Delta\nu)^2}\right]$$

Substituting the values we obtain $\delta\nu \approx 3.22$ GHz. Hence, the gain bandwidth would be 6.44 GHz.

The spacing between the modes of the given cavity is 300 MHz. Hence, the number of oscillating modes would be $6.44/0.3 \sim 21$ modes.

17.27 Since the transitions from level E_3 are very rapid we can assume that the population of that level is zero. At threshold the population of the level E_1 and E_2 would be equal and hence the net transition between these two levels will be just the spontaneous emissions from E_2 to E_1. Thus, the rate at which atoms have to be transported from level E_1 to the upper level should be equal to the rate at which atoms are dropping spontaneously from E_2 to E_1 which is just the inverse of the spontaneous life time of E_2. Hence the rate of pumping should be 10^6 per second.

17.28 (a) For continuous wave operation of the three level system, we should have $A_{32} > A_{21}$. Since in this case this condition is not satisfied, this cannot be used for continuous wave laser operation.

(b) The total spontaneous emission rate from level 3 is the sum of the spontaneous rates from 3 to 2 and 3 to 1. Hence, the total rate is given by $8 \times 10^7\ s^{-1}$. This corresponds to a life time of 12.5 ns.

(c) The power emitted spontaneously in the transition 3 -> 2 is $N_3 A_{32}$ $(E_3 - E_2)$. Substituting the values, we obtain a power of $2.24 \times 10^{10}\ W/m^3$.

Fiber Optics I: Basic Concepts and Ray Optics Considerations in Multimode Fibers

18

A Quick Review

18.1 STEP INDEX FIBER

The refractive index distribution (in the transverse direction) of a step index fiber is given by (see Fig. 18.1)

$$\left.\begin{aligned} n &= n_1 \qquad 0 < r < a \\ &= n_2 \qquad r > a \end{aligned}\right\} \tag{1}$$

where n_1 and $n_2(<n_1)$ represent respectively the refractive indices of core and cladding and a represents the radius of the core. We define a parameter Δ through the following equations

$$\Delta \equiv \frac{n_1^2 - n_2^2}{2n_1^2} \tag{2}$$

When $n_1 \approx n_2$, i.e., when $\Delta << 1$ (as is true for most silica fibers)

$$\Delta \approx \frac{n_1 - n_2}{n_2} \approx \frac{n_1 - n_2}{n_1} \tag{3}$$

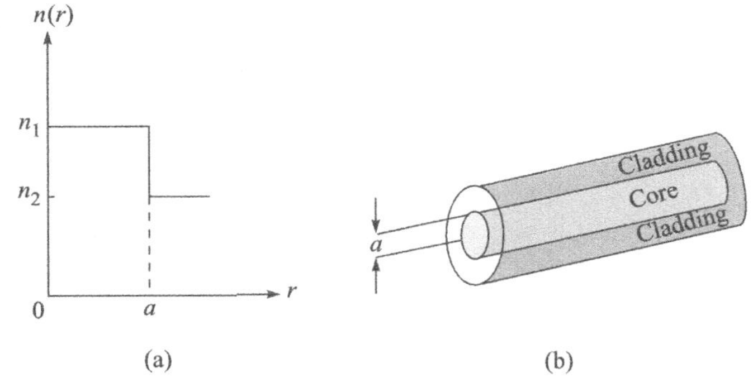

Fig. 18.1 (a) The refractive index variation of a step-index fiber. (b) The transverse cross-section of the fiber.

The numerical aperture of the fiber is given by the following equation:

$$\mathrm{NA} = \sin i_m = \sqrt{n_1^2 - n_2^2} \tag{4}$$

where i_m is the maximum value of the angle of incidence of the ray (that it makes with the z-axis) to be guided through the fiber (see Fig. 18.2).

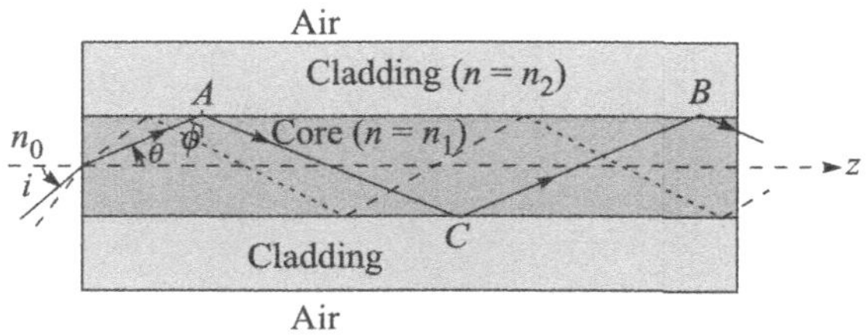

Fig. 18.2 If the angle of incidence (at the core-cladding interface) is greater than the critical angle, it will undergo total internal reflection and will guided through the optical fiber.

One defines the normalised waveguide parameter

$$V = \frac{2\pi}{\lambda_0} a\sqrt{n_1^2 - n_2^2} = \frac{2\pi}{\lambda_0} a n_1 \sqrt{2\Delta} \tag{5}$$

where λ_0 is the wavelength of operation. For $V \geq 10$, the total number of modes in a step index optical fiber is approximately given by

$$N \approx \frac{1}{2}V^2 \quad \text{(Number of modes for a multimode step-index fiber)} \tag{6}$$

Thus, for $V = 10$, the fiber will support approximately 50 modes. When the fiber supports such a large number of modes, the fiber is said to be a multimode fiber and one can use ray optics to calculate intermodal dispersion which is also known as ray dispersion which is given by:

$$\Delta\tau_i \cong \frac{n_1 L}{c}\Delta \quad \text{(Ray dispersion for a multimode step-index fiber)} \tag{7}$$

18.2 ATTENUATION

The attenuation in an optical fiber is usually measured in decibels (dB). If an input power P_1 results in an output power P_2, then the loss in decibels is given by

$$\alpha = 10 \log_{10}\left(\frac{P_{\text{input}}}{P_{\text{output}}}\right) \tag{8}$$

Thus, if the output power is only one hundredth of the input power, then the loss is = 20 dB etc. Further, the power level of a beam is measured in dBm which is defined as

$$P(\text{dBm}) = 10 \log_{10} P(\text{mW}) \tag{9}$$

Thus, e.g., 0.2 W = 200 mW $\Leftrightarrow \approx$ 23 dBm etc. Now,

$$P_{\text{output}}(\text{dBm}) = P_{\text{input}}(\text{dBm}) - \alpha(\text{dB}) \tag{10}$$

18.3 MATERIAL DISPERSION

For an optical pulse having a spectral width $\Delta\lambda_0$, the pulse broadening due to the dependence of the refractive index n on wavelength is given by [see Eq. (10) of Chapter 16].

$$\Delta\tau_m = -\frac{L\Delta\lambda_0}{\lambda_0 c}\left[\lambda_0^2 \frac{d^2n}{d\lambda_0^2}\right] \tag{11}$$

where L is the length of the fiber, $c \approx 3 \times 10^8$ m/s is the speed of light in free space, λ_0 is the free space wavelength and the quantity inside the square brackets is dimensionless. The material dispersion co-efficient (which is measured in ps/km-nm):

$$D_m = \frac{\Delta\tau_m}{L\Delta\lambda_0} = -\frac{10^4}{3\lambda_0}\left[\lambda_0^2 \frac{d^2n}{d\lambda_0^2}\right] \text{ ps/km.nm (Material Dispersion Co-efficient)} \tag{12}$$

where we have used $c \approx 3 \times 10^8$ m/s $= 3 \times 10^{-7}$ km/ps λ_0 is measured in μm and the quantity inside the square brackets is dimensionless. The quantity D_m is usually referred as the material dispersion coefficient (because it is due to the material properties of the medium) and hence the subscript m on D; it is tabulated (for pure silica) in Table 18.1. It represents the pulse broadening in picoseconds per kilometer length of the fiber per nanometer spectral width of the source. The total dispersion is given by

$$\Delta\tau = \sqrt{(\Delta\tau_i)^2 + (\Delta\tau_m)^2} \tag{13}$$

In Table 18.1 we have given the variation of $n(\lambda_0)$, $\frac{dn}{d\lambda_0}$, $\frac{d^2n}{d\lambda_0^2}$ and D_m for pure silica.

The maximum permissible bit rate (B_{max}), in one type of extensively used coding [known as NRZ (Non-Return to Zero)], is given by

$$B_{max} \approx \frac{0.7}{\Delta\tau} \tag{14}$$

Table 18.1 Values of n and D_m for pure silica[1]

λ_0 (μm)	$n(\lambda_0)$	$\frac{dn}{d\lambda_0}$ (μm^{-1})	$\frac{d^2n}{d\lambda_0^2}$ (μm^{-2})	D_m (ps/nm.km)
0.70	1.45561	−0.02276	0.0741	−172.9
0.75	1.45456	−0.01958	0.0541	−135.3
0.80	1.45364	−0.01725159	0.0400	−106.6
0.85	1.45282	−0.01552236	0.0297	−84.2
0.90	1.45208	−0.01423535	0.0221	−66.4
0.95	1.45139	−0.01327862	0.0164	−51.9
1.00	1.45075	−0.01257282	0.0120	−40.1

Contd.

1. The numerical values in the Table correspond to the refractive index variation as given in Ref. Pa1 (see Solution 16.5).

Table 18.1 Contd.

1.05	1.45013	−0.01206070	0.0086	−30.1
1.10	1.44954	−0.01170022	0.0059	−21.7
1.15	1.44896	−0.01146001	0.0037	−14.5
1.20	1.44839	−0.01131637	0.0020	−8.14
1.25	1.44783	−0.01125123	0.00062	−2.58
1.30	1.44726	−0.01125037	−0.00055	2.39
1.35	1.44670	−0.01130300	−0.00153	6.87
1.40	1.44613	−0.01140040	−0.00235	10.95
1.45	1.44556	−0.01153568	−0.00305	14.72
1.50	1.44498	−0.01170333	−0.00365	18.23
1.55	1.44439	−0.01189888	−0.00416	21.52
1.60	1.44379	−0.01211873	−0.00462	24.64

18.4 PARABOLIC INDEX FIBER

The refractive index distribution (in the transverse direction) of a parabolic index fiber (often abbreviated as PIF) is given by (see Fig. 18.3)

$$\left.\begin{aligned} n^2(r) &= n_1^2\left[1 - 2\Delta\left(\frac{r}{a}\right)^2\right] \quad 0 < r < a \\ &= n_2^2 \qquad\qquad\qquad\qquad r > a \end{aligned}\right\} \tag{15}$$

where n_1 represents the refractive index on the axis of the core and $n_2(< n_1)$ represents the refractive index of cladding and a represents the radius of the core. The parameter Δ, and the normalised waveguide parameter V is again defined by Eqs. (2) and (5) respectively. For $V \geq 10$, the total number of modes in a parabolic index optical fiber is approximately given by

$$N \approx \frac{1}{4}V^2 \tag{16}$$

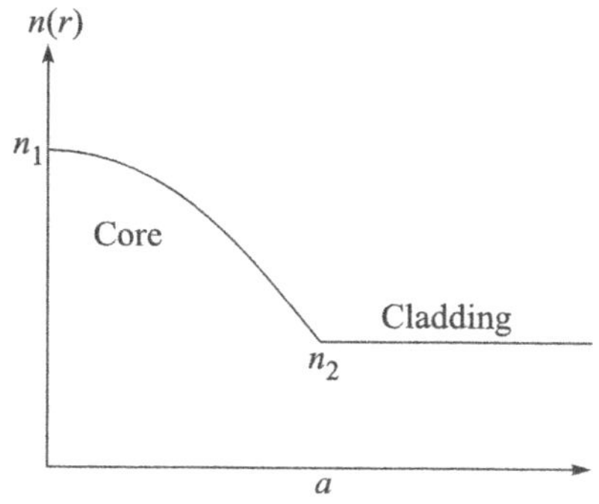

Fig. 18.3 The refractive index variation of a parabolic-index fiber.

For $V = 10$, the fiber will support approximately 25 modes. When the fiber supports such a large number of modes, one can use ray optics to calculate intermodal dispersion which is also known as ray dispersion which is given by:

$$\Delta\tau_i = \frac{n_2 L}{2c}\left(\frac{n_1 - n_2}{n_2}\right)^2 \quad \text{Pulse dispersion in multimode PIF} \tag{17}$$

When $\Delta << 1$, the above equation can be written as

$$\Delta\tau_i \approx \frac{n_2 L}{2c}\Delta^2 \tag{18}$$

18.5 POWER LAW PROFILE

A broad class of *multimoded* graded index fibers can be described by the following refractive index distribution

$$\begin{aligned} n^2(r) &= n_1^2\left[1 - 2\Delta\left(\frac{r}{a}\right)^q\right]; & 0 < r < a \\ &= n_2^2 = n_1^2(1 - 2\Delta); & r > a \end{aligned} \tag{19}$$

where r corresponds to a cylindrical radial coordinate, n_1 represents the value of the refractive index on the axis (i.e., at $r = 0$), n_2 represents the refractive index of the cladding and a represents the radius of the core. Equation (19) describes what is usually referred to as a power law profile or a q-profile; $q = 1$, $q = 2$ and $q = \infty$ correspond to the linear, parabolic, and step index profiles, respectively (see Fig. 18.4). The normalised waveguide parameter is again defined by Eq. (5). The total number of modes in a highly multimoded graded index optical fiber characterised by Eq. (19) are approximately given by

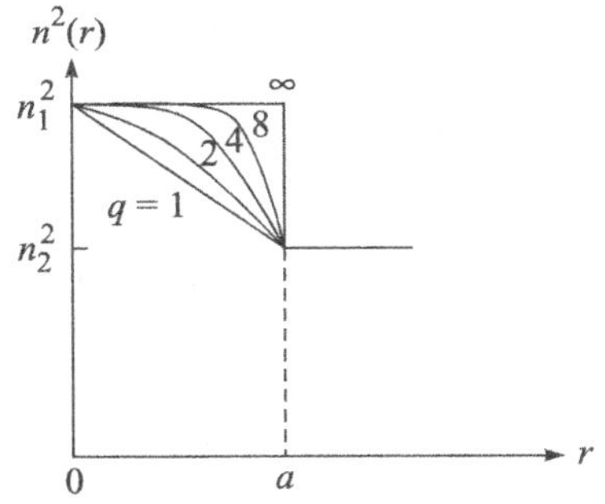

Fig. 18.4 Power law profiles for the refractive index distribution given by Eq. (19).

$$N \approx \frac{q}{2(2+q)}V^2 \tag{20}$$

Thus, a parabolic index fiber ($q = 2$) with $V = 10$ will support approximately 25 modes. Similarly, a step index fiber ($q = \infty$) with $V = 10$ will support approximately 50 modes. When the fiber supports such a large number of modes, the fiber is said to be a multimode fiber.

The time taken by a ray to propagate through a length L of a multimode fiber described by a q-profile (see Eq. 19) is given by [see Ref. An 1 and Chapter 5 of Gh 5]:

$$\tau(\tilde{\beta}) = \left(A\tilde{\beta} + \frac{B}{\tilde{\beta}} \right) L \tag{21}$$

where,

$$A = \frac{2}{c(2+q)}; \quad B = \frac{q n_1^2}{c(2+q)} \tag{22}$$

and for rays guided by the fiber $n_2 < \tilde{\beta} < n_1$.

PROBLEMS

18.1 Calculate the critical angle for (a) the glass-air interface ($n_1 = 1.5$, $n_2 = 1.0$) and (b) for the glass-water interface, $\left(n_1 = 1.5, n_2 = \frac{4}{3} \right)$.

18.2 Consider a step index fiber with $n_1 = 1.5$, $\Delta = 0.015$ and $a = 25$ μm placed in air. Calculate n_2, NA and the maximum acceptance angle (i_m).

[**Ans.** 1.477; 0.26; 15°]

18.3 In continuation of the previous problem, consider the same step index immersed in water of refractive index 1.33. Calculate the maximum acceptance angle.

[**Ans.** 11.3°]

18.4 Consider a 40 km fiber link (with a loss of 0.4 dB/km) having 3 connectors in its path and each connector has a loss of 1.8 dB. Calculate the total loss in dB.

18.5 The power of a 2 mW laser beam decreases to 15 μW after traversing through 25 km of a single mode optical fiber. Calculate the attenuation of the fiber.

[**Ans.** 0.85 dB/km]

18.6 A 5 mW laser beam passes through a 26 km fiber of loss 0.2 dB/km. Calculate the power at the output end.

[**Ans.** 1.5 mW]

18.7 Consider a 5 mW laser beam passing through a 40 km fiber link of loss 0.5 dB/km. Calculate the output in dBm and in mW.

18.8 For pure silica the refractive index variation in the wavelength domain 0.5 μm < λ_0 < 1.6 μm can be assumed to be given by the following approximate empirical formula

$$n(\lambda_0) \approx C_0 - a\lambda_0^2 + \frac{b}{\lambda_0^2} \tag{23}$$

where $C_0 \approx 1.451$, $a \approx 0.003\,(\mu\text{m})^{-2}$, $b \approx 0.003\,(\mu\text{m})^2$ and λ_0 is measured in μm. Calculate the group velocity at $\lambda_0 = 0.80$ μm and at $\lambda_0 = 0.85$ μm and show that the group velocity attains a maximum value at $\lambda_0 \approx 1.32$ μm.

18.9 Consider an LED operating at 0.85 μm with a spectral width of 50 nm. At this wavelength

$$\frac{d^2 n}{d\lambda_0^2} \approx 0.0297\,(\mu\text{m})^{-2}$$

Calculate the material dispersion in ns/km.

18.10 Consider again an LED operating at 1.3 μm with a spectral width of 20 nm. At this wavelength

$$\frac{d^2n}{d\lambda_0^2} \approx -0.00055\,(\mu\text{m})^{-2}$$

Calculate the material dispersion in ps/nm.

18.11 In the IV generation optical communication systems, one uses laser diodes operating at $\lambda_0 = 1.55$ μm where

$$\frac{d^2n}{d\lambda_0^2} \approx 0.0042\,(\mu\text{m})^{-2}$$

Assume a spectral width of $\Delta\lambda_0 \approx 2$ nm and calculate the material dispersion.

18.12 Consider a step index fiber with n_1 and n_2 representing the core and cladding refractive indices, respectively. Let θ be the angle that the rays make with the z-axis. Show that all rays for which

$$\theta < \theta_c = \cos^{-1}\left(\frac{n_2}{n_1}\right) \tag{24}$$

will get guided through the fiber. Thus, using simple ray optics, calculate the time taken by each ray to propagate through the length of the fiber and show that the ray dispersion will be given by:

$$\Delta\tau_i \cong \frac{n_1 L}{c}\Delta \tag{25}$$

18.13 Consider a multimode graded index fiber described by a q-profile (see Eq. 19). For rays guided by the fiber $n_2 < \tilde{\beta} < n_1$. Using Eq. (21), calculate the ray dispersion in fibers

(a) For the step profile, ($q = \infty$),
(b) For the parabolic profile, ($q = 2$), and
(c) For the optimum profile, ($q = 2 - 2\Delta$)

18.14 Consider a SIF with $n_1 = 1.5$, $a = 40$ μm and $\Delta = 0.015$ operating at 0.85 μm with a spectral width of 50 nm.

(a) Is this a single mode fiber or a multimode fiber?
(b) Calculate ray dispersion and using the value of material dispersion (from Problem 18.10), calculate total pulse dispersion.

18.15 In continuation of the previous problem, consider a parabolic index fiber with $n_1 = 1.5$, $a = 40$ μm and $\Delta = 0.015$ operating at 850 nm with a spectral width of 50 nm.

(a) Is this a single mode fiber or a multimode fiber?
(b) Calculate ray dispersion and using the value of material dispersion (from Problem 18.12), calculate total pulse dispersion and hence the maximum bit rate.

18.16 Consider fibers with step index profile, parabolic index profile and a triangular index profile. Show that in the step index fiber and parabolic index fiber, the fastest ray corresponds to the one propagating along the axis while for the triangular profile, the fastest ray corresponds to a ray making the largest angle with the axis.

SOLUTIONS

18.1 For the glass-air interface, $n_1 = 1.5$, $n_2 = 1.0$ and the critical angle is given by

$$\phi_c = \sin^{-1}\left(\frac{1.0}{1.5}\right) \approx 41.8°$$

On the other hand, for the glass-water interface, $n_1 = 1.5$, $n_2 = 4/3$ and

$$\phi_c = \sin^{-1}\left(\frac{4/3}{1.5}\right) \approx 62.7°.$$

18.2 $n_2 = n_1(1 - 2\Delta)^{1/2} = 1.5\sqrt{0.97} \approx 1.477$

$$\text{NA} = \sqrt{n_1^2 - n_2^2} = n_1\sqrt{2\Delta} = 1.5\sqrt{0.03} \approx 0.260$$

$$\sin i_m = 0.260 \quad \Rightarrow \quad i_m \approx 15°.$$

18.3 $n_{\text{water}} \sin i_m = \sqrt{n_1^2 - n_2^2} \Rightarrow \sin i_m \approx \dfrac{0.260}{1.33} \Rightarrow i_m \approx 11.3°.$

18.4 The total loss will be 0.4 dB/km × 40 km + 3 × 1.8 dB = 21.4 dB.

18.5 Loss in dB $= 10 \log_{10} \dfrac{2 \times 10^{-3}}{15 \times 10^{-6}} \approx 21.25 \Rightarrow$ Attenuation $= \dfrac{21.25}{25} \approx 0.85$ dB/km.

18.6 5 mW = 6.99 dBm; loss = 0.2 × 26 = 5.2 dB; Power at the output end = 6.99 – 5.20 = 1.79 dBm; Power in mW = $10^{0.179} \approx 1.5$ mW.

18.7 The input power is 6.99 dBm. The total loss is 20 dB. Thus the power at the output would be –13.01 dBm which is equal to 0.05 mW.

18.8 Since $n(\lambda_0) \approx C_0 - a\lambda_0^2 + \dfrac{b}{\lambda_0^2} \Rightarrow \lambda_0 \dfrac{dn}{d\lambda_0} = -2a\lambda_0^2 - \dfrac{2b}{\lambda_0^2}$

Thus, $$\tau = \frac{L}{v_g} = \frac{L}{c}\left[n(\lambda_0) - \lambda_0 \frac{dn}{d\lambda_0}\right] = \frac{L}{c}\left(C_0 + a\lambda_0^2 + \frac{3b}{\lambda_0^2}\right)$$

The group velocity attains a maximum value (or, the material dispersion becomes zero) when

$$\frac{d^2 n}{d\lambda_0^2} = 0 \quad \Rightarrow \quad -2a + \frac{6b}{\lambda_0^4} = 0 \quad \Rightarrow \quad \lambda_0 \approx 1.32\ \mu\text{m}$$

This is known as the zero material dispersion wavelength.

18.9 At $\lambda_0 \approx 0.85$ μm

$$\frac{d^2 n}{d\lambda_0^2} \approx 0.0297\ (\mu\text{m})^{-2}$$

giving $$D_m \approx -85 \text{ ps/km.nm}$$

the negative sign indicating that higher wavelengths travel faster than lower wavelengths. Thus for $\Delta\lambda_0 \approx 25$ nm, the broadening of the pulse (due to material dispersion) will be $\Delta\tau_m \approx 2.1$ ns/km.

18.11 In the IV generation optical communication systems, one uses laser diodes operating at $\lambda_0 = 1.55$ μm where,

$$\frac{d^2n}{d\lambda_0^2} \approx 0.0042\ (\mu\text{m})^{-2}$$

The corresponding material dispersion will be given by

$$D_m \approx +21.7 \text{ ps/km.nm}$$

the positive sign indicating that higher wavelengths travel slower than lower wavelengths. Thus for $\Delta\lambda_0 \approx 2$ nm, the broadening of the pulse (due to material dispersion) will be $\Delta\tau_m \approx 43$ ps/km.

18.12 If we assume that all rays lying between $\theta = 0$ and $\theta = \theta_c = \cos^{-1}\left(\frac{n_2}{n_1}\right)$ [see Fig. 18.2] are present, then the time taken by these extreme rays for a fiber of length L would be given by

$$t_{\text{min}} = \frac{n_1 L}{c} \quad \text{corresponding to } \theta = 0 \tag{26}$$

$$t_{\text{max}} = \frac{n_1^2 L}{c n_2} \quad \text{corresponding to } \theta = \theta_c = \cos^{-1}\left(\frac{n_2}{n_1}\right) \tag{27}$$

Hence, if all the input rays were excited simultaneously, the rays would occupy a time interval at the output end of duration

$$\Delta\tau_i = t_{\text{max}} - t_{\text{min}} = \frac{n_1 L}{c}\left[\left(\frac{n_1}{n_2}\right) - 1\right] \tag{28}$$

or,

$$\Delta\tau_i \cong \frac{n_1 L}{c}\Delta$$

18.13 For the step profile, $q = \infty$ and

$$A = 0; \quad \text{and} \quad B = \frac{n_1^2}{c} \quad \Rightarrow \quad \tau(\tilde{\beta}) = \frac{n_1^2}{c n_2} L \tag{29}$$

Thus, $$\tau_{\text{max}} = \tau(\tilde{\beta} = n_2) = \frac{n_1^2}{c n_2} L \quad \text{and} \quad \tau_{\text{min}} = \tau(\tilde{\beta} = n_1) = \frac{n_1}{c} L \tag{30}$$

giving $$\Delta\tau = \tau_{\text{max}} - \tau_{\text{min}} = \frac{n_1}{c}\frac{(n_1 - n_2)}{n_2} L \tag{31}$$

which is the same expression as given by Eq. (23). For the parabolic profile, $q = 2$ and

$$A = \frac{1}{2c}; \quad \text{and} \quad B = \frac{n_1^2}{2c} \quad \Rightarrow \quad \tau(\tilde{\beta}) = \frac{n_1^2}{c n_2} L \tag{32}$$

Thus, $$\tau_{\text{max}} = \tau(\tilde{\beta} = n_2) = \frac{1}{2c}\left[n_2 + \frac{n_1^2}{n_2}\right] L$$

and $$\tau_{\min} = \tau(\tilde{\beta} = n_1) = \frac{n_1}{c} L \tag{33}$$

giving $$\Delta\tau = \tau_{\max} - \tau_{\min} = \frac{n_2}{2c}\left(\frac{n_1 - n_2}{n_2}\right)^2 L \tag{34}$$

which is the same expression as given by Eq. (17). The calculation of the optimum value of q (which would give minimum ray dispersion) requires a plot of $\tau(\tilde{\beta})$ as a function of $\tilde{\beta}$ for different values of q. The details are given in References An 1 and Gh 5 and the minimum dispersion occurs for $q \approx 2 - 2\Delta$ where the pulse dispersion is given by:

$$\Delta\tau(\text{optimum profile}) = \frac{n_1}{8c}\left(\frac{n_1 - n_2}{n_2}\right)^2 L \tag{35}$$

18.14 $V = \dfrac{2\pi}{\lambda_0} a n_1 \sqrt{2\Delta} = \dfrac{2\pi}{0.85} \times 0.40 \times 1.5 \sqrt{0.03} \approx 77$

Since it is a step index fiber, the number of modes will be $\frac{1}{2}V^2 \approx 2965$; thus it is a highly multimode fiber. At 0.85 μm, $D_m \approx -85$ ps/km.nm (see Problem 18.10); thus for $\Delta\lambda_0 = 50$ nm, $|\Delta\tau_m| \approx 85 \times 50$ ps/km ≈ 4.2 ns/km. The ray dispersion (or the intermodal dispersion) is given by [see Eq. (17)]:

$$\Delta\tau_i = \frac{n_1 L}{c}\Delta \approx \frac{1.5 \times 10^3}{3 \times 10^8} \times 0.015 = 7.5 \times 10^{-8}\ \text{s}$$

where we have taken $L = 1$ km $= 1000$ m. Thus $\Delta\tau_i = 75$ ns/km.

Finally, $$(\Delta\tau) = \sqrt{(\Delta\tau_i)^2 + (\Delta\tau_m)^2} \approx 75.1\ \text{ns/km}$$

This gives a maximum bit rate of

$$B_{\max} \approx \frac{0.7}{75.1 \times 10^{-9}}\ \text{bits-km/s} \approx 9\ \text{Mbits-km/s}$$

Thus a 10 km link can at most support only 900 kbits/s.

18.15 (a) As in the previous problem

$$V = \frac{2\pi}{\lambda_0} a n_1 \sqrt{2\Delta} = \frac{2\pi}{0.85} \times 0.40 \times 1.5 \sqrt{0.03} \approx 77$$

Since it is a parabolic index fiber, the number of modes will be $\frac{1}{4}V^2 \approx 1482$; thus it is again a highly multimode fiber.

(b) Since the operating wavelength and the spectral width of the source are the same as in the previous problem, $\Delta\tau_m \approx 4.2$ ns/km. However, the ray dispersion (or the intermodal dispersion) is now given by [see Eq. (21)]:

$$\Delta\tau_i = \frac{n_1 L}{c}\Delta^2 \approx \frac{1.5 \times 10^3}{2 \times 3 \times 10^8} \times (0.015)^2 = 0.56 \times 10^{-9}\ \text{s}$$

which is much less than in the previous problem. The total dispersion is therefore given by

$$(\Delta\tau) = \sqrt{(\Delta\tau_i)^2 + (\Delta\tau_m)^2} = \sqrt{(4.2)^2 + (0.56)^2} \approx 4.2 \text{ ns/km}$$

This gives a maximum bit rate of

$$B_{\max} \approx \frac{0.7}{4.2 \times 10^{-9}} \text{ bits-km/s} \approx 167 \text{ Mbits-km/s}$$

Thus, a 10 km link can at most support about 17 Mbits/s.

18.16 For a step index fiber

$$\tau(\tilde{\beta}) = \frac{n_1^2}{c\tilde{\beta}} L$$

For a parabolic index fiber

$$\tau(\tilde{\beta}) = \left(2\tilde{\beta} + \frac{n_1^2}{\tilde{\beta}}\right)\frac{L}{2c}$$

For a triangular profile fiber

$$\tau(\tilde{\beta}) = \left(2\tilde{\beta} + \frac{n_1^2}{\tilde{\beta}}\right)\frac{L}{3c}$$

The axial ray corresponds to $\tilde{\beta} = n_1$ and the ray making largest angle with the axis corresponds to $\tilde{\beta} = n_2$. Calculating t for the two extreme rays for the different profiles, it can be shown that for the step index profile and the parabolic index profile, the axial ray takes less time to travel along the fiber than the rays making the largest angle with the axis while the contrary is true for the triangular profile fiber.

Basic Waveguide Theory and Concept of Modes

19

A Quick Review

We consider a waveguide with refractive index depending only on the x coordinate:

$$n^2 = n^2(x) \tag{1}$$

When the refractive index variation depends only on the x coordinate, we can always choose the z-axis along the direction of propagation of the wave and we may, *without any loss of generality*, write the solutions of Maxwell's equations in the form

$$\mathcal{E} = \mathbf{E}(x)e^{i(\omega t - \beta z)} \tag{2}$$

$$\mathcal{H} = \mathbf{H}(x)e^{i(\omega t - \beta z)} \tag{3}$$

The above equations ***define*** the modes of the system. Thus,

> ***modes represent transverse field distributions that suffer a phase change*** *only as they propagate through the waveguide along z.*

The quantity β represents the propagation constant of the mode. If we substitute the above solutions in Maxwell's equations, we will obtain two independent sets of equations. The first set of equations correspond to nonvanishing values of E_y, H_x and H_z with E_x, E_z and H_y vanishing, giving rise to what are known as TE modes because the electric field has only a transverse component. The second set of equations correspond to nonvanishing values of E_x, E_z and H_y with E_y, H_x and H_z vanishing, giving rise to what are known as TM modes because the magnetic field now has only a transverse component.

For TE modes, $E_y(x)$ satisfies the following differential equation

$$\frac{d^2E_y}{dx^2} + [k_0^2 n^2(x) - \beta^2]E_y = 0 \tag{4}$$

where,

$$k_0 = \omega\sqrt{\varepsilon_0\mu_0} = \frac{\omega}{c} \tag{5}$$

is the free space wave number and $c \left(= \dfrac{1}{\sqrt{\varepsilon_0\mu_0}} \right)$ is the speed of light in free space.

Once $E_y(x)$ is known, we can determine H_x and H_z from the following equations:

$$H_x = -\frac{\beta}{\omega\mu_0}E_y(x) \quad \text{and} \quad H_z = \frac{i}{\omega\mu_0}\frac{dE_y}{dx} \tag{6}$$

Since E_y and H_z represent tangential components,

$$E_y \quad \text{and} \quad \frac{dE_y}{dx} \text{ must be continuous everywhere} \tag{7}$$

Further, when the refractive index distribution is symmetric about $x = 0$; that is, when

$$n^2(-x) = n^2(x) \tag{8}$$

the solutions are either symmetric or antisymmetric functions of x (see Problem 19.11). Thus we must have,

$$E_y(-x) = E_y(x) \qquad \text{symmetric modes} \tag{9}$$

$$E_y(-x) = -E_y(x) \qquad \text{antisymmetric modes} \tag{10}$$

For TM modes, $H_y(x)$ satisfies the following equation:

$$n^2(x)\frac{d}{dx}\left[\frac{1}{n^2(x)}\frac{dH_y}{dx}\right] + [k_0^2 n^2(x) - \beta^2]H_y(x) = 0 \tag{11}$$

Once $H_y(x)$ is known, we can determine E_x and E_z from the following equations:

$$E_x = \frac{\beta}{\omega\varepsilon_0 n^2(x)}H_y \quad \text{and} \quad E_z = \frac{1}{i\omega\varepsilon_0 n^2(x)}\frac{dH_y}{dx} \tag{12}$$

Further, since $H_y(x)$ and $E_z(x)$ are tangential component,

$$H_y \quad \text{and} \quad \frac{1}{n^2}\frac{dH_y}{dx} \text{ must be continuous everywhere} \tag{13}$$

19.1 STEP INDEX SYMMETRIC PROFILE

The simplest planar optical waveguide consists of a thin dielectric film sandwiched between materials of slightly lower refractive indices and is characterised by the following refractive index variation (see Fig. 19.1).

$$n(x) = \begin{cases} n_1; & |x| < \dfrac{d}{2} \\ n_2; & |x| > \dfrac{d}{2} \end{cases} \tag{14}$$

with $n_1 > n_2$. The above equation describes what is usually referred to as a step-index profile. When

$$n_2^2 < \frac{\beta^2}{k_0^2} < n_1^2 \qquad \text{GUIDED MODES} \tag{15}$$

the solutions are oscillatory in nature in the region $|x| < \frac{d}{2}$ and exponential in nature in the region $|x| > \frac{d}{2}$. Only for certain discrete values of β, will we have decaying solutions in the region $x > \frac{d}{2}$ as well as in the region $x < -\frac{d}{2}$; these are the discrete guided modes of the waveguide. On the other hand, when $\beta^2 < k_0^2 n_2^2$, the solutions are also oscillatory in the region $|x| > \frac{d}{2}$ and they correspond to what are known as *radiation modes* of the waveguide. These radiation modes correspond to rays that undergo refraction (rather than total internal reflection) at the film-cover interface and when these are excited, they quickly leak away from the core of the waveguide.

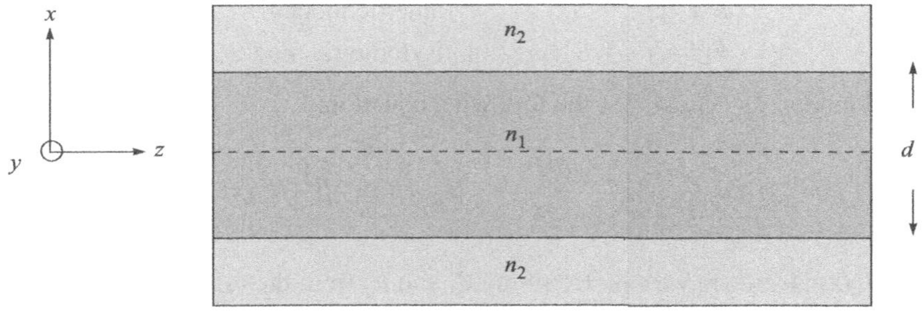

Fig. 19.1 A planar dielectric waveguide of thickness *d* (along *x* direction) but infinitely extended along the *y* direction. Light propagates along the *z* direction.

One often defines the effective index of the mode as

$$n_{\text{eff}} \equiv \frac{\beta}{k_0} \tag{16}$$

Thus, for guided modes

$$n_2 < n_{\text{eff}} < n_1 \tag{17}$$

19.2 TE MODES OF A SYMMETRIC STEP INDEX PLANAR WAVEGUIDE

If we solve Eq. (4) and apply the necessary continuity conditions, we will find that the propagation constants $\frac{\beta}{k_0}$ (for the TE modes) are determined by solving the following transcendental equations:

$$\xi \tan \xi = \sqrt{\left(\frac{V}{2}\right)^2 - \xi^2} \quad \text{for symmetric modes} \tag{18}$$

and

$$-\xi \cot \xi = \sqrt{\left(\frac{V}{2}\right)^2 - \xi^2} \quad \text{for antisymmetric modes} \tag{19}$$

where,

$$\xi = \frac{\kappa d}{2} \tag{20}$$

$$\kappa = \sqrt{k_0^2 n_1^2 - \beta^2} \tag{21}$$

and

$$V = k_0 d \sqrt{n_1^2 - n_2^2} \tag{22}$$

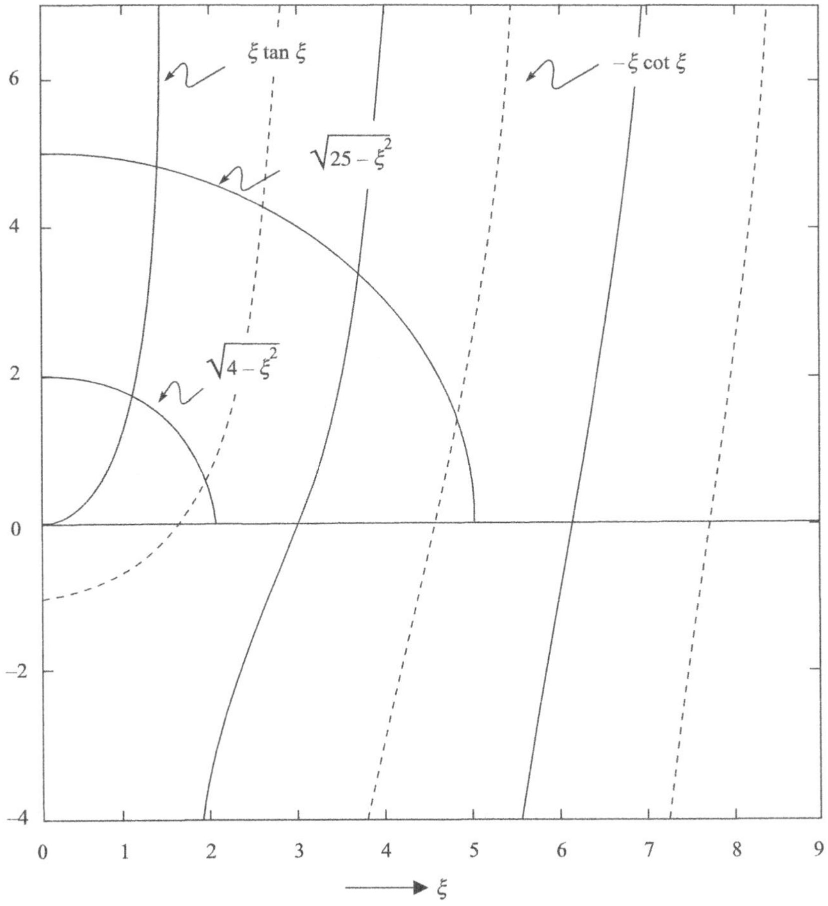

Fig. 19.2 Variation of $\xi \tan \xi$ (solid curve) and $-\xi \cot \xi$ (dashed curve) as a function of ξ. The points of intersection of these curves with the quadrant of a circle of radius $V/2$ determine the discrete propagation constants of the waveguide.

Since the equation

$$\eta = \sqrt{\left(\frac{V}{2}\right)^2 - \xi^2} \tag{23}$$

(for positive values of ξ) represents a circle (of radius $V/2$) in the first quadrant of the ξ–η plane[1], the numerical evaluation of the allowed values of ξ (and hence of the propagation constants) is quite simple. In Fig. 19.2 we have plotted the functions $\xi \tan \xi$ (solid curve) and $-\xi \cot \xi$ (dashed curve) as a function of ξ. For a given value

1. This follows from the fact that if we square Eq. (37) we would get $\eta^2 + \xi^2 = \left(\frac{V}{2}\right)^2$ which represents a circle of radius $V/2$.

of V, the points of intersection of these curves with the quadrant of the circle would determine the allowed (discrete) values of ξ. The two circles in Fig. 19.2 correspond to $V/2 = 2$ and $V/2 = 5$. Obviously, as can be seen from the figure, for $V = 4$ we will have one symmetric and one antisymmetric mode and for $V = 10$ we will have two symmetric and two antisymmetric modes. In general, when

$$(m - 1)\pi < V < m\pi \tag{24}$$

the waveguide will support a total of m modes. It is convenient to define the dimensionless propagation constant

$$b \equiv \frac{\dfrac{\beta^2}{k_0^2} - n_2^2}{n_1^2 - n_2^2} = \frac{n_{\text{eff}}^2 - n_2^2}{n_1^2 - n_2^2} = \frac{\gamma^2 d^2}{V^2} \tag{25}$$

where,
$$\gamma^2 = \beta^2 - k_0^2 n_2^2 \tag{26}$$

giving
$$\frac{\gamma d}{2} = \frac{1}{2} V \sqrt{b} \tag{27}$$

Further,
$$(\kappa^2 + \gamma^2)\frac{d^2}{4} = \frac{1}{4}[k_0^2 d^2 (n_1^2 - n_2^2)] = \frac{1}{4} V^2 \tag{28}$$

$$\xi = \frac{\kappa d}{2} = \sqrt{\left(\frac{1}{4}V^2 - \frac{\gamma^2 d^2}{4}\right)}$$

$$= \frac{1}{2} V \sqrt{1 - b} \tag{29}$$

Thus, Eqs (18) and (19) can be written in the form

$$\left(\frac{1}{2} V \sqrt{1-b}\right) \tan\left(\frac{1}{2} V \sqrt{1-b}\right) = \frac{1}{2} V \sqrt{b} \quad \text{for symmetric modes} \tag{30}$$

$$-\left(\frac{1}{2} V \sqrt{1-b}\right) \cot\left(\frac{1}{2} V \sqrt{1-b}\right) = \frac{1}{2} V \sqrt{b} \quad \text{for antisymmetric modes} \tag{31}$$

Obviously, because of Eq. (15), for guided modes we will have

$$0 < b < 1 \tag{32}$$

For a given value of V, solutions of Eqs (30) and (31) will give us discrete values of b; the mth solution ($m = 0, 1, 2, 3, \ldots$) is referred to as the TE_m mode. In Table 19.1 we have tabulated the discrete values of b for various values of V. For any given (step index) waveguide we just have to calculate V, and then obtain the corresponding value of b either by solving Eqs (30) and (31) or by using Table 19.1. From the values of b, one can obtain the propagation constants by using the following equation [see Eq. (25)]:

$$\frac{\beta}{k_0} = \sqrt{[n_2^2 + b(n_1^2 - n_2^2)]} \tag{33}$$

Table 19.1 Values of the normalised propagation constant (corresponding to TE modes) for a symmetric planar waveguide; the values are generated by using the software in Ref. Gh3. Notice that for $V < \pi$ we will have only one TE mode which will be symmetric in x and for $\pi < V < 2\pi$ we will have two TE modes one of them will be symmetric in x and the other anti-symmetric in x.

V	$b(TE_0)$	$b(TE_1)$
1.000	.189339	
1.125	.225643	
1.250	.261714	
1.375	.297049	
1.500	.331290	
1.625	.364196	
1.750	.395618	
1.875	.425479	
2.000	.453753	
2.125	.480453	
2.250	.505616	
2.375	.529300	
2.500	.551571	
2.625	.572502	
2.750	.592169	
2.875	.610649	
3.000	.628017	
3.125	.644344	
3.250	.659701	.002702
3.375	.674151	.011415
3.500	.687758	.024612
3.625	.700579	.041077
3.750	.712667	.059875
3.875	.724073	.080292
4.000	.734844	.101775

V	$b(TE_0)$	$b(TE_1)$	$b(TE_2)$
4.000	.734844	.101775	
4.125	.745021	.123903	
4.250	.754647	.146349	
4.375	.763756	.168864	
4.500	.772384	.191259	
4.625	.780563	.213390	
4.750	.788321	.235151	
4.875	.795686	.256461	
5.000	.802683	.277265	
5.125	.809335	.297523	
5.250	.815663	.317210	
5.375	.821689	.336310	
5.500	.827429	.354817	
5.625	.832902	.372731	
5.750	.838123	.390056	
5.875	.843107	.406800	
6.000	.847869	.422976	
6.125	.852420	.438596	
6.250	.856772	.453676	
6.375	.860938	.468231	.001845
6.500	.864926	.482278	.008819
6.625	.868748	.495834	.019189
6.750	.872412	.508916	.031806
6.875	.875926	.521541	.045942
7.000	.879298	.533727	.061106

For a guided (symmetric) mode, the complete field inside the film is given by

$$E_y(x) = A\cos \kappa x e^{i(\omega t - \beta z)}$$
$$= \frac{1}{2}Ae^{i(\omega t - \beta z - \kappa x)} + \frac{1}{2}Ae^{i(\omega t - \beta z + \kappa x)} \tag{34}$$

Thus, a guided mode can be considered to be a superposition of two plane waves with

$$k_x = \kappa, k_y = 0, k_z = \beta \tag{35}$$

and

$$k_x = -\kappa, k_y = 0, k_z = \beta \tag{36}$$

In each case $k_x^2 + k_y^2 + k_z^2 = \beta^2 + \kappa^2 = k_0^2 n_1^2$. Thus, the two plane waves make angles $+\theta$ and $-\theta$ with the z-axis where

$$\cos\theta = \frac{\beta}{\sqrt{\beta^2+\kappa^2}} = \frac{\beta}{k_0 n_1} \tag{37}$$

Since β takes discrete values, a guided mode can be considered to be superposition of two plane waves propagating at discrete angles with the z-axis (see Fig. 19.3).

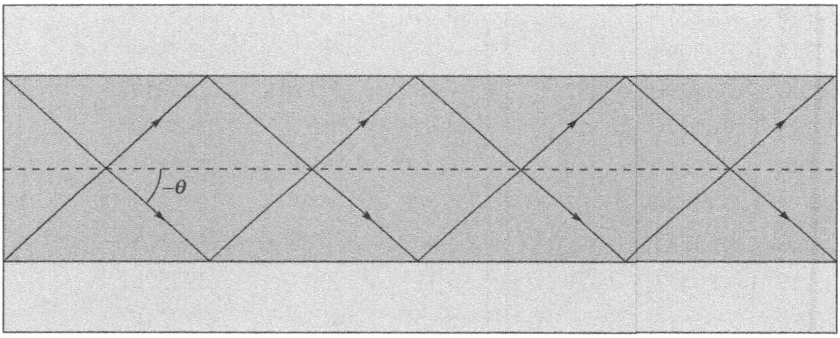

Fig. 19.3 A guided mode in a step index waveguide corresponds to the superposition of two plane waves propagating at particular angles $\pm\theta$ with the z-axis.

19.3 TM MODES OF A SYMMETRIC STEP INDEX PLANAR WAVEGUIDE

For TM modes, one has to solve Eq. (11) [with continuity conditions given by Eq. (13)] and one obtains the following transcendental equations which determine the discrete propagation constants β/k_0

$$\xi \tan \xi = \left(\frac{n_1}{n_2}\right)^2 \sqrt{\left(\frac{V}{2}\right)^2 - \xi^2} \quad \text{for symmetric TM modes} \tag{38}$$

and

$$-\xi \cot \xi = \left(\frac{n_1}{n_2}\right)^2 \sqrt{\left(\frac{V}{2}\right)^2 - \xi^2} \quad \text{for antisymmetric TM modes} \tag{39}$$

where ξ and V have been defined earlier. In terms of the parameters b and V, we have

$$\left(\frac{1}{2}V\sqrt{1-b}\right)\tan\left(\frac{1}{2}V\sqrt{1-b}\right) = \left(\frac{n_1}{n_2}\right)^2 \frac{1}{2}V\sqrt{b} \quad \text{for symmetric TM modes} \tag{40}$$

$$-\left(\frac{1}{2}V\sqrt{1-b}\right)\cot\left(\frac{1}{2}V\sqrt{1-b}\right) = \left(\frac{n_1}{n_2}\right)^2 \frac{1}{2}V\sqrt{b} \quad \text{for antisymmetric TM modes} \tag{41}$$

19.4 TE MODES IN A PARABOLIC INDEX MEDIUM

For a parabolic index medium characterised by the following refractive index distribution:

$$n^2(x) = n_1^2\left[1 - 2\Delta\left(\frac{x}{a}\right)\right]^2, \qquad |x| < a \text{ CORE} \tag{42}$$

$$= n_2^2 = n_1^2(1 - 2\Delta), \qquad |x| > a \text{ CLADDING}$$

the propagation constants are approximately given by

$$\beta = \beta_m \approx k_0 n_1 \sqrt{\left[1 - \frac{(2m+1)\sqrt{2\Delta}}{k_0 n_1 a}\right]}; \quad m = 0, 1, 2, 3, \ldots m_{\max} \tag{43}$$

where the maximum value of $m\,(= m_{\max})$ should be such that Eq. (15) is satisfied. The above expression for β is valid when

$$V \equiv k_0 a\sqrt{n_1^2 - n_2^2} = k_0 a n_1 \sqrt{2\Delta} \geq 10 \tag{44}$$

i.e., when the waveguide supports a large number of modes.

PROBLEMS

19.1 Consider a step index planar waveguide with d = 3 μm, n_1 = 1.5 and n_2 = 1.49153. The value of n_2 is chosen such that $\sqrt{n_1^2 - n_2^2} = \frac{1}{2\pi}$. Using Table 19.1, calculate the discrete values of b (and hence of β/k_0) for λ_0 = 1.5 μm, 1.0 μm and 0.6 μm. Show that in each case, the values of β/k_0 lie between n_1 and n_2.

19.2 In the above example, for λ_0 = 0.6 μm, calculate the values of θ that the component waves will make with the z-axis; show that the corresponding angles of incidence at the core-cladding interface is greater than the critical angle.

19.3 Using Eq. (43) and assuming $\frac{\sqrt{2\Delta}}{k_0 n_1 a} \ll 1$, show that the group velocity is approximately independent of the mode number.

19.4 Consider a step index planar waveguide with d = 2.5 μm, n_1 = 1.5 and n_2 = 1.47. Assume the operating wavelength λ_0 = 1.0 μm. Use Table 19.1 and linear interpolation, to determine the discrete values of b (and hence of β/k_0).

19.5 (a) Consider a symmetric step-index waveguide [see Eq. (14)] with n_1 = 1.5, n_2 = 1.46, d = 4 μm operating at λ_0 = 0.6328 μm. Calculate the number of TE and TM modes.

(b) Consider TE modes in a step index planar waveguide with n_1 = 1.5, d = 2 μm and the value of n_2 is chosen such that $\sqrt{n_1^2 - n_2^2} = \frac{1}{\pi}$. For λ_0 = 1 μm, 0.8 μm and 0.66667 μm calculate (using Table 19.1) the values of b and the corresponding value of β/k_0. Show that the values of β/k_0 lie between n_1 and n_2.

19.6 Consider now a parabolic index waveguide [see Eq. (43)] with $n_1 = 1.50$, $n_2 = 1.46$, $a = 2$ μm operating again at $\lambda_0 = 0.6328$ μm. Assuming the validity of Eq. (43) and that for discrete guided modes we must have $n_2^2 < \dfrac{\beta^2}{k_0^2} < n_1^2$, calculate the maximum value of m and the total number of TE modes.

19.7 Consider a step index symmetric waveguide with $n_1 = 1.50$, $n_2 = 1.48$, operating at $\lambda_0 = 0.6328$ μm. Calculate the value of d so that $V = 6$. Using Table 19.1, calculate the values of b, the corresponding propagation constants β/k_0 and the angles that the component waves make with the z-axis.

19.8 We consider the same waveguide as in the previous problem. At what wavelength will the value of V be equal to 3. Using Table 19.1, calculate the value of b and the corresponding propagation constant β/k_0.

19.9 (a) Consider a symmetric step-index waveguide [see Eq. (14)] with $n_1 = 1.49$, $n_2 = 1.46$, $d = 4$ μm operating at $\lambda_0 = 0.6328$ μm. Solve Eqs (18) and (19) to calculate the values of β/k_0.

(b) Calculate the corresponding values of θ_m.

19.10 (a) Consider a step index symmetric waveguide with $n_1 = 1.503$, $n_2 = 1.500$ and $d = 4$ μm. For $\lambda_0 = 1$ μm, calculate the value of V and use linear interpolation of the numbers given in Table 19.1 to calculate the value of β/k_0.

(b) If the operating wavelength is changed to 0.5 μm, show that $V = 4.771$ and by linear interpolation of the numbers given in Table 19.1 calculate the discrete values of β/k_0 ands the corresponding angles that the waves make with the z-axis.

19.11 In Eq. (4), make the transformation $x \to -x$ and assuming $n^2(x) = n^2(-x)$ show that $E_y(-x)$ satisfies the same equation as $E_y(x)$; hence we must have $E_y(-x) = \lambda E_y(x)$. Make the transformation $x \to -x$ again to prove that the solutions are either symmetric or antisymmetric functions of x [i.e., prove Eqs (9) and (10)].

19.12 Consider an infinitely extended parabolic index medium described by

$$n^2(x) = n_0^2 - \alpha x^2$$

Starting from the scalar wave equation and using the fact that the fundamental mode has a Gaussian field distribution of the form;

$$\psi(x) = Ae^{-x^2/2w_0^2}$$

Calculate w_0. Obtain the propagation constant of the fundamental mode.

19.13 Consider a symmetric planar waveguide with $n_1 = 2.3$, $n_2 = 2.2$ and $d = 2$ μm operating at $\lambda_0 = 1.0$ μm.

(a) How many guided TE and TM modes will the waveguide support?

(b) What are the minimum and maximum possible values of β of the TE_1 mode?

(c) In what range of λ_0 values will the waveguide be single moded (TE_0 and TM_0)?

19.14 Consider a planar waveguide with a refractive index profile given by

$$\begin{aligned} n^2(x) &= n_1^2; \quad |x| < d_1 \\ &= n_2^2; \quad d_1 < |x| < d_2 \\ &= n_3^2; \quad |x| > d_2 \end{aligned}$$

with $n_1 > n_3 > n_2$. Write down the range of propagation constant for guided modes in such a waveguide.

19.15 Consider a planar waveguide with the refractive index profile given in Problem 19.14 but with $n_1 = n_3$. Can such a waveguide support guided modes?

19.16 A symmetric single mode planar waveguide is excited by light which is polarised at 45° to the x-axis and lying in the x-y plane. Show that the state of polarisation will repeat itself after propagation through a certain distance and obtain this distance. Assume that the propagation constants of the TE and TM modes are β_{TE} and β_{TM}.

SOLUTIONS

19.1 For $d = 3$ µm, $n_1 = 1.5$ and $n_2 = 1.49153$, we get

$$\sqrt{n_1^2 - n_2^2} = \frac{1}{2\pi} \quad \text{and} \quad V = \frac{2\pi}{\lambda_0} d\sqrt{n_1^2 - n_2^2} = \frac{d}{\lambda_0} = \frac{3}{\lambda_0}$$

(where λ_0 is measured in µm) and

$$\frac{\beta}{k_0} = \sqrt{\left[n_2^2 + \frac{b}{4\pi^2}\right]}.$$

For $\lambda_0 = 1.5$ µm, V is equal to 2.0 and from Table 19.1 we see that there will be only one TE mode with $b = 0.453753$; the corresponding value of $\beta/k_0 \approx 1.49538$. The same waveguide operating at $\lambda_0 = 1.0$ µm will have $V = 3.0$ and from Table 19.1 we see that there will be again only one TE mode with $b = 0.628017$; the corresponding value of $\beta/k_0 \approx 1.49686$. However, for $\lambda_0 = 0.6$ µm, $V = 5.0$ and there will be two TE modes with $b = 0.802683$ (the TE_0 mode) and the other with $b = 0.277265$ (the TE_1 mode). The corresponding values of $\beta/k_0 \approx 1.49833$ and 1.49389. Finally, for $\lambda_0 = 0.4286$ µm, $V = 7.0$ and there will have 3 TE modes with $b = 0.879298$ (TE_0), 0.533727 (TE_1) and 0.061106 (TE_2). The corresponding values of β/k_0 are approximately given by

$$\frac{\beta}{k_0} \approx 1.4990,\ 1.49606 \quad \text{and} \quad 1.49205$$

respectively. Notice that all values of β/k_0 lie between n_1 and n_2. Further, in each case, the waveguide will support equal number of TM modes (see Fig. 19.2). Further, as the wavelength is made smaller, the waveguide will support larger number of modes and in the limit of the wavelength tending to zero, we will have a very large number of modes which is nothing but the ray-optics limit.

19.2 For $\lambda_0 = 0.6$ μm, V will be 5.0 and we will have two TE modes with $\beta/k_0 \approx 1.49833$ and 1.49389. Since $n_1 = 1.5$, the values of $\cos\theta$ will be 0.99889 and 0.99593 and therefore

$$\theta \approx 2.70° \quad \text{and} \quad 5.17°$$

corresponding to the symmetric TE_0 mode and the antisymmetric TE_1 mode respectively.

19.3 When $\frac{\sqrt{2\Delta}}{k_0 n_1 a} << 1$ and for not too large values of m, we may carry out a binomial expansion in Eq. (43) to obtain

$$\beta = \beta_m \approx k_0 n_1 - \left(m + \frac{1}{2}\right)\frac{\sqrt{2\Delta}}{a}$$

$$\approx \frac{\omega}{c} n_1 - \left(m + \frac{1}{2}\right)\frac{\sqrt{2\Delta}}{a}; \quad m = 0, 1, 2, 3, \ldots \tag{45}$$

Thus the group velocity v_g of the mode will be given by,

$$\frac{1}{v_g} = \frac{d\beta}{d\omega} \approx \frac{n_1}{c} \tag{46}$$

independent of the mode number. Thus, in this approximation, all modes travel with the same group velocity. Indeed, using ray optics, we had shown in Problem 2.8 that *all* rays take approximately the same time to propagate through a certain distance of a parabolic index waveguide. It is for this reason that parabolic index waveguides are often used in fiber-optic communication systems.

19.4 For $n_1 = 1.5$, $n_2 = 1.47$, $d = 2.5$ μm and $\lambda_0 = 1.0$ μm, we get $V = 4.6888$. If we carry out linear interpolation we would obtain for the TE_0 mode

$$b = 0.780563 + \frac{0.788321 - 0.780563}{0.125} \times 0.0638 \approx 0.78452$$

We therefore get $\frac{\beta}{k_0} \approx 1.49359$. Similarly for the TE_1 mode.

$$b = 0.213390 + \frac{0.235151 - 0.213390}{0.125} \times 0.0638 \approx 0.22450$$

and the corresponding value of β/k_0 will be $\approx$ 1.47679. Once again both values of β/k_0 lie between n_1 and n_2.

19.5 (a) We have a symmetric step-index waveguide [see Eq. (14)] with $n_1 = 1.50$, $n_2 = 1.46$, $d = 4$ μm operating at $\lambda_0 = 0.6328$ μm. Thus,

$$\frac{V}{2} = \frac{k_0 d}{2}\sqrt{n_1^2 - n_2^2} = \frac{\pi d}{\lambda_0}\sqrt{n_1^2 - n_2^2} = \frac{4\pi}{0.6328}\sqrt{1.5^2 - 1.46^2}$$

$$\approx 6.833$$

which is about 2.18π. From Fig. 19.2 we find that there will be 5 TE and 5 TM modes.

(b) Consider TE modes in a step index planar waveguide with $d = 2.0$ μm, $n_1 = 1.50$, and the value of n_2 is chosen such that $\sqrt{n_1^2 - n_2^2} = \dfrac{1}{\pi}$. Thus,

$$V = k_0 d\sqrt{n_1^2 - n_2^2} = \frac{2d}{\lambda_0} = \frac{4}{\lambda_0} \text{ (where } \lambda_0 \text{ is measured in microns)}$$

$$= 4.0, 5.0 \text{ and } 6.0$$

for $\lambda_0 = 1$ μm, 0.8 μm and 0.66667 μm respectively. Using Table 19.1, the corresponding values of b are

(0.734844 and 0.101775), (0.802683 and 0.277265) and (0.847869 and 0.422976)

Now, $\sqrt{n_1^2 - n_2^2} = \dfrac{1}{\pi} \Rightarrow n_2^2 = n_1^2 - \dfrac{1}{\pi^2} \approx 2.1486 \Rightarrow n_2 \approx 1.4658$

Thus, $\dfrac{\beta}{k_0} = \sqrt{\left[n_2^2 + \dfrac{b}{4\pi^2}\right]} = \sqrt{\left[2.1486 + \dfrac{b}{4\pi^2}\right]}$ and the corresponding value of β/k_0 are given by

(1.4721, 1.4667); (1.4727, 1.4682) and (1.4731, 1.4695)

for $\lambda_0 = 1$ μm, 0.8 μm and 0.66667 μm, respectively. As can be seen all values of β/k_0 lie between n_1 and n_2.

19.6 For a parabolic index waveguide, the allowed values of β^2 for discrete guided modes are given by

$$\beta^2 = \beta_m^2 = k_0^2 n_1^2 - (2m + 1)\frac{k_0 n_1 \sqrt{2\Delta}}{a}; \quad m = 0, 1, 2, 3, \ldots$$

which can be written as

$$m = \frac{1}{2}\left[\frac{k_0^2 n_1^2 - \beta_m^2}{k_0 n_1 \sqrt{2\Delta}/a} - 1\right]$$

For a guided mode, the minimum value of β is $k_0 n_2$; thus if m_{max} represents the maximum value of m; then m_{max} must be the integer less than

$$\frac{1}{2}\left[\frac{k_0^2 n_1^2 - k_0^2 n_2^2}{k_0 n_1 \sqrt{2\Delta}/a} - 1\right] = \frac{1}{2}\left[\frac{V^2/a^2}{V/a^2} - 1\right] = \frac{1}{2}[V - 1]$$

Now, $$V = k_0 a\sqrt{n_1^2 - n_2^2} = \frac{2\pi}{\lambda_0} a\sqrt{n_1^2 - n_2^2}$$

$$= \frac{2\pi}{0.6328} \times 2 \times \sqrt{1.5^2 - 1.46^2} \approx 6.8$$

Thus, m_{max} must be the integer less than 2.9; and therefore in this approximation there will be only 2 TE modes.

19.7 $V = k_0 d\sqrt{n_1^2 - n_2^2} = \dfrac{2\pi d}{\lambda_0}\sqrt{n_1^2 - n_2^2}$

Thus, $V = 6 = \dfrac{2\pi d}{0.6328}\sqrt{1.5^2 - 1.48^2}$ (where d is measured in microns)

$\Rightarrow \quad d = \dfrac{6 \times 0.6328}{2\pi\sqrt{1.5^2 - 1.48^2}} \approx 2.4752\ \mu\text{m}$

For $V = 6$, there are only 2 TE modes; the two values of b are: 0.847869 and 0.422976. Now,

$$\frac{\beta}{k_0} = \sqrt{\left[n_2^2 + \frac{b}{4\pi^2}\right]} = \sqrt{\left[2.1904 + \frac{b}{4\pi^2}\right]}$$

Thus the corresponding value of β/k_0 are 1.4872 $(= n_1 \cos\theta_1)$ and 1.4836 $(= n_1 \cos\theta_2)$ where θ_1 and θ_2 are the angles that the component waves make with the z-axis. Since, we get $\theta_1 \approx 7.49°$ and $\theta_2 \approx 8.48°$.

19.8 The wavelength (at which the value of V will be equal to 3) must be 1.2655 μm. For $V = 3$, there will be only one TE mode and the corresponding value of b will be: 0.628017. The corresponding value of β/k_0 will be:

$$\frac{\beta}{k_0} = \sqrt{\left[n_2^2 + \frac{b}{4\pi^2}\right]} = \sqrt{\left[2.1904 + \frac{b}{4\pi^2}\right]} \approx 1.485$$

which lies between n_1 and n_2.

19.9 $V = \dfrac{2\pi d}{\lambda_0}\sqrt{n_1^2 - n_2^2} = \dfrac{8\pi}{0.6328}\sqrt{1.49^2 - 1.46^2} \approx 11.815$

Thus, $V/2 \approx 5.91$. In Fig. 19.2, if we plot the quadrant of a circle of radius 5.91, we will have four points of intersection; two corresponding to symmetric modes and two corresponding to anti-symmetric modes. Now, the transcendental equation for symmetric modes [see Eq. (30)]

$$\left(\frac{1}{2}V\sqrt{1-b}\right)\tan\left(\frac{1}{2}V\sqrt{1-b}\right) = \frac{1}{2}V\sqrt{b}$$

can be written as

$$Q(b) = \left(\frac{1}{2}V\sqrt{1-b}\right)\sin\left(\frac{1}{2}V\sqrt{1-b}\right) - \frac{1}{2}V\sqrt{b}\cos\left(\frac{1}{2}V\sqrt{1-b}\right) = 0$$

One can use any program (like GNUPLOT or MATLAB or MATHEMATICA) to plot $Q(b)$ as a function of b in the region $0 < b < 1$. We have used GNUPLOT to plot $Q(b)$ as a function of b as shown in Fig. 19.4. The figure clearly shows that there are two (symmetric) TE modes one around $b \approx 0.95$ (TE_0 mode) and the other around $b \approx 0.55$ (TE_2 mode). In order to get greater accuracy, one may plot $Q(b)$ in the vicinity of the roots as we have done in Figs. 19.5 and 19.6. One finds that the values of b are close to 0.948 and 0.5474. In fact if we use the 'Find root' program in MATHEMATICA (or a similar program in MATLAB), we would obtain

$$b = 0.948421 \text{ (TE}_0 \text{ mode)} \quad \text{and} \quad b \approx 0.547390 \text{ (TE}_2 \text{ mode)}$$

```
GNU Program
set yrange[-10:10]
set xrange[0.:1.0]
set nokey
set ytics
set xtics
V=11.815
p0(x)=0.
p1(x)=0.5*V*sqrt(1-x)
p2(x)=0.5*V*sqrt(x)*cos(p1(x))
p3(x)=p1(x)*sin(p1(x))
Q(x)=p3(x)-p2(x)
plot p0(x) w l 8,Q(x) w l 8
```

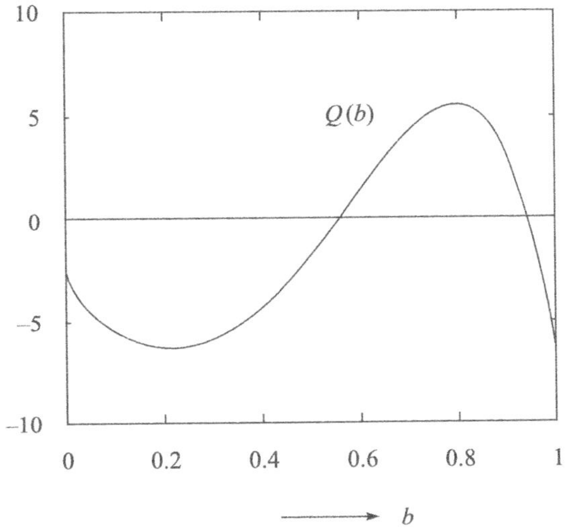

Fig. 19.4 The program for the GNU plot and (b) the corresponding variation of $Q(b)$

```
GNU Program
set yrange[-0.2:0.2]
set xrange[0.945:0.950]
set nokey
set ytics
set xtics
V=11.815
p0(x)=0.
p1(x)=0.5*V*sqrt(1-x)
p2(x)=0.5*V*sqrt(x)*cos(p1(x))
p3(x)=p1(x)*sin(p1(x))
Q(x)=p3(x)-p2(x)
plot p0(x) w l 8,Q(x) w l 8
```

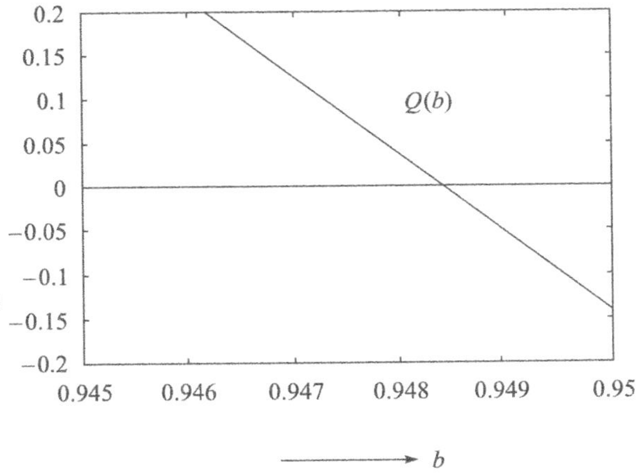

Fig. 19.5

GNU Program

```
set yrange[-0.05:0.05]
set xrange[0.547:0.548]
set nokey
set ytics
set xtics
V=11.815
p0(x)=0.
p1(x)=0.5*V*sqrt(1-x)
p2(x)=0.5*V*sqrt(x)*cos(p1(x))
p3(x)=p1(x)*sin(p1(x))
Q(x)=p3(x)-p2(x)
plot p0(x) w l8,Q(x) w l8
```

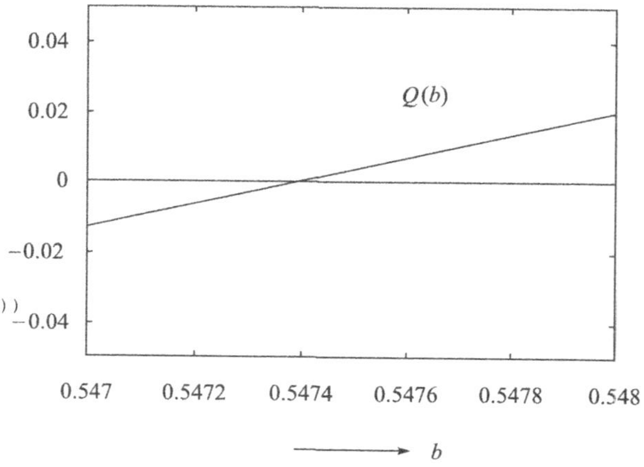

Fig. 19.6

Similarly, the transcendental equation for anti-symmetric modes [see Eq. (31)]

$$-\left(\frac{1}{2}V\sqrt{1-b}\right)\cot\left(\frac{1}{2}V\sqrt{1-b}\right)=\frac{1}{2}V\sqrt{b}$$

can be written as

$$R(b)=\left(\frac{1}{2}V\sqrt{1-b}\right)\cos\left(\frac{1}{2}V\sqrt{1-b}\right)+\frac{1}{2}V\sqrt{b}\,\sin\left(\frac{1}{2}V\sqrt{1-b}\right)=0$$

GNU Program

```
set yrange[-8.0:8.0]
set xrange[0.:1.0]
set nokey
set ytics
set xtics
V=11.815
p0(x)=0.
p1(x)=0.5*V*sqrt(1-x)
p2(x)=0.5*V*sqrt(x)*sin(p1(x))
p3(x)=p1(x)*cos(p1(x))
R(x)=p3(x)+p2(x)
plot p0(x) w l8,R(x) w l8
```

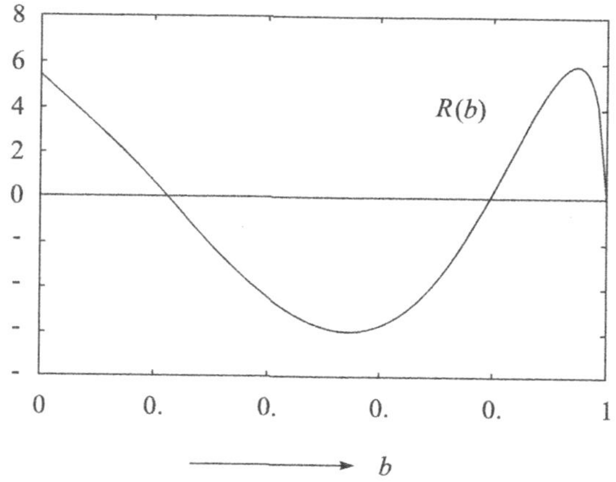

Fig. 19.7

One can again use any program to plot $Q(b)$ as a function of b in the region $0 < b < 1$ (see Fig. 19.7). The figure clearly shows that there are two TE (antisymmetric) modes one around $b \approx 0.8$ (TE_1 mode) and the other around $b \approx 0.2$ (TE_3 mode). In order to get greater accuracy, one may plot $Q(b)$ in the vicinity of the roots as we have done in Figs 19.8 and 19.9. One finds that the values of b are close to 0.800 and 0.224. In fact if we use the 'Find root' program in MATHEMATICA (or a similar program in MATLAB), we would obtain

$$b = 0.795397 \text{ (TE}_1\text{ mode)} \quad \text{and} \quad b \approx 0.223714 \text{ (TE}_3\text{ mode)}$$

```
set yrange[-0.3:0.3]
set xrange[0.790:0.80]
set nokey
set ytics
set xtics
V=11.815
p0(x)=0.
p1(x)=0.5*V*sqrt(1-x)
p2(x)=0.5*V*sqrt(x)*sin(p1(x))
p3(x)=p1(x)*cos(p1(x))
R(x)=p3(x)+p2(x)
plot p0(x) w l8,R(x) w l8
```

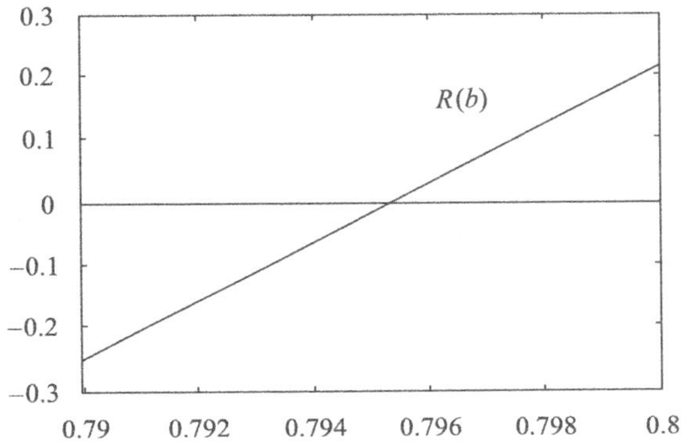

Fig. 19.8

```
set yrange[-0.3:0.3]
set xrange[0.22:0.23]
set nokey
set ytics
set xtics
V=11.815
p0(x)=0.
p1(x)=0.5*V*sqrt(1-x)
p2(x)=0.5*V*sqrt(x)*sin(p1(x))
p3(x)=p1(x)*cos(p1(x))
R(x)=p3(x)+p2(x)
plot p0(x) w l8,R(x) w l8
```

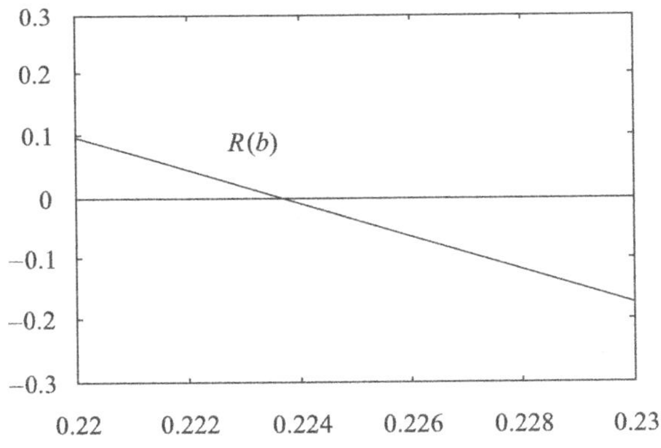

Fig. 19.9

The corresponding values of $\frac{\beta}{k_0}$ can easily be obtained by using the equation

$$\frac{\beta}{k_0} = \sqrt{[n_2^2 + b(n_1^2 - n_2^2)]}$$

and we get, $\beta/k_0 = 1.4885$ for the first symmetric mode which is denoted as the TE_0 mode,

$= 1.4839$ for the first anti-symmetric mode which is denoted as the TE_1 mode,

= 1.4765 for the second symmetric mode which is denoted as the TE_2 mode,

and = 1.4668 for the second anti-symmetric mode which is denoted as the TE_3 mode.

Now, $\beta/k_0 = n_1 \cos\theta = 1.49 \cos\theta$. Thus,

$$\theta = \cos^{-1}\left(\frac{\beta/k_0}{n_1}\right) = \cos^{-1}\left(\frac{1.4885}{1.49}\right) \approx 2.6^\circ \text{ for the TE}_0 \text{ mode,}$$

Similarly,

$$\theta = \cos^{-1}\left(\frac{1.4839}{1.49}\right) \approx 5.2^\circ \text{ for the TE}_1 \text{ mode,}$$

$$\theta = \cos^{-1}\left(\frac{1.4765}{1.49}\right) \approx 7.7^\circ \text{ for the TE}_2 \text{ mode, and}$$

$$\theta = \cos^{-1}\left(\frac{1.4668}{1.49}\right) \approx 10.1^\circ \text{ for the TE}_3 \text{ mode.}$$

19.10 (a) $n_1 = 1.503$, $n_2 = 1.500$ and $d = 4$ μm. and $\lambda_0 = 1$ μm. Thus,

$$V = \frac{2\pi}{\lambda_0} d\sqrt{n_1^2 - n_2^2} = \frac{2\pi}{1} \times 4 \times \sqrt{1.503^2 - 1.500^2} \approx 2.3855$$

Now, $b(V = 2.375) \approx 0.529300$ and $b(V = 2.500) \approx 0.551571$. Thus,

$$b(V = 2.3855) \approx 0.529300 + (0.551571 - 0.529300) \times \frac{0.0105}{0.125}$$
$$\approx 0.5312$$

$$\text{Now, } \frac{\beta}{k_0} = \sqrt{[n_2^2 + b(n_1^2 - n_2^2)]} = \sqrt{1.5^2 + 0.5312 \times (1.503^2 - 1.5^2)}$$
$$\approx 1.5016$$

(b) $V = 4.771$

Now, $b(V = 4.750) \approx 0.788321$ and 0.235151

and $b(V = 4.875) \approx 0.795686$ and 0.256461.

Thus, for the first mode

$$b(V = 4.771) \approx 0.788321 + (0.795686 - 0.788321) \times \frac{0.021}{0.125}$$
$$\approx 0.7896$$

$$\frac{\beta}{k_0} = \sqrt{[n_2^2 + b(n_1^2 - n_2^2)]} = \sqrt{1.5^2 + 0.7896 \times (1.503^2 - 1.5^2)}$$
$$\approx 1.5024$$

Similarly, for the second mode

$$b(V = 4.771) \approx 0.235151 + (0.256461 - 0.235151) \times \frac{0.021}{0.125}$$
$$\approx 0.2387$$

$$\frac{\beta}{k_0} = \sqrt{[n_2^2 + b(n_1^2 - n_2^2)]} = \sqrt{1.5^2 + 0.2387 \times (1.503^2 - 1.5^2)}$$
$$\approx 1.5007$$

19.11 We will show that if

$$n^2(x) = n^2(-x) \tag{47}$$

i.e., if the refractive index distribution function is symmetric about $x = 0$ then the eigen functions of the wave equation are either symmetric functions of x [i.e., $\psi(x) = \psi(-x)$] or antisymmetric functions[2] of x [i.e., $\psi(x) = -\psi(-x)$]; we are representing $E_y(x)$ by $\psi(x)$. In order to prove this, we first rewrite the wave equation [Eq. (4)] in the form

$$\frac{d^2\psi(x)}{dx^2} + k_0^2 n^2(x)\,\psi(x) = \beta^2\psi(x) \tag{48}$$

Making the transformation $x \to -x$ we get

$$\frac{d^2\psi(-x)}{dx^2} + k_0^2 n^2(x)\,\psi(-x) = \beta^2\psi(-x) \tag{49}$$

where we have used the fact that $n^2(x) = n^2(-x)$. Comparing the above two equations, we see that $\psi(x)$ and $\psi(-x)$ are eigenfunctions belonging to the same eigenvalue β^2. Thus, $\psi(-x)$ must be a multiple of $\psi(x)$:

$$\psi(-x) = \lambda\psi(x)$$

Clearly,

$$\psi(x) = \lambda\psi(-x) = \lambda^2\psi(x)$$

so that $\lambda^2 = 1$ or $\lambda = \pm 1$. Hence,

$$\psi(-x) = \pm\psi(x) \tag{50}$$

proving the theorem.

19.12 Substituting the solution in Eq. (4) we obtain the following equation:

$$-\frac{1}{w_0^2} + \frac{x^2}{w_0^4} + k_0^2 n_0^2 - k_0^2\alpha x^2 - \beta^2 = 0$$

In order that the above equation be satisfied for all values of x, the coefficient of x^2 on both sides must be equal and so also with the x-independent term. Thus, we obtain

$$w_0^2 = \frac{1}{k_0\sqrt{\alpha}}$$

and

$$\beta^2 = k_0^2 n_0^2 - k_0\sqrt{\alpha}$$

19.13 (a) For the given waveguide parameters, $V = k_0 d\sqrt{(n_1^2 - n_2^2)} \approx 2.683\pi$. Thus, the waveguide will support three TE modes and three TM modes.

---

2. The theorem is strictly true for nondegenerate states only. By a nondegenerate state, we imply that there is only one wave function for a particular value of β^2. If for the same value of β^2, there are more than one linearly independent wave function, we have what is known as a degenerate state. For degenerate states the wave functions need not be symmetric or antisymmetric functions of x. However, even for degenerate states one can always construct appropriate linear combinations which are either symmetric or antisymmetric functions of x.

(b) The maximum and minimum values of b of the TE1 mode would be $2.3k_0$ and $2.2k_0$ respectively.

(c) The waveguide will be single moded in the range $0 < d < 0.745$ μm.

19.14 For guided modes, the field distribution should decay in the cladding region. Hence, the guided modes will have their propagation constants in the range $n_1 > \tilde{\beta} > n_3$.

19.15 No, such a waveguide cannot support any guided modes.

19.16 For exciting the TE mode of the waveguide, the incident polarisation must be along the y-direction while for exciting the TM mode of the waveguide, the incident polarisation must be along the x-direction. When light polarised at 45° to the x-axis and lying in the x-y plane is incident on the waveguide, then the incident light is a linear combination of the x- and y-polarisations and thus it will excite both the TE and the TM modes of the waveguide. Since the propagation constants of the TE and TM modes are unequal they will develop a phase difference as they propagate and this will change the polarisation state of the propagating wave. The polarisation will repeat itself when the phase difference between the two modes becomes equal to 2π.

Thus, if L is the distance for the polarisation to return to the incident polarisation, then we have

$$(\beta_{TE} - \beta_{TM})L = 2\pi \quad \text{or} \quad L = \frac{2\pi}{(\beta_{TE} - \beta_{TM})}$$

Fiber Optics II: Single Mode Fibers

20

A Quick Review

Most of the discussion in this chapter will correspond to a step index fiber which is characterised by the following refractive index distribution (see Fig. 20.1):

$$\begin{aligned} n(r) &= n_1 \qquad 0 < r < a \qquad \text{core} \\ &= n_2 \qquad r > a \qquad \text{cladding} \end{aligned} \tag{1}$$

where we are using the cylindrical system of coordinates (r, ϕ, z). In actual fibers,

$$\frac{n_1 - n_2}{n_2} \le 0.01 \tag{2}$$

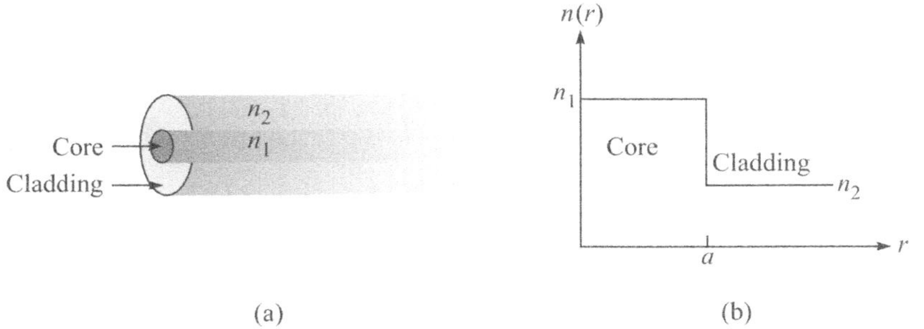

Fig. 20.1 (a) A step index fiber is a cylindrical structure in which the refractive index is n_1 for $0 < r < a$ and n_2 for $r > a$ (b) The refractive index variation of a step index fiber.

When the above condition is satisfied we have what is usually referred to as the weakly guiding approximation in which, the transverse component of the electric field (E_x or E_y) satisfies the scalar wave equation

$$\nabla^2 \Psi = \varepsilon_0 \mu_0 n^2 \frac{\partial^2 \Psi}{\partial t^2} = \frac{n^2}{c^2} \frac{\partial^2 \Psi}{\partial t^2} \tag{3}$$

where $c \left(= \frac{1}{\sqrt{\varepsilon_0 \mu_0}} \right) \approx 3 \times 10^8$ m/s is the speed of light in free space. In most practical fibers n^2 depends only on the cylindrical coordinate r and therefore it is convenient to use the cylindrical system of coordinates (r, ϕ, z) and the solution of Eq. (3) can be written in the form

$$\Psi(r, \phi, z, t) = R(r)\begin{Bmatrix}\cos l\phi \\ \sin l\phi\end{Bmatrix} e^{i(\omega t - \beta z)}; \quad l = 0, 1, 2, \ldots \tag{4}$$

where ω is the angular frequency, β is known as the propagation constant and $R(r)$ satisfies the radial part of the equation

$$r^2 \frac{d^2 R}{dr^2} + r\frac{dR}{dr} + \{[k_0^2 n^2(r) - \beta^2] r^2 - l^2\} R = 0 \tag{5}$$

where,

$$k_0 = \frac{\omega}{c} = \frac{2\pi}{\lambda_0} \tag{6}$$

is the free space wave number. As in the previous chapter, Eq. (4) defines the modes of the system. For each value of l there can be two independent states of polarisation; modes with $l \geq 1$ are four-fold degenerate (corresponding to two orthogonal polarisation states and to the ϕ dependence being cos $l\phi$ or sin $l\phi$). Modes with $l = 0$ are ϕ independent and have two-fold degeneracy[1]. For an arbitrary cylindrically symmetric profile having a refractive index that decreases monotonically from a value n_1 on the axis to a constant value n_2 beyond the core-cladding interface $r = a$ [see Fig. 20.2], we can make the general observation that the solutions of Eq. (5) can be divided into two distinct classes [compare with the discussions in the previous chapter]; the first class of solutions correspond to

$$n_2^2 < \frac{\beta^2}{k_0^2} < n_1^2 \qquad \textbf{GUIDED MODES} \tag{7}$$

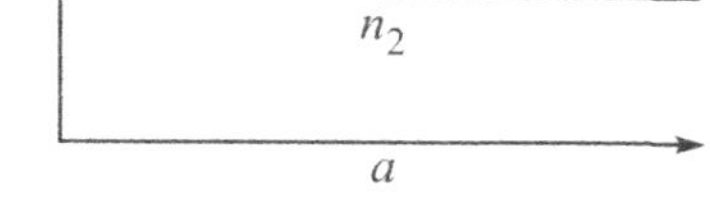

Fig. 20.2 A cylindrically symmetric refractive index profile having a refractive index that decreases monotonically from a value n_1 on the axis to a constant value n_2 beyond the core-cladding interface $r = a$.

---

1. The word degeneracy means that for the same value of the propagation constant there are more than one field profiles. For $l = 0$ we will have two independent state of polarisation; thus the mode is said to be 2 fold degenerate. On the other hand, for $l = 1, 2, 3\ldots$ the mode will be 4 fold degenerate because (for the same value of β^2) we will have two field profiles: one proportional to cos $l\phi$ and the other to sin $l\phi$ and for each field profile, we will again have two independent states of polarisation.

For β^2 lying in the above range, the field $R(r)$ are oscillatory in the core and decay in the cladding and β^2 assumes only discrete values; these are known as the *guided modes* of the waveguide. For a given value of l, there will be a finite number of guided modes, these are designated as LP_{lm} modes ($m = 1, 2, 3, \ldots$).

The second class of solutions correspond to

$$\beta^2 < k_0^2 n_2^2 \qquad \textbf{RADIATION MODES} \tag{8}$$

For such β values, the fields are oscillatory even in the cladding and β can assume a continuum of values. These are known as the *radiation modes*[2].

If we solve Eq. (5) corresponding to a step index fiber [see Eq. (1)], we would obtain the following expressions for the complete field pattern:

$$\Psi(r, \phi, z, t) = \begin{cases} \dfrac{A}{J_l(U)} J_l\left(\dfrac{Ur}{a}\right)\begin{bmatrix}\cos l\phi \\ \sin l\phi\end{bmatrix} e^{i(\omega t - \beta z)}; & r < a \\ \dfrac{A}{K_l(W)} K_l\left(\dfrac{Wr}{a}\right)\begin{bmatrix}\cos l\phi \\ \sin l\phi\end{bmatrix} e^{i(\omega t - \beta z)}; & r > a \end{cases} \tag{9}$$

where $J_l(x)$ is the Bessel function and $K_l(x)$ is the modified Bessel function; A is a constant and we have assumed the continuity of ψ at the core-cladding interface ($r = a$). Further,

$$U \equiv a\sqrt{k_0^2 n_1^2 - \beta^2} \tag{10}$$

and

$$W \equiv a\sqrt{\beta^2 - k_0^2 n_2^2} \tag{11}$$

Because of Eq. (7), both U and W are real. The normalised waveguide parameter V is defined by

$$V = \sqrt{U^2 + W^2} = k_0 a\sqrt{n_1^2 - n_2^2} \tag{12}$$

In terms of the wavelength,

$$V = \frac{2\pi}{\lambda_0} a\sqrt{n_1^2 - n_2^2} \tag{13}$$

Continuity of $\partial\psi/\partial r$ at $r = a$ and use of identities involving Bessel functions [see, e.g., Ref. Ir 1 and Gh 5] give us the following transcendental equations which determine the allowed discrete values of the normalised propagation constant b of the guided LP_{lm} modes:

$$V\sqrt{1-b}\,\frac{J_{l-1}\left[V\sqrt{1-b}\right]}{J_l\left[V\sqrt{1-b}\right]} = -V\sqrt{b}\,\frac{K_{l-1}\left[V\sqrt{b}\right]}{K_l\left[V\sqrt{b}\right]}; \quad l \geq 1 \tag{14}$$

and

$$V\sqrt{1-b}\,\frac{J_1\left[V\sqrt{1-b}\right]}{J_0\left[V\sqrt{1-b}\right]} = V\sqrt{b}\,\frac{K_1\left[V\sqrt{b}\right]}{K_0\left[V\sqrt{b}\right]}; \quad l = 0 \tag{15}$$

---

2. For more details about radiation modes and also excitation of leaky modes, see, e.g. Ref. Gh 5 and Sn 1.

The solution of the above transcendental equations will give us universal curves describing the dependence of b (and therefore of U and W) on V. For a given value of l, there will be a finite number of solutions and the mth solution ($m = 1, 2, 3, \ldots$) is referred to as the LP_{lm} mode. The variation of b with V form a set of universal curves, which are plotted in Fig. 20.3. Table 20.1 gives the numerical values of b (corresponding to the LP_{0m} mode) for values of V lying between 1.0 and 2.5. It is convenient to define the normalised propagation constant

$$b = \frac{\dfrac{\beta^2}{k_0^2} - n_2^2}{n_1^2 - n_2^2} = \frac{W^2}{V^2} \tag{16}$$

Thus,

$$W = V\sqrt{b} \tag{17}$$

$$U = V\sqrt{1-b} \tag{18}$$

and

$$\frac{\beta}{k_0} = \sqrt{\left[n_2^2 + b\left(n_1^2 - n_2^2\right)\right]} \approx n_2\sqrt{1 + (2\Delta)b} \tag{19}$$

From Eq. (7) we find that for guided modes $0 < b < 1$.

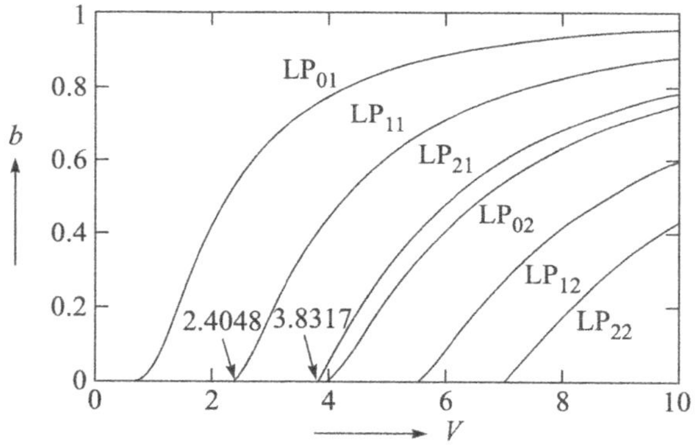

Fig. 20.3 Variation of the normalised propagation constant b with normalised waveguide parameter V corresponding to a few lower order modes [Calculations courtesy Ms. Triranjita Srivastava].

20.1 CUT-OFF FREQUENCIES AND NUMBER OF MODES

From Fig. 20.3 we see that the value of b decreases as we decrease the value of V. For every mode, there is a value of V when b becomes zero (i.e., when β/k_0 becomes equal to n_2) and the mode ceases to be a guided mode. The value of V for which b becomes zero is known as the *cutoff frequency* of the mode. Now, for a given step-index fiber, the value of V decreases as we increase the wavelength [see Eq. (13)] and the value of the wavelength at which b becomes zero is known as the *cutoff wavelength* for that mode.

We can see from Eq. (15) that the cutoff frequencies of the LP_{0m} modes will occur at the zeroes of $J_1(V)$, i.e., when $V=0$ (LP_{01}), 3.8317 (LP_{02}), 7.0156 (LP_{03}), 10.1735 (LP_{04}),...

Similarly, we can see from Eq. (14) that the cut-off frequencies of the LP_{1m} modes will occur at the zeroes of $J_0(V)$, i.e., when $V=2.4048$ (LP_{11}), 5.5201 (LP_{12}), 8.6537 (LP_{13}), 11.7915 (LP_{14}),...; cut-off frequencies of the LP_{2m} modes occur at the zeroes of $J_1(V)$ (excluding the value $V = 0$), i.e., when $V = 3.8317$ (LP_{21}), 7.0156 (LP_{22}), 10.1735 (LP_{23}),....

For $l \geq 1$, cutoff frequencies of the LP_{lm} modes will occur at the zeroes of $J_{l-1}(V)$ (excluding the value $V = 0$); thus[3] cut-off frequencies of the LP_{3m} modes occur when $V = 5.1356$ (LP_{31}), 8.4172 (LP_{32}), 11.6198 (LP_{33}); cutoff frequencies of the LP_{4m} modes occur when $V = 6.3802$ (LP_{41}), 9.7610 (LP_{42}), 13.015 (LP_{43}); cutoff frequencies of the LP_{5m} modes occur when $V=7.5883$ (LP_{51}), 11.0647 (LP_{52}); cutoff frequencies of the LP_{6m} modes occur when $V=8.7715$ (LP_{61}), 12.3386 (LP_{62}),...

Thus, as can also be seen from the figure:

For $0 < V < 2.4048$ we will only have the LP_{01} mode (which is referred to as the fundamental mode); $V=2.4048$ represents the cutoff of the LP_{11} mode where (for the LP_{11} mode) b becomes 0, i.e., β/k_0 becomes equal to n_2.

For $2.4048 < V < 3.8317$ we will only have LP_{01} and LP_{11} modes; $V = 3.8317$ represents the cutoff of the LP_{02} and the LP_{21} modes where (for the LP_{02} and the LP_{21} modes) b becomes 0, i.e., β/k_0 becomes equal to n_2.

For $3.8317 < V < 5.1356$ we will only have LP_{01}, LP_{02}, LP_{11} and LP_{21} modes; $V=5.1356$ represents the cutoff of the LP_{31} mode.

Thus at a particular V value, the fiber can support only a finite number of modes. We must mention here that each LP_{0m} mode is two fold degenerate; i.e., there are two independent modes with the same value of b, corresponding to two independent states of polarisation. Further each LP_{lm} mode ($l > 1$) is 4 fold degenerate; i.e., there are four independent modes with the same value of b, corresponding to ϕ-dependence of $\cos l\phi$ and $\sin l\phi$ with each mode having two independent states of polarisation. The total number of modes in a highly multimoded ($V \geq 10$) step index fiber is approximately given by

$$N \approx \frac{1}{2}V^2 \tag{20}$$

20.2 EMPIRICAL FORMULA FOR THE NORMALISED PROPAGATION CONSTANT

For a single mode step index fiber, a convenient empirical formula for $b(V)$ is given by

$$b(V) = \left(A - \frac{B}{V}\right)^2; \quad 1.5 \lesssim V \lesssim 2.5 \tag{21}$$

3. The values of the zeros of the Bessel functions are taken from Ref. Ab1.

with $A \approx 1.1428$ and $B \approx 0.996$. The above formula gives values of b which are within about 0.2% of the exact values (see Table 20.1).

Table 20.1 Values of b, $(bV)'$ and $V(bV)''$ vs V for a step index fiber; the values in the second, fourth and fifth columns are generated by solving Eq. (15) for a step index fiber using the software given in Ref. Gh3 and Gh4.

V	b	b [using Eq. (30)]	$\frac{d}{dV}(bV)$	$V(bV)''$
1.5	0.229248	0.229249	0.849	1.063
1.6	0.270063	0.270712	0.913	0.919
1.7	0.309467	0.310157	0.965	0.785
1.8	0.347068	0.347471	1.006	0.664
1.9	0.382660	0.382653	1.039	0.556
2.0	0.416163	0.415767	1.065	0.462
2.1	0.447581	0.446911	1.086	0.380
2.2	0.476969	0.476200	1.102	0.309
2.3	0.504416	0.503754	1.114	0.248
2.4	0.530026	0.529693	1.124	0.195
2.5	0.553915	0.554131		

20.3 SPOT SIZE OF THE FUNDAMENTAL MODE

For most single mode fibers, the fundamental mode field distributions can be well-approximated by a Gaussian function, which may be written in the form

$$\psi(x, y) = A e^{-\frac{x^2 + y^2}{w^2}} = A e^{-\frac{r^2}{w^2}} \tag{22}$$

where w is referred to as the spot size of the mode field pattern and $2w$ is called the mode field diameter (MFD). MFD is a very important characteristic of a single mode optical fiber. For a step index fiber one has the following empirical expression for w (see Ref. Ma1):

$$\frac{w}{a} \approx 0.65 + \frac{1.619}{V^{3/2}} + \frac{2.879}{V^6}; \quad 0.8 \le V \le 2.5 \tag{23}$$

where a is the core radius. Many data sheets describing a commercially available single mode fiber would not always give the actual refractive index profile. They would instead give the MFD may be at more than one wavelength. They would also give the cutoff wavelength (see for example Ref. My1). For example, the standard single mode fiber designated as G.652 fiber when operating at 1.3 μm has a MFD of 9.2 ± 0.4 μm; the same fiber when operating at 1.55 μm has a MFD of 10.4 ± 0.8 μm.

20.4 SPLICE LOSS DUE TO TRANSVERSE MISALIGNMENT

The most common misalignment at a joint between two similar fibers is the transverse misalignment similar to that shown in Fig. 20.4. Corresponding to a transverse misalignment of u the loss in decibels is given by

$$\alpha(\text{dB}) \approx 4.34\,(u/w)^2 \tag{24}$$

Thus a larger value of w will lead to a greater tolerance to transverse misalignment. For a single mode fiber operating at 1300 nm, $w = 5$ μm, and if the splice loss is to be below 0.1 dB, then from Eq. (24) we obtain $u < 0.76$ μm. Thus, for a low-loss joint, the transverse alignment is very critical and connectors for single-mode fibers require precision matching and positioning for achieving low loss. For $w \approx 5$ μm, and a transverse offset of 1 μm the loss at the joint will be approximately 0.17 dB; on the other and, for $w \approx 3$ μm, a transverse offset of 1 μm will result in a loss of about 0.5 dB.

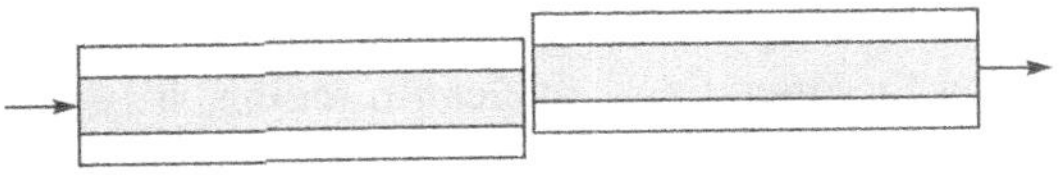

Fig. 20.4 A transverse alignment between two fibers would result in a loss of the optical beam.

20.5 PULSE DISPERSION IN SINGLE MODE FIBERS

In single-mode fibers since there is only one mode and there is no intermodal dispersion. However, we have (in addition to material dispersion) waveguide dispersion which is characteristic of the transverse refractive index variation. The total dispersion is given by the sum of material and waveguide dispersions:

$$D_{\text{total}} = D_m + D_w \tag{25}$$

In Chapters 16 and 18 we have discussed material dispersion which arises because of the wavelength dependence of the refractive index and had obtained the following expression for the material dispersion coefficient:

$$D_m = \frac{\Delta\tau_m}{L\Delta\lambda_0} = -\frac{10^4}{3\lambda_0}\left[\lambda_0^2\frac{d^2n}{d\lambda_0^2}\right] \text{ps/km.nm (Material Dispersion Coefficient)} \tag{26}$$

where λ_0 is measured in μm and the quantity inside the square brackets is dimensionless. Now, even if n_1 and n_2 are independent of wavelength (i.e., even if there is no material dispersion), the group velocity of a particular mode will depend on the wavelength; physically this arises because of the dependence of the spot size on wavelength. This leads to what is known as the *waveguide dispersion*. The waveguide dispersion coefficient D_w is approximately given by

$$D_w \approx -\frac{n_2\Delta}{3\lambda_0}\times 10^7 f(V) \quad \text{ps/km.nm} \tag{27}$$

where λ_0 is measured in nanometers and we have assumed $c = 3 \times 10^{-4}$ m/ps [meters per picosecond] and

$$f(V) \equiv V \frac{d^2}{dV^2}(bV) \tag{28}$$

For a step index fiber, b as a function of V is an universal curve; therefore the variation of $f(V)$ with V will also be universal (see Table 20.1). A convenient empirical formula for a step index fiber is given by [Ref. Ma 2]

$$f(V) \approx 0.080 + 0.549(2.834 - V)^2; \quad 1.3 < V < 2.4 \tag{29}$$

A comparison between the above empirical values with the exact values have been made in Ref. Gh 5. Thus the waveguide dispersion coefficient is approximately given by

$$D_w \equiv \frac{\Delta\tau_w}{L\Delta\lambda_0} \approx -\frac{n_2\Delta}{3\lambda_0} \times 10^7 [0.080 + 0.549\,(2.834 - V)^2] \text{ ps/km.nm} \tag{30}$$

where λ_0 is measured in nanometers. In the single-mode regime, the quantity within the bracket in Eq. (30) is usually positive; hence the waveguide dispersion is negative indicating that longer wavelengths travel faster. Since the sign of material dispersion depends on the operating wavelength region, it is possible that the two effects namely, material and waveguide dispersions cancel each other at a certain wavelength. Such a wavelength, which is a very important parameter of single-mode fibers, is referred to as the zero-dispersion wavelength (λ_{ZD}).

PROBLEMS

20.1 Consider a step index fiber with $n_1 = 1.474$, $n_2 = 1.470$ and having a core radius $a = 4.5$ μm Determine the cut-off wavelength.

20.2 Consider a step index fiber with $n_1 = 1.5$, $n_2 = 1.48$ and having a core radius $a = 6.0$ μm. Determine the operating wavelength λ_0 for which $V = 8$.

20.3 In continuation of the previous problem, (a) calculate the total number of modes for $V = 8$ and (b) compare with the approximate value given by Eq. (20).

20.4 Consider a step index fiber with $n_1 = 1.474$, $n_2 = 1.470$ and having a core radius $a = 3.0$ μm operating at a wavelength 0.889 μm. Calculate the spot size of the fundamental mode.

20.5 We consider a step index fiber with $n_1 = 1.5$, $n_2 = 1.49$ and the core radius $a = 3.0$ μm. Calculate the range of the values of λ_0 for which LP_{01}, LP_{11}, LP_{21} and LP_{02} modes will exist.

20.6 Consider a step index fiber with $n_1 = 1.5$, $n_2 = 1.48$ and core radius $a = 6.0$ μm. Assuming the operating wavelength $\lambda_0 = 1.3$ μm calculate the number of modes and compare with that obtained by using the approximate formula [Eq. (20)].

20.7 We consider a step index fiber with $n_2 = 1.447$, $\Delta = 0.003$ and $a = 4.2$ μm. Calculate the domain of single mode operation. Find the value of λ_0 for which $V = 2.0$ and therefore use Table 20.1 to determine b and then the values of β/k_0 and of β.

20.8 In continuation of the previous problem, we consider the same step index fiber [$n_2 = 1.447$, $\Delta = 0.003$ and $a = 4.2$ μm] now operating at $\lambda_0 = 1.55$ μm. Use Table 20.1 and linear interpolation to determine b and then the values of β/k_0 and of β.

20.9 Fibers used in IV generation optical communication systems (operating at 1.55 μm) have a small value of core radius and a large value of Δ. Consider a fiber with $n_2 = 1.444$, $\Delta = 0.0075$ and $a = 2.3$ μm. Assuming the operating wave length to be $\lambda_0 = 1.55$ μm, calculate the values of b and β/k_0.

20.10 Consider a step index fiber (operating at 1300 nm) with $n_2 = 1.447$, $\Delta = 0.003$ and $a = 4.2$ μm (see Problem 20.3). Using the empirical formula [Eq. (23)], calculate the spot size of the fundamental mode at $\lambda_0 = 1300$ nm and at $\lambda_0 = 1550$ nm.

20.11 For a step index fiber with $n_2 = 1.444$, $\Delta = 0.0075$ and $a = 2.3$ μm (see Problem 20.5). Using the empirical formula [Eq. (23)], calculate the spot size of the fundamental mode at $\lambda_0 = 1300$ nm and at $\lambda_0 = 1550$ nm and show that the spot size increases with wavelength.

20.12 Assume the single mode fiber to have a Gaussian spot size $w = 4.5$ μm. Calculate the splice loss at a joint between two such identical fibers with a transverse misalignment of 1, 2 and 3 μm.

20.13 Consider a step index fiber with $n_2 = 1.447$, $\Delta = 0.003$ and $a = 4.2$ μm (see Problem 20.7).

Calculate and plot D_m, D_w and D_{total} and determine the wavelength corresponding to zero total dispersion.

20.14 We next consider the fiber discussed in Problem 20.9 for which $n_2 = 1.444$, $\Delta = 0.0075$ and $a = 2.3$ μm. Calculate and plot D_m, D_w and D_{total} and determine the wavelength corresponding to zero total dispersion.

20.15 The modal field is said to be normalised if

$$\iint |\psi(x, y)|^2 \, dx\, dy = 1$$

Show that the normalised Gaussian field is given by

$$\psi(x, y) = \sqrt{\frac{2}{\pi}}\frac{1}{w} e^{-\frac{x^2 + y^2}{w^2}} = \sqrt{\frac{2}{\pi}}\frac{1}{w} e^{-\frac{r^2}{w^2}}$$

20.16 Consider two identical single mode fibers joined together with a transverse misalignment of u (along the x-axis). The fractional power that is coupled to the fundamental mode of the second fiber is given by the overlap integral

$$T = \left| \iint \psi_1(x, y)\, \psi_2(x, y)\, dx\, dy \right|^2$$

Show that $T = \exp\left(-\frac{u^2}{w^2}\right)$. Thus,

$$\text{Loss in dB} = 10 \log T = 4.34 \left(\frac{u}{w}\right)^2$$

20.17 Consider a parabolic index fiber characterised by the following refractive index variation

$$n^2(r) = n_1^2\left[1 - 2\Delta\left(\frac{r}{a}\right)^2\right] = n_1^2\left[1 - 2\Delta\frac{x^2+y^2}{a^2}\right] \quad 0 < r < a \quad \text{core}$$

$$= n_2^2 \qquad\qquad r > a \quad \text{cladding}$$

The corresponding propagation constants for guided modes are approximately given by

$$\beta^2 = \beta_{mn}^2 \approx k_0^2 n_1^2 - 2(m+n+1)\gamma k_0;\ m, n = 0, 1, 2, 3, \ldots \text{ where } \gamma = \frac{n_1\sqrt{2\Delta}}{a}$$

(a) Show that the group velocity is independent of the mode number.
(b) Assuming Eq. (7), calculate approximately the number of modes for a given value of V.

20.18 For a singlemode fiber operating at 1300 nm, the mode field diameter is approximately 10 μm. For far field measurements, at what distance from the fiber tip should the detector be placed?

20.19 Two different step index fibers have the same value of V at some wavelength. How will their spots sizes be related?

20.20 Show that the excitation efficiency of the LP_{11} mode in a step index fiber is zero for an incident Gaussian beam given by

$$\Psi(r, \phi) = Ae^{-r^2/w^2}$$

20.21 Identify the following transverse modal field patterns from a fiber

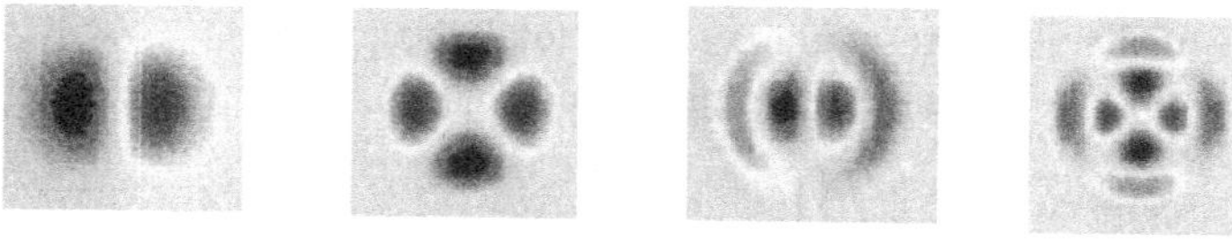

(a) (b) (c) (d)

SOLUTIONS

20.1 $V = \dfrac{2\pi}{\lambda_0} a\sqrt{n_1^2 - n_2^2} = \dfrac{3.068}{\lambda_0}$

Thus,

$$V = \frac{2\pi}{\lambda_0} a\sqrt{n_1^2 - n_2^2} = \frac{3.068}{\lambda_0} \Rightarrow 2.4045 = \frac{3.068}{\lambda_c} \Rightarrow \lambda_c \approx 1.28\ \mu\text{m}.$$

20.2 $V = \dfrac{2\pi}{\lambda_0} a\sqrt{n_1^2 - n_2^2} = \dfrac{9.204}{\lambda_0}$. Thus, $V = 8$ would imply $\lambda_0 \approx 1.15\ \mu$m.

20.3 There will be

2 LP_{01} modes
2 LP_{02} modes

2 LP_{03} modes
4 LP_{11} modes
4 LP_{12} modes
4 LP_{21} modes
4 LP_{22} modes
4 LP_{31} modes
4 LP_{41} modes
and 4 LP_{51} modes

In above we have used the fact that the $l = 0$ modes are two fold degenerate because of two independent states of polarisation; on the other hand, for $l \geq 1$, modes are four fold degenerate because for each independent state of polarisation the ϕ dependence can be either cos $l\phi$ or sin $l\phi$. Thus, there will be a total of 34 modes. The approximate formula [Eq. (20)] will give

$$N \approx \frac{1}{2}V^2 \approx 32.$$

20.4 $V = \dfrac{2\pi}{\lambda_0} a\sqrt{n_1^2 - n_2^2} \approx 2.3$. For a step index fiber, the empirical expression for w is given by

$$\frac{w}{a} \approx 0.65 + \frac{1.619}{V^{3/2}} + \frac{2.879}{V^6}; \quad 0.8 \leq V \leq 2.5$$

where a is the core radius. Thus,

$$\frac{w}{a} \approx 0.65 + 0.464 + 0.0194 = 1.13$$

and since $a \approx 3$ μm, we get $w \approx 3.4$ μm.

20.5 $V = \dfrac{2\pi}{\lambda_0} a\sqrt{n_1^2 - n_2^2} = \dfrac{3.2594}{\lambda_0}$ where λ_0 is measured in μm. Thus,

the cutoff wavelength of the LP_{11} mode will be 1.355 μm,
cutoff wavelengths of the LP_{21} and LP_{02} modes will be 0.8506 μm,
cutoff wavelength of the LP_{31} mode will be 0.6347 μm,...

The LP_{01} mode has no cutoff. Thus for $\lambda_0 > 1.355$ μm, we will only have the LP_{01} mode and for 0.8506 μm $< \lambda_0 <$ 1.355 μm, we will have LP_{01} and LP_{11} modes. For 0.6347 μm $< \lambda_0 <$ 0.8506 μm, we will have LP_{01}, LP_{11}, LP_{21} and LP_{02} modes.

20.6 $V = \dfrac{2\pi}{\lambda_0} a\sqrt{n_1^2 - n_2^2} = \dfrac{9.2035}{\lambda_0} \approx 7.080.$

Thus we will have two each of LP_{01}, LP_{02} and LP_{03} modes, four each of LP_{11}, LP_{12}, LP_{21}, LP_{22}, LP_{31} and LP_{41} modes and we will have a total of 30 modes. For $V = 7.0796$, we get $N \approx 25$. For higher value of V the values given by Eq. (20) will become closer to the exact value.

20.7 $V \approx \dfrac{2.958}{\lambda_0}$, where λ_0 is measured in μm. Thus for $\lambda_0 > 1.23$ μm, the fiber will be single moded. The cutoff wavelength λ_c (for which $V = 2.4045$) is

1.23 μm. We assume the operating wavelength $\lambda_0 = 1.479$ μm so that $V = 2.0$ and therefore (from Table 20.1)

$$b \approx 0.4162 \Rightarrow \frac{\beta}{k_0} \approx n_2 \sqrt{1+(2\Delta)b} \approx 1.4488 \Rightarrow \beta \approx 6.1549 \times 10^6 \text{ m}^{-1}.$$

20.8 $V \approx 1.908$ and we have a single mode fiber. Using Table 20.1 and linear interpolation we get

$$b \approx 0.382660 + \frac{0.416163 - 0.382660}{0.1} \times 0.008 \approx 0.38534$$

$$\Rightarrow \qquad \frac{\beta}{k_0} \approx n_2 \sqrt{1+(2\Delta)b} \approx 1.4487 \Rightarrow \beta \approx 5.8725 \times 10^6 \text{ m}^{-1}.$$

20.9 At $\lambda_0 = 1.55$ μm

$$V = \frac{2\pi}{1.55} \times 2.3 \times 1.444 \times \sqrt{0.015} \approx 1.649$$

Thus, the fiber will be single moded at 1.55 μm and

$$b \approx 0.270063 + \frac{0.309467 - 0.270063}{0.1} \times 0.049 \approx 0.28937$$

$$\Rightarrow \qquad \frac{\beta}{k_0} \approx n_2 \sqrt{1+(2\Delta)b} \approx 1.447$$

Further, for the given fiber we may write $V = \dfrac{2.556}{\lambda_0}$ and therefore the cut off wavelength will be $\lambda_c = 2.556/2.4045 \approx 1.06$ μm.

20.10 At $\lambda_0 = 1300$ nm, $V \approx 2.28$ giving $w \approx 4.8$ μm. The same fiber will have a V value of 1.908 at $\lambda_0 = 1550$ nm giving a value of the spot size ≈ 5.5 μm. *Thus the spot size increases with wavelength.*

20.11 At $\lambda_0 = 1300$ nm, $V \approx 1.97$ giving $w \approx 3.0$ μm. The same fiber, at $\lambda_0 = 1550$ nm, will have a V value of 1.65 giving a value of the spot size $w \approx 3.6$ μm. Thus, we again obtain the result that the spot size increases with wavelength.

20.12 $T = \exp\left(-\dfrac{u^2}{w^2}\right)$ and therefore

$$\text{Splice loss in dB} = -10 \log_{10} T = 10 (\log_{10} e) \times \frac{u^2}{w^2} = 4.34 \frac{u^2}{w^2}$$

Thus, splica loss in dB ≈ 0.21 dB for $u = 1$ μm

≈ 0.86 dB for $u = 2$ μm

≈ 1.93 dB for $u = 3$ μm

20.13 $V \approx \dfrac{2958}{\lambda_0}$, where λ_0 is measured in nanometers. Substituting in Eq. (30), we get

$$D_w = -\frac{1.447 \times 10^4}{\lambda_0}\left[0.080 + 0.549\left(2.834 - \frac{2958}{\lambda_0}\right)^2\right] \text{ ps/km.nm}$$

Elementary calculations show that at $\lambda_0 \approx 1300$ nm, $D_w = -2.8$ ps/km.nm. The variations of D_m, D_w and D_{total} with λ_0 are shown in Fig. 20.5; the variation of D_m is taken from Table 20.1 of Chapter. The total dispersion passes through zero around $\lambda_0 \approx 1300$ nm which is the zero total dispersion wavelength. These fibers are usually referred to as G 652 (or SMF 28) fibers and are extensively used in optical communication systems.

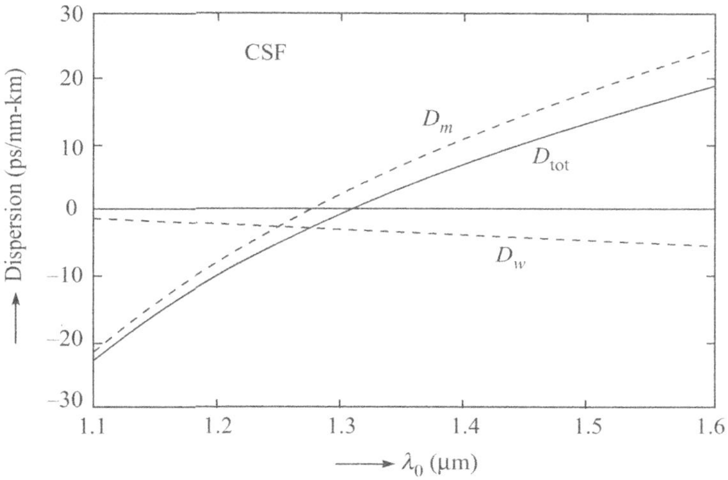

Fig. 20.5 The wavelength dependence of D_m, D_w and D_{tot} for a typical conventional single mode fiber (CSF) with parameters as given in Problem 20.13. The total dispersion passes through zero around $\lambda_0 \approx 1300$ nm which is known as zero dispersion wavelength.

20.14 $V \approx \dfrac{2556}{\lambda_0}$, where λ_0 is measured in nanometers. Substituting in Eq. (30), we get

$$D_w = -\frac{3.61 \times 10^4}{\lambda_0}\left[0.080 + 0.549\left(2.834 - \frac{2556}{\lambda_0}\right)^2\right] \text{ ps/km.nm}$$

Elementary calculations show that at $\lambda_0 \approx 1550$ nm, $D_w = -20$ ps/km.nm. On the other hand, the material dispersion at this wavelength is given by [see Table 20.1]

$$D_m = +20 \text{ ps/km.nm}$$

We therefore see that the two expressions are of opposite sign and almost cancel each other. Physically, because of waveguide dispersion, longer wavelengths travel slower than shorter wavelengths and because of material dispersion, longer wavelengths travel faster than shorter wavelengths–and the two effects compensate each other resulting in zero total dispersion around 1550 nm. The corresponding variation of D_m, D_w and D_{tot} with wavelength is shown in Fig. 20.6. As can be seen from the figure, we have been able to shift the zero dispersion wavelength by changing the fiber parameters; these are known as the ***dispersion shifted fibers***. Thus, dispersion shifted fibers are those fibers whose total dispersion becomes zero at a shifted wavelength. Commercially available dispersion shifted fibers (which are abbreviated as DSF and referred to as G 653 fibers) do not usually have a step variation of refractive index; the refractive index variation is bit complicated and is such that the total dispersion passes through zero around 1550 nm wavelength.

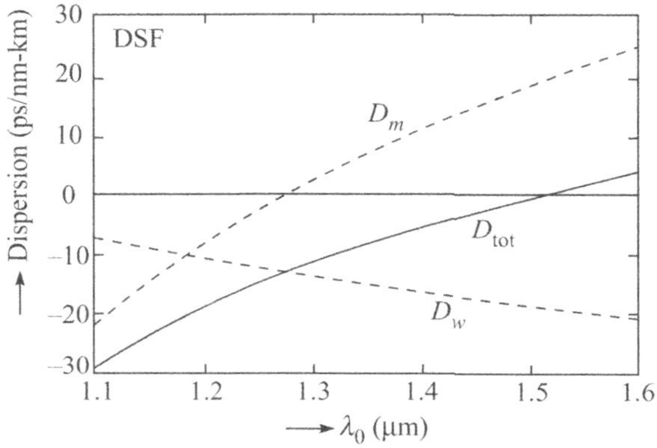

Fig. 20.6 The wavelength dependence of D_m, D_w and D_{tot} for a typical dispersion shifted fiber (DSF) with parameters as given in Problem 20.14. The zero dispersion wavelength is around 1550 nm.

20.15 Let the normalised Gaussian field be given by

$$\psi(x, y) = N e^{-\frac{x^2+y^2}{w^2}} = N e^{-\frac{r^2}{w^2}}$$

Thus, the normalisation condition

$$\iint |\psi(x, y)|^2 \, dx\, dy = 1$$

will give us

$$1 = N^2 \left[\int_{-\infty}^{+\infty} dx\, e^{-\frac{2x^2}{w^2}} \right] \times \left[\int_{-\infty}^{+\infty} dy\, e^{-\frac{2y^2}{w^2}} \right] = N^2 \left[\sqrt{\frac{\pi w^2}{2}} \right]^2$$

where we have used the integral

$$\int_{-\infty}^{+\infty} e^{-\alpha x^2 + \beta x} dx = \sqrt{\frac{\pi}{\alpha}} \exp\left[\frac{\beta^2}{4\alpha}\right]$$

Thus, $$N = \frac{1}{w}\sqrt{\frac{2}{\pi}}$$

The normalisation condition can also be written as

$$\iint |\psi(r,\phi)|^2\, r\,dr\,d\phi = 1$$

which will give us

$$1 = N^2\left[\int_0^{+\infty} r\,dr\; e^{-\frac{2r^2}{w^2}}\right]\times\left[\int_0^{2\pi} d\phi\right] = N^2\left[-\frac{w^2}{4}e^{-\frac{2r^2}{w^2}}\right]_0^{\infty}[2\pi]$$

$$= N^2\left[\frac{\pi w^2}{2}\right]$$

Thus, $$N = \frac{1}{w}\sqrt{\frac{2}{\pi}}$$

20.16 Consider two identical single mode fibers joined together with a transverse misalignment of u (along the x-axis). The fractional power that is coupled to the fundamental mode of the second fiber is given by the overlap integral

$$T = \left|\iint \psi_1(x,y)\,\psi_2(x,y)\,dx\,dy\right|^2$$

where $\psi_1(x,y)$ and $\psi_2(x,y)$ are the normalised field patterns of the fundamental mode of the two fibers. Thus,

$$\psi_1(x,y) = Ne^{-\frac{x^2+y^2}{w^2}}$$

Since for the second fiber, we have a transverse misalignment of u (along the x-axis), we will have

$$\psi_2(x,y) = Ne^{-\frac{(x-u)^2+y^2}{w^2}}$$

Thus, $$T = N^4\left|\iint e^{-\frac{x^2+y^2}{w^2}}\,e^{-\frac{(x-u)^2+y^2}{w^2}}\,dx\,dy\right|^2$$

$$= N^4 e^{-\frac{2u^2}{w^2}}\left|\int_{-\infty}^{+\infty} e^{-\frac{2x^2}{w^2}}e^{\frac{2xu}{w^2}}dx\int_{-\infty}^{+\infty} e^{-\frac{2y^2}{w^2}}dy\right|^2$$

$$= \left(\frac{1}{w^2}\frac{2}{\pi}\right)^2 e^{-\frac{2u^2}{w^2}}\left|\sqrt{\frac{\pi w^2}{2}}\,e^{\frac{u^2}{2w^2}}\times\sqrt{\frac{\pi w^2}{2}}\right|^2$$

where we have used the integral

$$\int_{-\infty}^{+\infty} e^{-\alpha x^2+\beta x}\,dx = \sqrt{\frac{\pi}{\alpha}}\exp\left[\frac{\beta^2}{4\alpha}\right]$$

Thus, $$T = \exp\left(-\frac{u^2}{w^2}\right).$$

20.17 $\beta^2 = \beta_{mn}^2 \approx k_0^2 n_1^2 - 2(m+n+1)\gamma k_0; \quad m, n = 0, 1, 2, 3, \ldots$

Thus,
$$\beta = \beta_{mn} \approx k_0 n_1 \left[1 - 2(m+n+1)\frac{\gamma}{k_0 n_1^2}\right]^{1/2}$$
$$\approx k_0 n_1 - (m+n+1)\frac{\gamma}{n_1} \approx \frac{\omega}{c} n_1 - (m+n+1)\frac{\gamma}{n_1}$$

Thus,
$$\frac{1}{v_g} = \frac{d\beta_{mn}}{d\omega} \approx \frac{n_1}{c}$$

independent of the mode number and, in this approximation, all modes travel with the same group velocity.

20.18 For far field, we must have
$$z >> \frac{d^2}{\lambda} \approx 77\ \mu\text{m}$$

The detector is usually placed at a much greater distance than 77 μm.

20.19 Using the approximate expression for the spot size dependence on V-number (see Eq. 23), we see that if the core radii of the two fibers are a_1 and a_2 and their spot sizes are w_1 and w_2 then
$$\frac{w_1}{w_2} = \frac{a_1}{a_2}$$

20.20 The excitation efficiency is proportional to the overlap integral between the field distribution of the exciting field and the modal field distribution. The field distribution of the LP_{11} modal field has a ϕ dependence of the form $\cos\phi$ or $\sin\phi$. Since the incident field distribution has no ϕ dependence when the overlap integral is evaluated over the range from 0 to 2π, the integral will vanish. Thus, the excitation efficiency would be zero.

20.21 (a) LP_{11} (b) LP_{12} (c) LP_{21} (d) LP_{22}.

Integrated Optics

21

A Quick Review

Integrated optics deals with optical devices on a planar substrate much like integrated electronic circuits. By having the optical components like source, modulators, switches, taps, etc. the integrated optical circuit is to process optical signals in an efficient way and also have lower power consumption. Using waveguides for guided propagation of light among different devices on the same integrated optical chip, it is possible to achieve very efficient functions of modulation, switching, filtering, etc.

Figure 21.1 shows a planar waveguide in which a high index planar region is surrounded by lower index regions. The high index region is usually fabricated on a substrate and the upper region is referred to as cover. If the refractive indices of the substrate and the cover are the same, then such a waveguide is referred to as a symmetric planar waveguide. If the two refractive indices are different then it corresponds to an asymmetric planar waveguide.

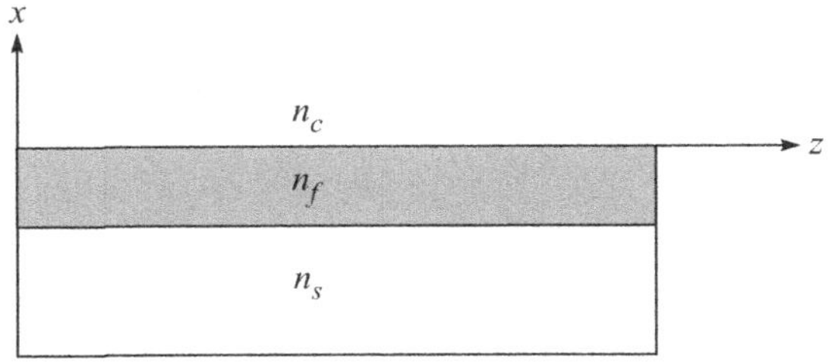

Fig. 21.1

In a planar waveguide the modes split into two groups namely TE and TM modes. In the case of TE modes the electric field is purely transverse (along y-direction in Fig. 21.1) while the magnetic field has both x and z components. Similarly for the TM mode the magnetic field is purely transverse (along the y-direction) and the electric field has components along x and z. For a given refractive indices of the film, substrate and cover and a film thickness and a wavelength of operation the waveguide would support only a finite number of guided TE and TM modes. In a symmetric planar waveguide the fundamental modes TE_0 and TM_0 have no cutoffs while in an asymmetric waveguide even the fundamental modes have a finite cutoff.

The guided modes of a planar waveguide have to satisfy boundary conditions on the interface and also the fields should tend to zero at large distances from the waveguide (see Chapter 19). For a step index planar waveguide with a film index of n_f, substrate index of n_s and cover index of n_c, the propagation constant of TE modes is obtained as solutions of the following transcendental equation:

$$\tan \kappa_f d = \frac{\dfrac{\gamma_c}{\kappa_f} + \dfrac{\gamma_s}{\kappa_f}}{1 - \dfrac{\gamma_c \gamma_s}{\kappa_f^2}} \tag{1}$$

where,

$$\kappa_f = k_0 \sqrt{(n_f^2 - n_{\text{eff}}^2)}$$
$$\gamma_s = k_0 \sqrt{(n_{\text{eff}}^2 - n_s^2)} \tag{2}$$
$$\gamma_c = k_0 \sqrt{(n_{\text{eff}}^2 - n_c^2)}$$

Here n_{eff} represents the effective index of the mode. We also define the V-parameter by the following equation:

$$V = k_0 d \sqrt{(n_f^2 - n_s^2)} \tag{3}$$

The electric field distribution of the TE mode is given by[1]

$$E_y = A e^{-\gamma_c x}; \qquad x \geq 0$$
$$= A\left[\cos \kappa_f x - \frac{\gamma_c}{\kappa_f} \sin \kappa_f x\right]; \qquad 0 \geq x \geq -d$$
$$= A\left[\cos \kappa_f d + \frac{\gamma_c}{\kappa_f} \sin \kappa_f d\right] e^{\gamma_s (x+d)}; \quad x \leq -d \tag{4}$$

where A is a constant and is determined by the power carried by the mode.

The propagation constant of TM modes is determined from the following eigenvalue equation:

$$\tan \kappa_f d = \frac{\dfrac{1}{\gamma_1}\dfrac{\gamma_c}{\kappa_f} + \dfrac{1}{\gamma_2}\dfrac{\gamma_s}{\kappa_f}}{1 - \dfrac{1}{\gamma_1 \gamma_2}\dfrac{\gamma_c \gamma_s}{\kappa_f^2}} \tag{5}$$

where,

$$\gamma_1 = \frac{n_s^2}{n_f^2}; \quad \gamma_2 = \frac{n_c^2}{n_f^2} \tag{6}$$

The field profiles of the TM modes are given by

$$H_y = A e^{-\gamma_c x}; \qquad x \geq 0$$

1. TE and TM modes have been defined in Chapter 19.

$$= A\left[\cos\kappa_f x - \frac{n_f^2}{n_c^2}\frac{\gamma_c}{\kappa_f}\sin\kappa_f x\right]; \qquad 0 \geq x \geq -d$$

$$= A\left[\cos\kappa_f d + \frac{n_f^2}{n_c^2}\frac{\gamma_c}{\kappa_f}\sin\kappa_f d\right]e^{\gamma_s(x+d)}; \quad x \leq -d \tag{7}$$

The cut off of the modes are given by

$$V_c^{\mathrm{TE}} = \tan^{-1}\sqrt{a} + m\pi; \quad \text{TE modes}$$

$$V_c^{\mathrm{TM}} = \tan^{-1}(\sqrt{a}/\gamma_2) + m\pi; \quad \text{TM modes} \tag{8}$$

A directional coupler consists of two closely lying waveguides in which there is an interaction between the two waveguides through the evanescent fields. If the amplitudes of the modes of the two waveguides are given by A and B, then they satisfy the following coupled equations (neglecting self coupling terms):

$$\frac{dA}{dz} = -i\kappa B e^{i\Delta\beta z} \tag{9}$$

$$\frac{dB}{dz} = -i\kappa A e^{-i\Delta\beta z} \tag{10}$$

where κ is the coupling co-efficient which depends on the waveguide parameters, the wavelength of operation and the distance between the two waveguides. The quantity $\Delta\beta$ represents the difference in the propagation constant of the two interacting modes.

If at the input power is coupled into one of the waveguides, then the fractional power at any value of z in the coupled waveguide is given by

$$P_2(z) = \frac{\kappa^2}{\kappa^2 + \frac{\Delta\beta^2}{4}}\sin^2\left[\sqrt{\left(\kappa^2 + \frac{\Delta\beta^2}{4}\right)}\, z\right] \tag{11}$$

The fractional power remaining in the input waveguide is $1 - P_2(z)$.

When $\Delta\beta = 0$, i.e., when the two modes have the same propagation constant then there is complete power exchange between the two waveguides. If $\Delta\beta \neq 0$ then the power transfer is incomplete. This is used in realising optical switches using the electro optic effect.

A periodic variation in the waveguide property such as refractive indices of core or cladding or the thickness of the waveguide can lead to coupling of light between a forward propagating and a backward propagating mode. If the effective index of the mode is n_{eff}, then the spatial period Λ required for efficient coupling to the same mode propagating in the backward direction is

$$\Lambda = \frac{\lambda_B}{2n_{\mathrm{eff}}} \tag{12}$$

where λ_B is called the *Bragg wavelength*. Such a device behaves as a wavelength filter. The peak reflectivity of such a periodic structure is given by

$$R_p = \tanh^2\left(\frac{\pi\Delta n L I}{\lambda_B}\right) \tag{13}$$

where $I(0 < I < 1)$ is an overlap integral between the interacting modes and the periodic perturbation, Δn represents the peak value of refractive index modulation of the periodic variation and L represents the length of the periodic medium.

Assuming the modes to have an approximate Gaussian transverse field profile we obtain for the overlap integral

$$I \approx 1 - e^{-2a^2/w_0^2} \tag{14}$$

where a is the width of the region having the periodic variation (assumed to be symmetrically placed in the core of the waveguide) and w_0 is the Gaussian spot size of the mode.

If the length of the grating is L and the coupling coefficient for the coupling is denoted by κ then the bandwidth of the filter is given by

$$\Delta\lambda = \frac{\lambda_B^2}{n_{\text{eff}} L}\left[1 + \left(\frac{\Delta n L I}{\lambda_B}\right)^2\right] \tag{15}$$

PROBLEMS

21.1 Consider an x-cut and a z-cut $LiNbO_3$ crystal. Planar waveguides are grown on these by proton exchange techniques. What modes can the waveguides support when light propagates along the 'y' and 'z' directions in the x-cut waveguide and along the 'x' and 'y' directions in the z-cut waveguide?

21.2 A directional coupler with two identical waveguides is designed to have a coupling length of 2 mm. What is the permissible variation in the fabrication of the length of the coupler so that in the cross-state if unit power is incident on waveguide 1, the power emanating from the same waveguide is less than –20 dB.

21.3 Consider two directional couplers, the cross-sections of which are shown in Fig. 21.2;

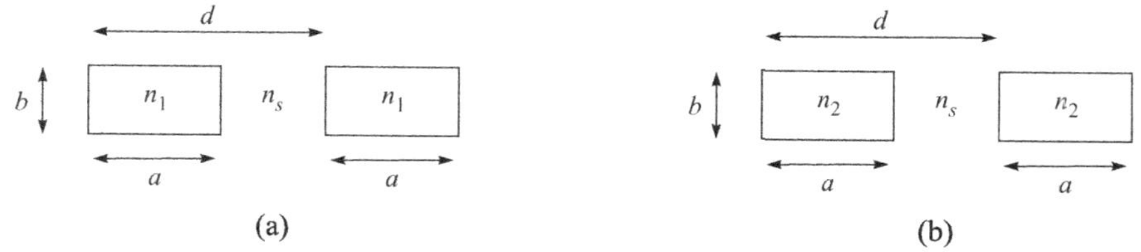

Fig. 21.2

If $n_1 > n_2$, which coupler has a larger coupling length and why?

21.4 A directional coupler with 50% coupling ratio is to be fabricated so that it is least sensitive to fabrication errors in the coupler length. What value of $\Delta\beta/\kappa$ would you chose and why?

21.5 Consider a symmetric planar waveguide with $n_1 = 2.3$, $n_2 = 2.2$ and $d = 2$ μm operating at $\lambda_0 = 1.0$ μm.

(a) How many guided TE and TM modes will the waveguide support?

(b) What are the minimum and maximum possible values of n_{eff} of the TE_1 mode?

(c) In what range of λ_0 values will the waveguide be single moded (TE_0 and TM_0)?

21.6 Consider a symmetric planar waveguide with the following refractive index profile:

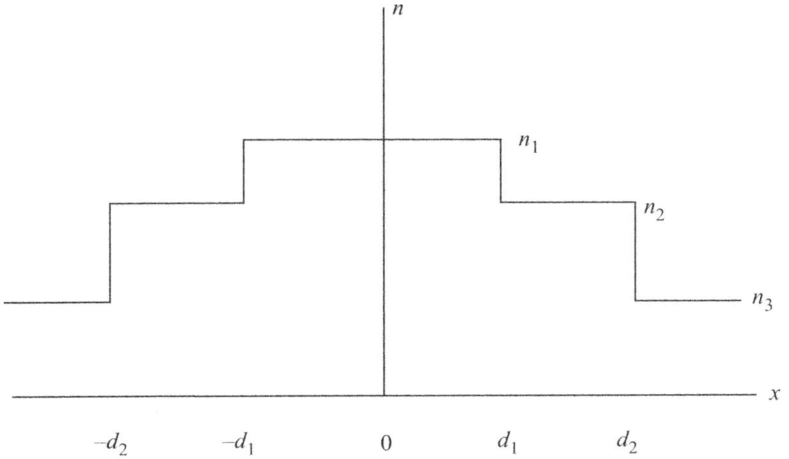

Fig. 21.3

For $n_1 > n_{\text{eff}} > n_2$, write down the solution for E_y for symmetric TE modes in various regions defining all the quantities.

21.7 Consider an asymmetric planar waveguide with $n_f = 1.5$, $n_c = 1.0$ and $n_s = 1.495$. If the waveguide is operated at 1 μm, then

(a) In what range should the thickness lie so that only TE_0 mode can propagate.

(b) What is the penetration depth of the mode in the cover when it is just at the cutoff?

21.8 Consider a planar inhomogeneous waveguide described by a refractive index variation;

$$n(x) = 1.48 + 0.02\,e^{-x/10}; \quad x > 0$$
$$= 1.0; \quad x < 0$$

where, x is measured in μm.

(a) A ray ($\lambda = 1$ μm) enters the waveguide horizontally at $x = 5$ μm.

(i) Calculate the propagation constant of the ray.

(ii) The angle at which the ray will strike the surface at $x = 0$.

(b) What is the range of n_{eff} for guided modes in the waveguide?

(c) A mode with $\beta/k_0 = 1.49$ is propagating in the waveguide. What are the turning points of the rays corresponding to this mode?

21.9 Consider a symmetric step index planar waveguide with a film index n_f and a substrate and cover index of n_s. If the effective index of the fundamental mode is n_{eff}, obtain an expression for the approximate mode size if the wavelength is λ_0.

21.10 In the figure below is shown the cross-section of a strip waveguide with electrodes for fabricating a phase modulator using $LiNbO_3$.

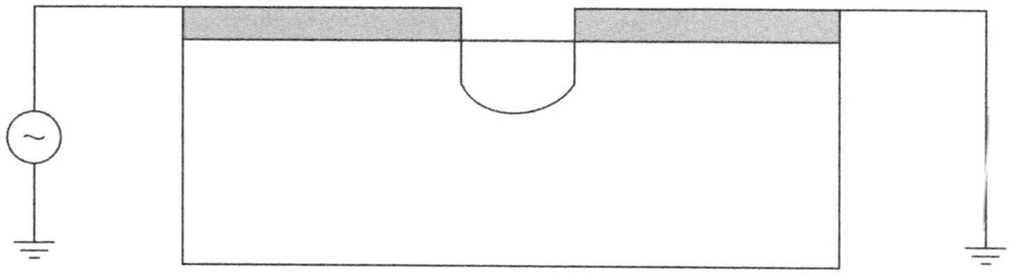

Fig. 21.4

What should be the direction of the optic axis of the crystal and the state of polarisation of the mode for minimum voltage for π phase shift. (Electro-optic effect is discussed in Chapter 22).

21.11 In the figure below is shown a *Y*-branch with input as shown. If the power in the input waveguide is P_0 what would be the output power? Assume all branches to be single mode waveguides.

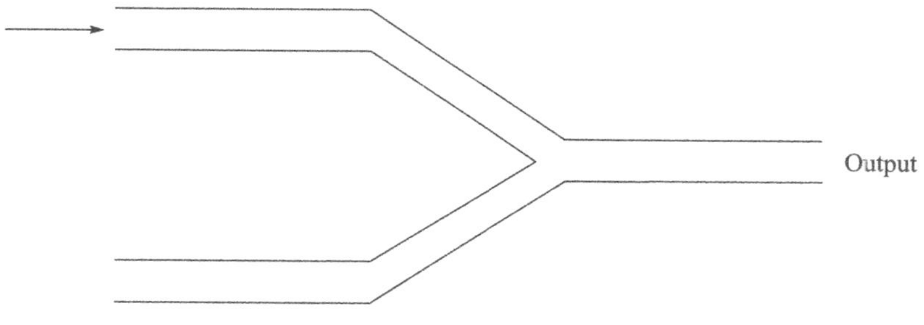

Fig. 21.5

21.12 Obtain the variation of power in the two waveguides of a directional coupler (with $\Delta\beta = 0$) with the following initial conditions;

(a) $a(0) = 1, \quad b(0) = 1$
(b) $a(0) = 1, \quad b(0) = -1$
(c) $a(0) = 1, \quad b(0) = i$
(d) $a(0) = 1, \quad b(0) = -i$

21.13 An acousto-optic tunable filter is to be made in $LiNbO_3$ with an effective index difference between the TE and TM modes of 0.07 at $\lambda_0 = 1.3$ μm. What should be the acoustic frequency if the acoustic velocity is 4 km/s? (Acousto-optic effect is discussed in Chapter 23).

21.14 The coupled mode equations for periodic coupling between two modes propagating in the same direction (with a period satisfying the phase matching condition, $\Gamma = 0$) are given as;

$$\frac{dA}{dz} = -i\kappa B; \quad \frac{dB}{dz} = -i\kappa A$$

where A and B represent the amplitude of the two modes and κ represents the coupling co-efficient. For efficient coupling between two modes with effective indices 1.5086 and 1.5046 at $\lambda_0 = 0.6$ μm, what periodicity would you choose?

21.15 In a Mach-Zehnder interferometer with unequal arm lengths, as the input wavelength is varied, the output power reaches a maximum at 1.53 μm and becomes zero at 1.56 μm (with no other zero between 1.53 μm and 1.56 μm). Assuming the mode effective index to be 1.5 (at both wavelengths) calculate the difference in arm lengths.

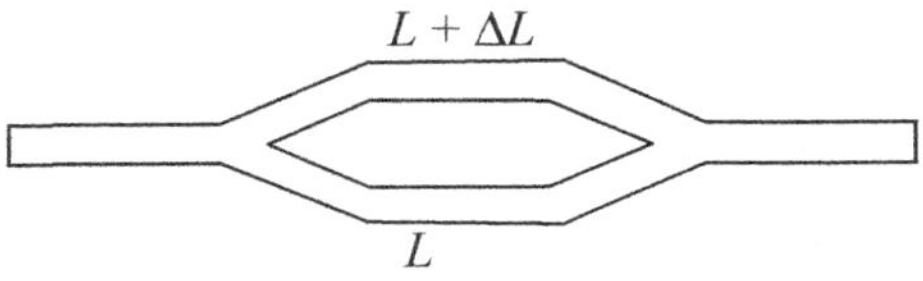

Fig. 21.6

21.16 Shown below is $P_2(z)$ in a directional coupler when unit power is incident at $z = 0$ in waveguide #1. The maximum coupled power is 0.81 and the coupling length is 5 mm.

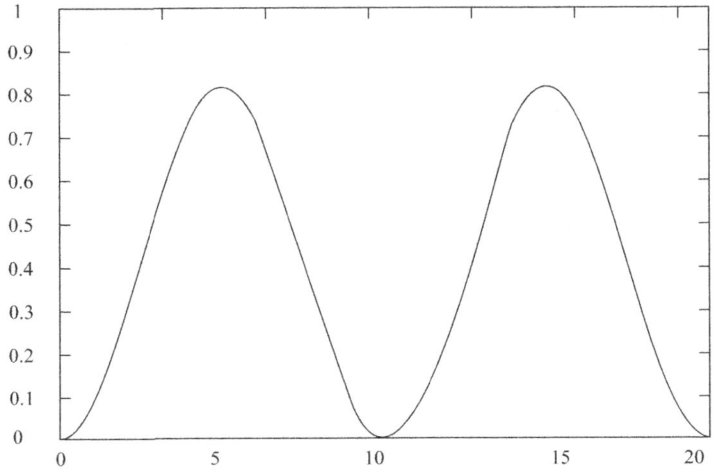

Fig. 21.7

(a) Obtain the values of κ and $\Delta\beta$ of the directional coupler.
(b) What is the distance from the input end where the power is equally divided between the two waveguides?

SOLUTIONS

21.1 In proton exchange technique only the extraordinary index increases in the exchanged region. Hence x-cut y-propagating crystal will support only TE modes while x-cut z-propagating crystal no modes would be supported. Similarly, in the z-cut x-propagating or y-propagating crystals, only TM modes will be supported.

21.2 Power remaining in the input waveguide at any length L is given by

$$P_1 = \cos^2\left(\frac{\pi L}{2L_c}\right)$$

If the length of the coupler is exactly L_c, then the power exiting the first waveguide would be zero. If the length is $L_c \pm \Delta L$ then the power exiting the first waveguide would be

$$P_1 = \cos^2\left(\frac{\pi}{2}\left[1 + \frac{\Delta L}{L_c}\right]\right) = \sin^2\left(\frac{\pi}{2}\frac{\Delta L}{L_c}\right)$$

If the power exiting the first waveguide should be less than 0.01 then assuming $\Delta L << L$, we obtain

$$\Delta L < 0.2\,\frac{L_c}{\pi} \approx 0.13 \text{ mm}$$

21.3 Since $n_1 > n_2$, the individual waveguide modes in the first coupler are more confined to the core than in the second coupler. This implies that the coupling co-efficient of the first coupler would be smaller than that of the second coupler. Hence, the first coupler would have larger coupling length.

21.4 Since the coupler is required for a 3 dB splitting, if we fabricate a coupler for which the maximum energy transfer corresponds to 3 dB, then the variation of coupled power with length will reach a maximum value of 50% and hence such a coupler would be least sensitive to errors in length. For such a coupler we need to have $\Delta\beta = 2\kappa$.

21.5 (a) The V value of the waveguide is $2.68\,\pi$. Hence, the number of guided TE and TM modes supported by the waveguide would each be 3.

(b) The maximum and minimum values of n_{eff} of the TE_1 mode are 2.3 and 2.2 respectively.

(c) The waveguide will be single moded in the region $0 < V < \pi$. Thus for $\lambda >$ 2.683 μm, the waveguide would be single moded.

21.6 In the region $-d_1 < x < d_1$, the solution would be

$$E_y = A\sin\kappa x + B\cos\kappa x$$
$$\kappa^2 = k_0^2(n_1^2 - n_{\text{eff}}^2)$$

For symmetric solution $A = 0$ and for antisymmetric solution $B = 0$.
For $d_1 < x < d_2$

$$E_y = Ce^{\delta x} + De^{-\delta x}$$

$$\delta^2 = k_0^2 (n_{\text{eff}}^2 - n_2^2)$$

For $d_2 < x$

$$E_y = Ee^{-\gamma x}$$
$$\gamma^2 = k_0^2 (n_{\text{eff}}^2 - n_3^2)$$

21.7 (a) For the given waveguide, the asymmetry parameter is $a = 82.47$. Hence, for the waveguide to support only the fundamental TE mode, the V number must be less than 1.461 which implies that the thickness should be less than 1.9 μm.

(b) The depth of penetration is given by $1/\gamma_c$ and at cutoff since the effective index of the mode is n_s, the depth of penetration is 0.143 μm.

21.8 (a) (i) $\beta = k_0 n(x_0) \approx 9.375\ \mu\text{m}^{-1}$

(ii) The ray will make an angle of 5.9° with the interface of the waveguide.

(b) The range of effective indices of the guided modes would be $1.48 < n_{\text{eff}} < 1.5$.

(c) The turning points would be $x = 6.93$ μm and $x = 0$.

21.9 The modal size would be approximately given by the sum of the waveguide width and the depths of penetration of the mode into the surrounding cladding regions. The depth of penetration is the distance from the interface where the field drops to $1/e$ of its value on the interface and is given by $1/\gamma$ where

$$\gamma = k_0 \sqrt{(n_{\text{eff}}^2 - n_s^2)}$$

Hence, the modal width would be approximately

$$w = d + \frac{2}{\gamma} = d + \frac{2}{k_0\sqrt{(n_{\text{eff}}^2 - n_s^2)}}$$

21.10 Since the largest electro-optic co-efficient in lithium niobate is r_{33}, for maximum electro-optic change the electric field should be applied along the z-direction and the polarisation should be oriented along the z-direction. Since the electrodes are oriented such that the electric field on the waveguide is horizontal, the crystal must be either x-cut, y-propagating or y-cut, x-propagating and the incident light should be polarised parallel to the surface of the waveguide.

21.11 Since all waveguides are single moded, the output power would be half of the input power. The other half of the power will get radiated away into the substrate.

21.12 Cases (a) and (b) correspond to the excitation of the fundamental mode and the first excited mode of the directional coupler. Hence, in these two cases there will be no exchange of power between the waveguides.

In cases (c) and (d), the two waveguides have equal power but the amplitudes of the electric field are $\pi/2$ out of phase. In one of the cases, as the waves propagate the power will start to get transferred to the first waveguide and in the other case to the second waveguide. In both cases after complete transfer the power will exchange periodically between the waveguides.

21.13 The acoustic wave provides for periodic coupling between the two modes. For efficient coupling the following condition needs to be satisfied:

$$K = \frac{2\pi}{\Lambda} = \Delta\beta = \frac{2\pi}{\lambda_0}\Delta n_{\text{eff}}$$

where K is the propagaton constant of the acoustic wave and Λ is the acoustic wavelength. Substituting for the values we obtain the required acoustic frequency as 215 MHz.

21.14 The required periodicity is 150 μm.

21.15 If the wavelength corresponding to maximum intensity is λ_1 and the wavelength corresponding to the adjacent minimum intensity is λ_2, then we have

$$\frac{2\pi}{\lambda_1} n_{\text{eff}}\,\Delta l = 2m\pi;$$

$$\frac{2\pi}{\lambda_2} n_{\text{eff}}\,\Delta l = (2m+1)\pi;$$

Eliminating m from the above two equations we obtain the required value of $\Delta l = 26.52$ μm.

21.16 The maximum power transfer is given by

$$P_{2,\,\max} = \frac{\kappa^2}{\kappa^2 + \dfrac{\Delta\beta^2}{4}}$$

This happens when,

$$\sqrt{\left(\kappa^2 + \frac{\Delta\beta^2}{4}\right)}\,L = \frac{\pi}{2}$$

Using the above two equations and the given parameters from the figure, we obtain $\kappa = 90\pi\ \text{m}^{-1}$ and $\Delta\beta = 87.18\ \text{m}^{-1}$.

The length at which the power gets equally divided among the two waveguides is given by

$$P_2(L) = \frac{\kappa^2}{\kappa^2 + \dfrac{\Delta\beta^2}{4}}\sin^2\left[\sqrt{\left(\kappa^2 + \frac{\Delta\beta^2}{4}\right)}\,L'\right] = 0.5$$

Using the various values obtained earlier, we get $L' = 2.88$ mm.

Electro-Optic Effect 22

A Quick Review

1. When an external electric field is applied to a medium, then this applied field can change the optical properties such as refractive index, anisotropy, etc., of the medium. This effect is referred to as the electro-optic effect.
2. If changes in the refractive index are proportional to the applied field, then this is referred to as the *linear electro-optic effect* or *Pockels effect*. If the changes are quadratic with respect to the applied electric field, then this is referred to as the *Kerr effect*. Only crystals possessing no inversion symmetry posses the Pockels effect; however Kerr effect is present in all materials.
3. The linear electro-optic effect is defined by the changes in the co-efficient of the index ellipsoid equation:

$$\Delta\left(\frac{1}{n_{ij}^2}\right) = \sum_{k=1,2,3} r_{ijk} E_k; \quad i, j = 1, 2, 3 \tag{1}$$

The co-efficient r_{ijk} is the electro-optic tensor and is a characteristic of the material. Since i and j can be interchanged, we can contract the two indices using the following standard convention:

11->1, 22->2, 33->3, 23,32->4, 13,31->5, 12,21->6

Hence Eq. (1) can be written as

$$\Delta\left(\frac{1}{n_i^2}\right) = \sum_{k=1,2,3} r_{ik} E_k; \quad i = 1, 2, 3, \ldots 6 \tag{2}$$

4. In the absence of the electric field, the index ellipsoid equation in the principal axis system is given by

$$\frac{x^2}{n_x^2} + \frac{y^2}{n_y^2} + \frac{z^2}{n_z^2} = 1 \tag{3}$$

where n_x, n_y and n_z are the principal refractive indices of the medium.

5. In the presence of the applied electric field, the equation for the index ellipsoid in the principal axis system becomes

$$x^2\left(\frac{1}{n_x^2}+r_{11}E_x+r_{12}E_y+r_{13}E_z\right)+y^2\left(\frac{1}{n_y^2}+r_{21}E_x+r_{22}E_y+r_{23}E_z\right)$$
$$+y^2\left(\frac{1}{n_z^2}+r_{31}E_x+r_{32}E_y+r_{33}E_z\right)+2yz\,(r_{41}E_x+r_{42}E_y+r_{43}E_z)$$
$$+2xz\,(r_{51}E_x+r_{52}E_y+r_{53}E_z)+2xy\,(r_{61}E_x+r_{62}E_y+r_{63}E_z)=1 \quad (4)$$

6. The r tensor for KDP and ADP is

$$[r]=\begin{pmatrix}0&0&0\\0&0&0\\0&0&0\\r_{41}&0&0\\0&r_{52}&0\\0&0&r_{63}\end{pmatrix} \quad (5)$$

The values of the coefficients for KDP are $r_{41}=r_{52}=8.77\times10^{-12}$ m/V, $r_{63}=10.5\times10^{-12}$ m/V.

7. When an electric field is applied along the z-direction in KDP, then the index ellipsoid equation becomes

$$\frac{x^2+y^2}{n_o^2}+\frac{z^2}{n_e^2}+2r_{63}xyE_z=1 \quad (6)$$

Since the index ellipsoid contains a cross term, in the presence of the electric field, the orientation of the principal axes have changed. Since the equation is symmetric in x and y, the new principal axes must be rotated by 45° about the z-axis. The new principal refractive indices are

$$n_x'=n_{x'}\approx n_o-\frac{n_o^3r_{63}E_z}{2}$$
$$n_y'=n_{y'}\approx n_o+\frac{n_o^3r_{63}E_z}{2} \quad (7)$$
$$n_z'=n_z\approx n_e$$

Thus, the crystal becomes biaxial in the presence of the electric field along the z-direction.

8. For lithium niobate and lithium tantalite the electro-optic tensor is

$$[r]=\begin{pmatrix}0&-r_{22}&r_{13}\\0&+r_{22}&r_{13}\\0&0&r_{33}\\0&r_{51}&0\\r_{51}&0&0\\-r_{22}&0&0\end{pmatrix} \quad (8)$$

9. Lithium niobate is a uniaxial crystal and hence the index ellipsoid in the absence of an electric field is given by

$$\frac{x^2 + y^2}{n_o^2} + \frac{z^2}{n_e^2} = 1 \tag{9}$$

In the presence of an electric field applied along the z-direction, the index ellipsoid equation becomes

$$x^2\left(\frac{1}{n_o^2} + r_{13}E_z\right) + y^2\left(\frac{1}{n_o^2} + r_{13}E_z\right) + x^2\left(\frac{1}{n_e^2} + r_{33}E_z\right) = 1 \tag{10}$$

In the presence of the electric field along the z-direction, the direction of the principal axes remains the same. The new principal indices are given by

$$\begin{aligned} n_x' &\approx n_o - \frac{n_o^3 r_{13} E_z}{2} \\ n_y' &\approx n_o - \frac{n_o^3 r_{13} E_z}{2} \\ n_z' &\approx n_e - \frac{n_e^3 r_{33} E_z}{2} \end{aligned} \tag{11}$$

Thus, the crystal remains uniaxial but all the three principal indices have changed.

10. The values of the various co-efficients for lithium niobate are

$$\begin{aligned} r_{33} &= 30.8 \times 10^{-12} \text{ m/V} \\ r_{13} &= 8.6 \times 10^{-12} \text{ m/V} \\ r_{51} &= 28 \times 10^{-12} \text{ m/V} \\ r_{22} &= 3.4 \times 10^{-12} \text{ m/V} \end{aligned}$$

11. The longitudinal configuration corresponds to the case when the applied electric field is along the propagation direction of the light beam. In this case, the required voltages for inducing a phase shift are independent of the length of the crystal.

12. The transverse configuration is one in which the electric field is applied perpendicular to the direction of the propagation of the light wave. In this case, the phase shift is determined by the length of propagation while the electric field for a given voltage is determined by the spacing between the electrodes which is nothing but the thickness of the crystal. Hence, in this case, the voltage required for a given phase change depends on the ratio of thickness to the length of the crystal.

13. In the longitudinal configuration, the half wave voltage for a modulator based on KDP is given by

$$V_\pi = \frac{\lambda_0}{2 n_o^3 r_{63}}$$

14. In the transverse configuration the half wave voltage for a modulator based on KDP is given by

$$V_\pi = \frac{\lambda_0}{n_o^3 r_{63}}\left(\frac{d}{L}\right)$$

Here d is the thickness of the crystal and L the length of the modulator.

15. In the case of lithium niobate operating in the transverse configuraration with the electric field applied along the optic axis, the half wave voltage is given by

$$V_\pi = \frac{\lambda_0}{(n_e^3 r_{33} - n_o^3 r_{13})}\left(\frac{d}{L}\right)$$

16. In the transverse configuration, the length and the thickness are not independent due to diffraction. For a given length of the crystal, in order that the beam fit into the crystal the minimum thickness is given by

$$d = 2\sqrt{\frac{\lambda L}{\pi}}$$

where λ is the wavelength within the medium and L is the length of the crystal, d is the thickness of the crystal (see Fig. 22.1)

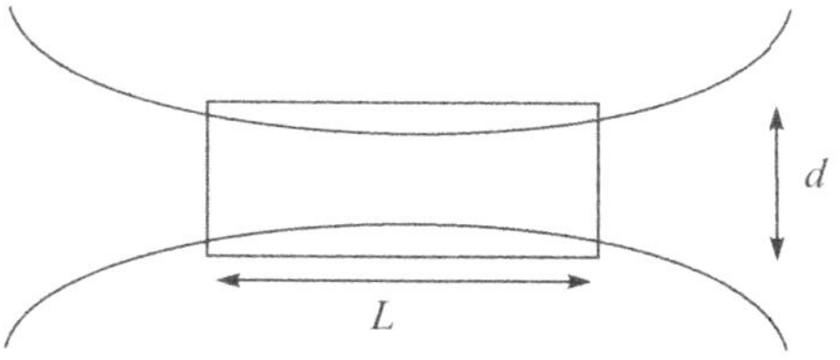

Fig. 22.1

PROBLEMS

22.1 Linearly polarised light (polarised along x) at a wavelength of 500 nm propagates along the z-direction in a KDP crystal of length 2 cm in which an external electric field is applied along the z-direction. State whether the intensity of the output will change as the applied voltage is changed.

22.2 Consider a modulator shown in Fig. 22.2 and state in which cases there would be phase modulation and in which cases there would be amplitude modulation of the output light

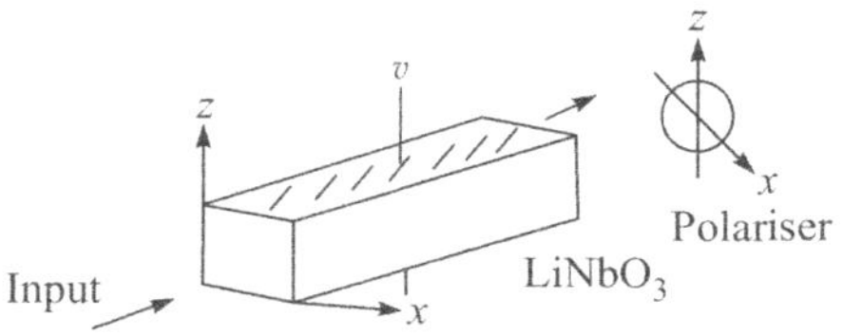

Fig. 22.2

(a) Input light LP along z, output polariser pass axis at 45° to z-axis.
(b) Input light LP at 30° to z, and output polariser pass axis along z.
(c) Input light LP at 30° to z, and output polariser pass axis at 30° to z.
(d) Input light CP, and output polariser at 45° to z axis.
(LP: Linearly Polarised, CP: Circularly Polarised)

22.3 For $BaTiO_3$ (a uniaxial medium) the only nonzero electro optic coefficients are

$$r_{13}, r_{23} = r_{13}, r_{33}, r_{42}, r_{51} = r_{42}. \tag{12}$$

Write down the equation of the index ellipsoid in the presence of an electric field along the y-direction.

22.4 Consider a uniaxial medium with the following electro-optic tensor:

$$[r] = \begin{pmatrix} 0 & -r_{22} & r_{13} \\ 0 & +r_{22} & r_{13} \\ 0 & 0 & r_{33} \\ 0 & r_{51} & 0 \\ r_{51} & 0 & 0 \\ -r_{22} & 0 & 0 \end{pmatrix} \tag{13}$$

(a) Obtain the equation of the index ellipsoid in the presence of an electric field along y-direction.
(b) Which polarisations directions are coupled by the field?
(c) If a uniform electric field is applied will it lead to efficient coupling between the polarisations? Give reasons. Suggest a solution for increasing the efficiency.

22.5 Consider an isotropic crystal having the following nonzero electro-optic co-efficients:

$$r_{41}, r_{52} = r_{41}, r_{63} = r_{41}. \tag{14}$$

(a) Obtain the equation of the index ellipsoid in the absence and in the presence of an electric field along the x-direction.
(b) What are the new principal axes and the corresponding principal refractive indices?
(c) For propagation along x direction what are the eigen polarisations?

22.6 Consider a KDP electro-optic intensity modulator as shown below:

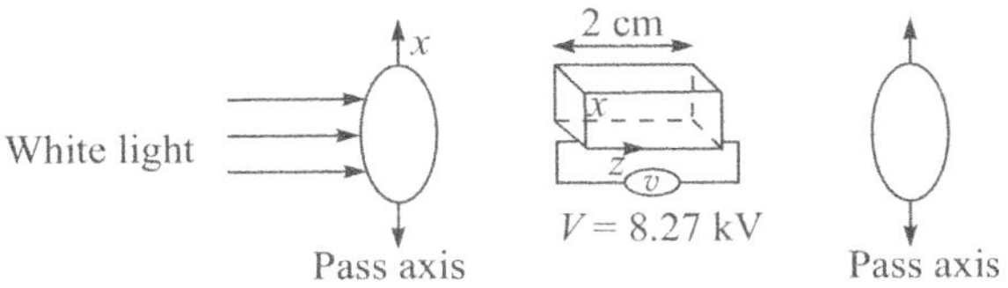

Fig. 22.3

Which wavelengths in the visible region (400 to 800 nm) will be absent in the output? Given that for KDP $n_0 = 1.512$, $r_{63} = 10.5 \times 10^{-12}$ m/V.

22.7 Consider the following arrangement with lithium niobate crystal. What will be the output intensity variation with the applied voltage V if the pass axis of the analyser is (a) parallel to x and (b) is parallel to the pass axis of the input polariser. Given $n_o = 2.30$, $n_e = 2.21$, $\lambda_0 = 1$ μm, length of the crystal = 10 mm.

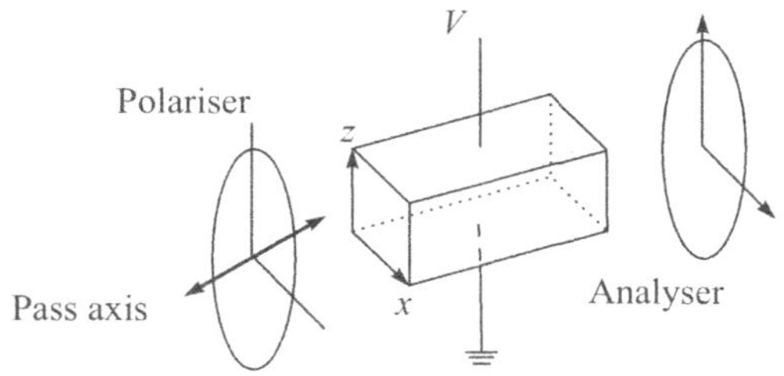

Fig. 22.4

22.8 Consider a KDP phase modulator operating in the longitudinal configuration with an applied voltage given by $V = 100 \sin(2\pi \times 10^9 t)$ volts and an input light beam with a wavelength of 0.6 μm.

(a) What frequency components would be present in the output beam?

(b) Calculate the approximate power in the first upper side band frequency.

You may use the following parameters of KDP: $n_o = 1.51$, $n_e = 1.47$, $r_{63} = 10^{-11}$ m/V.

22.9 Suppose you wish to fabricate a KDP transverse electro optic modulator having a modulation bandwidth of 200 MHz. Calculate the values of l, d and V_π of the modulator. Given $n_o = 1.51$, $r_{63} = 10 \times 10^{-12}$ m/V, $\lambda_0 = 1$ μm, $R_s = 100\ \Omega$, $\kappa = 21$, $\varepsilon_0 = 8.85 \times 10^{-12}$ mks units, safety factor $S = 2$.

22.10 An electro-optic phase modulator (in the transverse configuration) in lithium niobate is 25 mm long. Given that the refractive index of lithium niobate is 2.2, at what modulation frequency will one have zero depth of modulation?

22.11 A laser beam ($\lambda_0 = 1$ μm) passes through an electro-optic phase modulator operating at 1 GHz in the longitudinal configuration in ADP. (Given that for ADP, $n_0 = 1.53$, $n_e = 1.48$ and the only non-zero terms in the r matrix are r_{41}, r_{52} and r_{63}.)

(a) What is the required wavelength resolution of a Fabry-Perot interferometer that can resolve the different frequency components in the phase modulated beam?

(b) For a peak applied voltage of 1 kV, the ratio of the intensity of the first side band to the fundamental frequency is found to be 2.25×10^{-3}. Calculate the value of r_{63}.

22.12 Consider an electro-optic intensity modulator in the transverse configuration in KDP with a crystal length of 25 mm operating at 1500 nm.

(a) What cross sectional dimension can one choose to obtain minimum half wave voltage (assume a safety factor of unity).

(b) What will be the corresponding half wave voltage?

(c) If the total resistance of the modulator circuit is 1000 Ω, calculate the modulator bandwidth. (For KDP, $n_o = 1.51$, $\kappa = 21$. $\varepsilon_0 = 8.85 \times 10^{-12}$ SI units)

22.13 Consider an electro-optic modulator made of lithium niobate consisting of two parts of lengths $L_1 = 3$ mm and $L_2 = 1$ mm with their optic axes pointing in opposite directions as shown. Calculate the electro optically induced phase change suffered by a wave polarised along the z-axis, when a voltage of 100 V is applied.

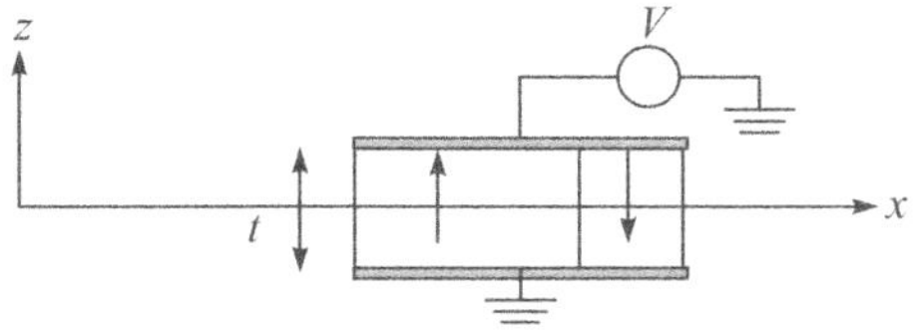

Fig. 22.5

$$n_o = 2.28$$
$$n_e = 2.20$$
$$r_{33} = 30 \times 10^{-12} \text{ m/V}$$
$$r_{13} = 8 \times 10^{-12} \text{ m/V}$$
$$\lambda_0 = 1 \ \mu\text{m}$$
$$t = 2 \text{ mm}$$

SOLUTIONS

22.1 No the intensity will not change. Only the state of polarisation will change. For converting this to an intensity change, the output needs to be sent through an analyser.

22.2 (a) Phase modulation

(b) Phase modulation

(c) Amplitude modulation

(d) Amplitude modulation

22.3 For the given crystal

$$[r] = \begin{pmatrix} 0 & 0 & r_{13} \\ 0 & 0 & r_{13} \\ 0 & 0 & r_{33} \\ 0 & r_{51} & 0 \\ r_{51} & 0 & 0 \\ 0 & 0 & 0 \end{pmatrix} \tag{15}$$

The change in the co-efficients of the index ellipsoid will be given by

$$\begin{pmatrix} \Delta\left(\dfrac{1}{n_{xx}^2}\right) \\ \Delta\left(\dfrac{1}{n_{yy}^2}\right) \\ \Delta\left(\dfrac{1}{n_{zz}^2}\right) \\ \Delta\left(\dfrac{1}{n_{yz}^2}\right) \\ \Delta\left(\dfrac{1}{n_{xz}^2}\right) \\ \Delta\left(\dfrac{1}{n_{xy}^2}\right) \end{pmatrix} = \begin{pmatrix} 0 & 0 & r_{13} \\ 0 & 0 & r_{13} \\ 0 & 0 & r_{33} \\ 0 & r_{51} & 0 \\ r_{51} & 0 & 0 \\ 0 & 0 & 0 \end{pmatrix} \begin{pmatrix} 0 \\ E_y \\ 0 \end{pmatrix} = \begin{pmatrix} 0 \\ 0 \\ 0 \\ r_{51}\,E_y \\ 0 \\ 0 \end{pmatrix} \tag{16}$$

Hence, the equation of the index ellipsoid in the presence of the electric field would be

$$\frac{x^2 + y^2}{n_o^2} + \frac{z^2}{n_e^2} + 2yzr_{51}E_y = 1 \tag{17}$$

22.4 (a) Equation of the index ellipsoid:

$$x^2\left(\frac{1}{n_o^2} - r_{22}E_y\right) + y^2\left(\frac{1}{n_o^2} + r_{22}E_y\right) + \frac{z^2}{n_e^2} + 2yzr_{51}E_y = 1 \tag{18}$$

(b) Since the cross term contains the product yz, the components along the y-direction and z-direction will get coupled.

(c) No a uniform field will not induce efficient coupling since the phase velocities of the waves polarised along y and z are not equal. One would have to use periodic electrodes to overcome this problem.

22.5 (a) In the absence of the applied electric field the index ellipsoid equation is given by

$$\frac{x^2 + y^2 + z^2}{n^2} = 1 \tag{19}$$

where n is the refractive index of the medium.

In the presence of an applied electric field along the x-direction, the index ellipsoid equation becomes

$$\frac{x^2 + y^2 + z^2}{n^2} + 2r_{41}E_x yz = 1 \tag{20}$$

(b) Since the equation is symmetric in y and z, the new principal axes (x, y', z') are rotated by 45° from the x-y-z directions about the x-axis.

The new principal indices of refraction are given by

$$\begin{aligned} n_x &= n \\ n_{y'} &\approx n - \frac{n^3 r_{41} E_x}{2} \\ n_{z'} &\approx n + \frac{n^3 r_{41} E_x}{2} \end{aligned} \tag{21}$$

(c) For propagation along the x-direction, the eigen polarisations are oriented along y' and z'.

22.6 The phase difference introduced between the two eigen polarisation states is given by

$$\Delta\phi = \frac{2\pi}{\lambda_0} n_0^3 r_{63} V$$

Since the polariser and analyser pass axes are parallel to each other, those wavelengths for which $\Delta\phi$ is $m\pi$ will be absent in the output. Thus, the wavelengths that will be absent will be given by

$$\lambda_0 = \frac{0.6}{m}\,\mu\text{m}$$

where m is an integer. Within the visible spectrum, the only wavelength that will be completely absent from the output would correspond to $m = 1$ which gives 0.6 μm.

22.7 (a) If the pass axis of the analyser is along x, then since the eigenmodes of propagation in the absence or presence of the electric field are along x and z, the applied voltage would induce a phase difference between the two components. If the analyser is along the x direction, then the analyser passes only the x-component and hence there would be no change in intensity as the voltage is changed.

(b) If the analyser pass axis is parallel to the input polariser, then analyser passes components of both x- and z-eigen modes and depending on the phase difference between these two components, the output intensity will change.

22.8 (a) The phase modulator leads to a sinusoidal phase modulation of the input light wave. Phase modulation creates side bands and hence the frequencies in the output phase modulated light will be $(\nu_0 \pm m \times 10^9)$ Hz where $m = 1, 2, 3\ldots$

(b) The power in the first side band would be $J_1^2(\zeta)$ where $\zeta \sim 1.8 \times 10^{-3}$.

22.9 The bandwidth of the modulator is given by

$$\Delta\nu = \frac{1}{2\pi RC}$$

where C is the capacitance of the modulator and R is the resistance in the circuit. Assuming the transverse cross section of the modulator to be a square of side d and the length of the modulator to be L, the capacitance of the modulator would be given by $C = \varepsilon_0 \kappa dL/d = \varepsilon_0 \kappa L$. Using the expression for C in the bandwidth expression and substituting the numerical values we obtain $L \sim 43$ mm. The transverse dimension, assuming a safety factor of 2 would be $d \sim 1.9$ mm. Substituting in the expression for the half wave voltage in the transverse configuration we obtain $V_\pi \sim 1.28$ kV.

22.10 $\simeq 5.4$ GHz.

22.11 (a) The phase modulator produces side bands at the frequency of modulation. Since the frequency of modulation is 1 GHz, the output would have side bands separated by 1 GHz. At the given wavelength of 1 μm, this corresponds to a wavelength spacing of 3.3 pm. Hence, the Fabry–Perot should have a resolution of better than 3.3 pm.

(b) The ratio of the intensity of the first side band to the fundamental frequency is given by

$$R = \frac{J_1^2(\varsigma)}{J_0^2(\varsigma)} \tag{22}$$

where,

$$\zeta = \frac{\pi n_0^3 r_{63} V_0}{\lambda_0} \tag{23}$$

Substituting the values of various parameters the ratio can be evaluated.

22.12 (a) Due to diffraction effects, the transverse width d of the crystal and its length L are related through the following equation:

$$d = 2\sqrt{\frac{\lambda L}{\pi}} \tag{24}$$

where λ is the wavelength in the medium. Since the length of the crystal given is 25 mm, we can substitute all the parameter values and obtain $d \sim 0.18$ mm.

(b) The corresponding half wave voltage will be given by

$$V_\pi = \frac{2\lambda_0}{n_0^3 r_{63}} \sqrt{\frac{\lambda_0}{\pi n_o L}} \approx 3.7 \text{ kV}$$

(c) The capacitance of the modulator would be $\varepsilon L = 4.65$ pF. Hence, the RC time constant would be about 215 MHz.

22.13 Since the directions of the optic axes are opposite in the two parts of the crystal, for an applied electric field, the refractive index changes would be opposite in sign in the two parts. Hence, the total phase change suffered by the wave would be

$$\Delta\phi = \frac{2\pi}{\lambda_0} \frac{n_e^3 r_{33} V}{d} (L_1 - L_2) \approx 0.064\,\pi$$

Acousto-Optic Effect 23

A Quick Review

When an acoustic wave is launched into a medium, it creates a periodic strain variation which propagates alongwith the acoustic wave. The strain causes a change in the refractive indices of the medium through the strain optic effect. *Acousto-optic effect* refers to the change in the optical properties of a medium in the presence of an acoustic wave. The presence of the acoustic wave creates a periodic refractive index variation in the medium and this periodic variation leads to diffraction of an incident light beam. In the regime of *Raman Nath diffraction*, the medium behaves as a thin phase grating and one observes multiple order diffraction while in the regime of *Bragg diffraction*, the medium behaves as a volume phase grating and one observes a single order diffraction.

1. If L is the length of the medium, k is the propagation constant of the light wave and K is the propagation constant of the acoustic wave, then Raman Nath diffraction occurs when

$$L << \frac{k}{K^2} \tag{1}$$

2. For Bragg diffraction we must satisfy

$$L >> \frac{k}{K^2} \tag{2}$$

3. The angle of diffraction (within the medium) corresponding to the mth order of Raman Nath diffraction is given by

$$\sin\theta_m = m\frac{\lambda_0}{n_0\Lambda}; \quad m = 0, \pm 1, \pm 2, \ldots \tag{3}$$

where λ_0 is the free space optical wavelength, n_0 is the refractive index of the medium in the absence of the acoustic wave and Λ is the wavelength of the acoustic wave.

4. The diffraction efficiency into the mth order Raman Nath diffraction is given by

$$\eta_m = J_m^2(\zeta) \tag{4}$$

where $\zeta = k_0 \Delta n L$ with Δn is the peak refractive index modulation due to the presence of the acoustic wave and L is the length of interaction with the acoustic wave.

5. Coupled mode equations for small Bragg angle diffraction

$$\frac{d\tilde{A}_0}{dx} = \kappa \tilde{A}_+ e^{i\Delta\alpha x}; \quad \frac{d\tilde{A}_+}{dx} = -\kappa \tilde{A}_0 e^{-i\Delta\alpha x} \tag{5}$$

where,
$$\kappa \approx \frac{\pi \Delta n}{\lambda_0} \tag{6}$$

$$\Delta\alpha = \alpha - \alpha_+ \tag{7}$$

$$\beta_+ = \beta + K \tag{8}$$

$$k^2 \approx k_+^2 = (\alpha^2 + \beta^2) = (\alpha_+^2 + \beta_+^2) \tag{9}$$

The powers carried by the incident and diffracted waves are $|\tilde{A}_0|^2$ and $|\tilde{A}_-|^2$ respectively.

6. Power coupled in small angle Bragg diffraction

$$P_+ = \frac{\kappa^2}{\kappa^2 + \dfrac{\Delta\alpha^2}{4}} \sin^2\left[\sqrt{\kappa^2 + \frac{\Delta\alpha^2}{4}}\, x\right] \tag{10}$$

7. Coupled mode equations for large Bragg angle diffraction:

$$\frac{d\tilde{A}_0}{dx} = \frac{\beta}{|\beta|} \sigma \tilde{A}_+ e^{i\Delta\beta z}; \quad \frac{d\tilde{A}_+}{dx} = -\frac{\beta_+}{|\beta_+|} \sigma \tilde{A}_0 e^{-i\Delta\beta z} \tag{11}$$

where,
$$\sigma \approx \frac{\pi \Delta n}{\lambda_0} \tag{12}$$

$$\Delta\beta = \beta + K - \beta_+ \tag{13}$$

$$\alpha_+ = \alpha \tag{14}$$

$$k^2 \approx k_+^2 = (\alpha^2 + \beta^2) = (\alpha_+^2 + \beta_+^2) \tag{15}$$

and $|\tilde{A}_0|^2$ and $|\tilde{A}_+|^2$ represent the power carried by the incident light wave and the +1 order diffracted light wave.

8. For codirectional coupling, the signs of β and β_+ are the same while in contra directional coupling they have opposite signs
9. Power coupled in codirectional coupling

$$P_+(L) = \frac{\sigma^2}{\sigma^2 + \dfrac{\Delta\beta^2}{4}} \sin^2\left(\sqrt{\sigma^2 + \frac{\Delta\beta^2}{4}}\, L\right) \tag{16}$$

10. Power coupled in contradirectional coupling with Bragg condition satisfied

$$P_+ = \tanh^2 \sigma L \tag{17}$$

11. Bandwidth for codirectional interaction

$$\Delta\lambda = \frac{\sqrt{3}}{2} \frac{\Lambda}{L} \lambda_0 \tag{18}$$

12. Bandwidth of reflection in contra directional coupling

$$\delta\lambda = \frac{\lambda_B^2}{\pi n L} \sqrt{(\sigma^2 L^2 + \pi^2)} \tag{19}$$

13. Relationship between acoustic intensity and strain $\bar{s}$ in the medium

$$I_a = \frac{1}{2}\rho v_a^3 \bar{s}^2 \tag{20}$$

PROBLEMS

23.1 Light undergoes Raman Nath diffraction from acoustic waves propagating in water. State what will happen to the diffracted light waves as (a) the frequency of the acoustic wave is changed (b) as the amplitude of the acoustic wave is changed.

23.2 Consider Raman Nath diffraction of a light wave at $\lambda_0 = 1$ μm from acoustic waves of frequency 10 MHz propagating in water (n = 1.33). If the cell width is 1 cm and Δn produced by the acoustic waves is 2×10^{-6}, obtain the approximate diffraction efficiency into the first order.

23.3 A light beam consisting of two wavelength components at 0.6 and 0.61 μm falls normally on a Raman Nath cell filled with water and undergoes diffraction. In the first order, the angular separation between the diffracted beams is 0.1434×10^{-2} degrees. Assuming the angles of diffraction to be small, calculate the frequency of the acoustic wave. Assume v_a = 1500 m/s and refractive index of water 1.33.

23.4 Raman Nath diffraction of light at 500 nm occurs from a standing acoustic wave at frequency 10 MHz propagating in water. What time dependent intensity variation, if any, of light do you expect to observe along the direction of the first order? [v_a = 1500 m/s, n = 1.33].

23.5 Consider a thin medium (of thickness d along x) with a refractive index variation given by

$$n(z) = n_0 - \alpha z;$$

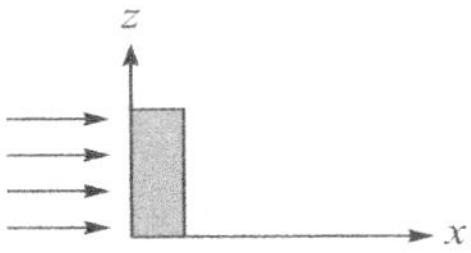

A plane light wave is incident normally as shown in the figure. Obtain the field at the output plane ($x = d$) and interpret the result.

23.6 A laser beam at a wavelength of 1500 nm is propagating in an isotropic medium of refractive index 1.5 and v_a = 4 km/s. An acoustic wave at a specific frequency is launched in such a direction that a frequency upshifted diffracted light wave satisfying the Bragg condition appears at an angle (within the medium) making 2° to the incident light wave. Calculate the frequency difference between the incident and the diffracted light waves.

23.7 Suppose you wish to design a Bragg fiber reflector with the following characteristic:

Center wavelength λ_c = 1550 nm

Peak reflectivity R = 81%

(a) If the fiber grating has $\Delta n = 5 \times 10^{-5}$, what should be the length of the grating?

(b) What would you do to reduce the spectral width of the grating without changing the peak reflectivity?

23.8 When light at $\lambda_0 = 1.06$ μm is passed through flint glass with $n = 1.92$ and $v_a = 3.1$ km/s, we observe an acousto optically diffracted beam appearing at right angles. What is the frequency and direction of propagation of the acoustic wave in the medium?

23.9 Consider a glass slide of refractive index 1.5 with a thickness varying as

$$t = t_0 + \Delta t \sin 600\,\pi z \text{ (}z\text{ in m)}$$

with $t_0 = 1$ mm, $\Delta t = 0.03$ μm. Light wave at a wavelength of 600 nm is incident normally on the slide and undergoes Raman Nath diffraction

(a) Obtain the angle or which you will observe the *I* order in air.

(b) Calculate the approximate power of the *I* order.

23.10 Consider Bragg reflection under normal incidence from a medium with

$$n = 1.5 + 0.001 \sin 5\pi z; \quad (z \text{ is measured in μm})$$

Obtain the wavelength corresponding to peak reflectivity.

23.11 Light of wavelength $\lambda_0 = 1$ μm undergoes +1 order Bragg diffraction by acoustic waves of frequency f_a in a medium with $n_0 = 2.0$ and $v_a = 5$ km/s. The figure below shows the dependence of P_+ on the angle of incidence θ for a length of interaction of 10 mm.

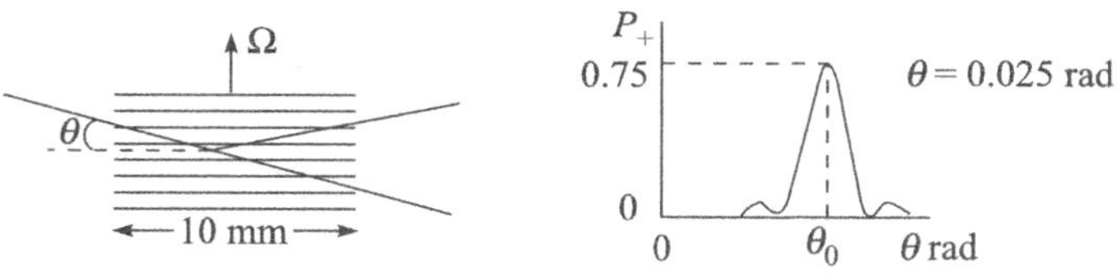

Fig. 23.1

(a) Calculate the acoustic frequency.

(b) Obtain the coupling coefficient κ.

23.12 An acousto-optic tunable filter is to be made with collinear codirectional interaction in $LiNbO_3$ for which $n_0 = 2.28$, $n_e = 2.20$ and $v_a = 3.6$ km/s and with both the light beam and the acoustic beam propagating perpendicular to the optic axis along the same direction.

(a) For $\lambda_0 = 1500$ nm, what should be the acoustic frequency for maximum conversion?

(b) If the filter is to be used for filtering 1500 nm and 1500.8 nm, with negligible cross talk, what should be the minimum length of the device?

(c) If the input light has ordinary polarisation, will the diffracted light correspond to +1 or −1 order?

23.13 A particular Bragg reflector has a peak reflectivity of 64%. By what factor the length of the periodic refractive index region needs to be increased/decreased for the peak reflectivity to be 81%. (**Ans.** 1.34)

23.14 A medium is characterised by a refractive index variation of the form

$$n(z) = 1.5 + 0.001 \sin Kz \qquad (z \text{ is measured in } \mu\text{m})$$

has maximum reflectivity at $\lambda_0 = 1.5$ μm when illuminated normally.

(a) What is the value of K?

(b) If the peak reflectivity is 0.25 what is the length of the interaction?

23.15 The figure below shows the measured reflectivity under normal incidence as a function of wavelength from a Bragg reflector made in a medium with $n_0 = 1.5$.

(a) Calculate the period, the length and the peak refractive index modulation of the grating.

(b) What modification would one do to decrease the bandwidth of the reflector without changing the peak reflectivity?

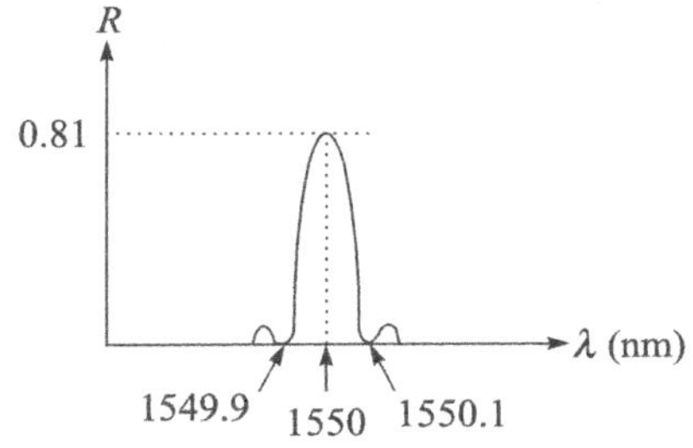

Fig. 23.2

23.16 Consider large angle Bragg diffraction when Bragg condition is satisfied. Show that in the case of codirectional coupling

$$\frac{d}{dz}(|\tilde{A}_0|^2 + |\tilde{A}_+|^2) = 0.$$

23.17 Consider large Bragg angle diffraction when Bragg condition is satisfied. Show that in the case of contra directional coupling

$$\frac{d}{dz}(|\tilde{A}_0|^2 - |\tilde{A}_+|^2) = 0.$$

SOLUTIONS

23.1 (a) As the frequency of the acoustic wave changes, the angle of diffraction of all orders will change and the diffracted beams would scan angularly.

(b) As the amplitude of the acoustic wave changes, the diffraction efficiency of all orders will change leading to an intensity modulation of all orders.

23.2 The diffraction efficiency of the first order is given by $J_1^2(\zeta)$ where $\zeta = (2\pi/\lambda_0)\Delta nL$. For small diffraction efficiency, we can approximate $J_1(x)$ by $x/2$. Substituting the values of various parameters we obtain a diffraction efficiency of 3.95×10^{-4}.

23.3 The angle of emergence (in air) of the first order in Raman Nath diffraction is given approximately by

$$\theta_d = \frac{\lambda_0 f_a}{v_a}$$

For two closely lying wavelengths the difference in deflection angles would be

$$\Delta\theta_d = \frac{(\Delta\lambda_0) f_a}{v_a}$$

Substituting the values of various parameters we obtain the acoustic frequency of 3.75 MHz.

23.4 A standing acoustic wave consists of acoustic waves propagating in both directions. Each of the acoustic waves would induce Raman Nath diffraction along the same set of directions. However, since the two acoustic waves are propagating in opposite directions, the frequency shift produced by the two acoustic waves will be of opposite signs. Hence, the diffracted beams in each order will beat with each other producing sinusoidal intensity modulation. For the acoustic frequency of 10 MHz, the intensity in the first order will be modulated at a frequency of 20 MHz.

23.5 On the plane $x = 0$, the incident plane wave propagating along the x-direction will be given by

$$E = A$$

The thin medium induces phase changes at different values of z depending on the refractive index at z. Thus, assuming the medium to be thin, the field distribution at the output of the medium at $x = d$ will be given by

$$E = Ae^{-ikn(z)d} = A\exp[-ikn_0 d + ikd\alpha z]$$

The above expression represents a plane wave propagating with a z-component of propagation vector $-k\alpha$. Thus, the emerging wave is a plane wave propagating along a direction making an angle of $\sin^{-1}(\alpha d)$ with the x-axis and propagating along the $-z$-direction.

23.6 When small angle Bragg diffraction takes place, the angular deviation between the incident and diffracted beam is $2\theta_B$ where θ_B is the Bragg angle. In the given problem the angle of deviation is given as 2°. Hence, the Bragg angle must be 1°. Using the formula for Bragg angle, we obtain the frequency of the acoustic wave to be 139.6 MHz. Hence, the frequency of the diffracted beam is upshifted by 139.6 MHz with respect to the incident beam.

23.7 (a) Using Eq. (17) for the peak reflectivity, we find that for a peak reflectivity of 0.81, we must have $\sigma L \sim 1.472$. Using the expression of σ, we get the required length of the grating as 1.4 cm.

(b) This can be achieved by increasing the grating length and decreasing Δn such that $\Delta n L$ remains constant.

23.8 Since the diffracted light wave makes an angle of 90° with the incident light wave, the acoustic wave must be propagating at an angle of 45° with respect to the light wave. Using the Bragg condition we obtain for the frequency of the acoustic wave

$$f_a = \frac{\sqrt{2} n v_a}{\lambda_0}$$

Substituting the various values we obtain for the acoustic frequency 7.94 GHz.

23.9 (a) First order Raman Nath diffraction occurs at an angle given by

$$\theta = \sin^{-1}\left(\frac{\lambda_0}{\Lambda}\right) \approx 0.01°$$

(b) The power in the first order is given by $J_1^2(\zeta)$ where $\zeta = k_0 n \Delta t$. Substituting the values we obtain $\zeta \approx 0.47$. Since for small ζ, $J_1(\zeta) \approx \zeta/2$, the diffraction efficiency is approximately given by 0.055.

23.10 The wavelength corresponding to peak reflectivity is given by $\lambda_0 = 2n\Lambda$. Using the various values we obtain the peak wavelength as 1.2 μm.

23.11 (a) Since the Bragg angle is 0.025 rad, the frequency of the acoustic wave is 500 MHz.

(b) The maximum diffraction efficiency is 0.75. Hence we have $\sin \kappa L = \sqrt{3}/2$. Using $L = 10$ mm, we obtain $\kappa \approx 1.05 \times 10^{-2}\ \text{m}^{-1}$.

23.12 (a) In the acousto-optic wavelength filter, the incident and diffracted waves have orthogonal polarisations. Thus, the required Bragg condition is given by:

$$K = k_o - k_e = \frac{2\pi}{\lambda_0}(n_o - n_e)$$

Using the various values, the frequency of the acoustic wave is 192 MHz.

(b) Using the expression for the bandwidth, we obtain $L > 3.04$ cm.

(c) Since the incident light wave and the acoustic wave are propagating in the same direction and the incident wave is an ordinary wave, the diffraction corresponds to –1 order diffraction.

23.13 If the reflectivity is R, then $\tanh \sigma L = \sqrt{R}$. If L_1 and L_2 are the two lengths with reflectivities R_1 and R_2, then

$$\tanh \sigma L_1 = \sqrt{R_1}$$

$$\tanh \sigma L_2 = \sqrt{R_2}$$

Using the given values, we obtain the values of σL_1 and σL_2 and ratio of the two lengths as 1.34.

23.14 (a) $K = 4\pi\ \mu\text{m}^{-1}$.

(b) For the given parameters, $\sigma = 2.09 \times 10^{-3}\ \mu\text{m}^{-1}$. Since the peak reflectivity is 0.25, the length of interaction is 0.26 mm.

23.15 From the figure, we can see that the center wavelength is 1550 nm. Hence, using the Bragg condition, we obtain the period of index modulation as 0.517 μm. Since the peak reflectivity is 0.81, we use Eq. (17), and obtain

$\sigma L \sim 1.4722$. Since the bandwidth of the reflection spectrum is 0.1 nm, we can use Eq. (19) to obtain the length of the grating as 1.768 cm.

Using various values in Eq. (12), we obtain $\Delta n \sim 7.26 \times 10^{-5}$.

23.16 For large Bragg angle diffraction with Bragg condition satisfied and co-directional coupling,

$$\frac{d\tilde{A}_0}{dx} = \sigma\tilde{A}_+; \quad \frac{dA_+}{dx} = -\sigma\tilde{A}_0$$

assuming that both β and β_+ are positive. Now,

$$\frac{d}{dz}(|\tilde{A}_0|^2 + |\tilde{A}_+|^2) = \tilde{A}_0^*\frac{d\tilde{A}_0}{dz} + \tilde{A}_0\frac{d\tilde{A}_0^*}{dz} + \tilde{A}_+^*\frac{d\tilde{A}_+}{dz} + \tilde{A}_+\frac{d\tilde{A}_+^*}{dz} = 0$$

23.17 For large Bragg angle diffraction with Bragg condition satisfied and contra directional coupling,

$$\frac{d\tilde{A}_0}{dx} = \sigma\tilde{A}_+; \quad \frac{dA_+}{dx} = \sigma\tilde{A}_0$$

assuming β to be positive and β_+ to be negative. Now,

$$\frac{d}{dz}(|\tilde{A}_0|^2 - |\tilde{A}_+|^2) = \tilde{A}_0^*\frac{d\tilde{A}_0}{dz} + \tilde{A}_0\frac{d\tilde{A}_0^*}{dz} - \tilde{A}_+^*\frac{d\tilde{A}_+}{dz} - \tilde{A}_+\frac{d\tilde{A}_+^*}{dz} = 0$$

Nonlinear Optics 24

A Quick Review

1. For large intensities of light, matter behaves in a nonlinear fashion and we can describe the electric polarisation of the medium by the following equation:

$$\mathcal{P} = \varepsilon_0 \chi \mathcal{E} + 2\varepsilon_0 d \mathcal{E}^2 + \varepsilon_0 \chi^{(3)} \mathcal{E}^3 + \ldots \tag{1}$$

   Here the first term on the right hand side is the linear polarisation term, χ represents the linear susceptibility and the second and third terms represent nonlinearities. The co-efficients d and $\chi^{(3)}$ represent second order and third order susceptibilities respectively.
2. The second term corresponds to second order nonlinearity and is present only in media possessing no inversion symmetry. The third term corresponds to third order nonlinearity and is found in all media.
3. The second term leads to second harmonic generation, sum and difference frequency generation and parametric amplification. The third term leads to self phase modulation, cross phase modulation and four wave mixing. This is very important in optical fiber communications.
4. The coupled equations describing second harmonic generation are given by

$$\frac{dE_1}{dz} = -i\kappa E_2 E_1^* e^{-i\Delta kz};$$

$$\frac{dE_2}{dz} = -i\kappa E_1^2 e^{i\Delta kz}; \tag{2}$$

$$\kappa = \frac{\omega d}{cn_1}$$

$$\Delta k = k_2 - 2k_1 \tag{3}$$

   where E_1 and E_2 represent the electric field amplitudes of the fundamental wave at frequency ω and the second harmonic at frequency 2ω and n_1 and n_2 represent the refractive indices of the media at these frequencies respectively. k_1 and k_2 represent the propagation constants at frequencies ω and 2ω respectively.
5. The second harmonic generation efficiency is given by

$$\eta = \frac{P_2(L)}{P_1(0)} = \frac{2\mu_0 \omega^2}{cn_1^2 n_2} d^2 L^2 \frac{P_1}{A} \operatorname{sinc}^2\left(\frac{\Delta k L}{2}\right) \tag{4}$$

where P_1 and P_2 represent the powers at frequencies ω and 2ω, respectively, A is the area of cross section of the beams and L is the length of interaction.

6. In vectorial form the nonlinear polarisation for second harmonic generation is given by

$$P_i^{(2\omega)} = \varepsilon_0 d_{ijk} E_j^{(\omega)} E_k^{(\omega)}; \quad i, j, k = 1, 2, 3 \tag{5}$$

where $E_j^{(\omega)}$ represents the jth component of the electric field at ω frequency and similarly $E_k^{(\omega)}$ represents the kth component of the electric field at ω frequency. d_{ijk} is a tensor of rank three. The last two indices can be contracted and the tensor in contracted form is represented by d_{ij} with i = 1, 2, 3 and j = 1.2 … 6.

7. The components of nonlinear polarisation at the second harmonic and the electric field components of the fundamental frequency are related through the following equation:

$$\begin{pmatrix} \tilde{P}_x \\ \tilde{P}_y \\ \tilde{P}_z \end{pmatrix} = \varepsilon_0 \begin{pmatrix} d_{11} & d_{12} & d_{13} & d_{14} & d_{15} & d_{16} \\ d_{21} & d_{22} & d_{23} & d_{24} & d_{25} & d_{26} \\ d_{31} & d_{32} & d_{33} & d_{34} & d_{35} & d_{36} \end{pmatrix} \begin{pmatrix} E_x^{(\omega)} E_x^{(\omega)} \\ E_y^{(\omega)} E_y^{(\omega)} \\ E_z^{(\omega)} E_z^{(\omega)} \\ 2E_y^{(\omega)} E_z^{(\omega)} \\ 2E_x^{(\omega)} E_z^{(\omega)} \\ 2E_x^{(\omega)} E_y^{(\omega)} \end{pmatrix} \tag{6}$$

where, $P_i^{(2\omega)} = \dfrac{1}{2}[\tilde{P}_i e^{2(\omega t - k_1 z)} + c.c.]$ (7)

8. The nonlinearity tensor of lithium niobate is given by

$$[d] = \begin{pmatrix} 0 & 0 & 0 & 0 & d_{15} & -d_{22} \\ -d_{22} & d_{22} & 0 & d_{15} & 0 & 0 \\ d_{31} & d_{31} & d_{33} & 0 & 0 & 0 \end{pmatrix} \tag{8}$$

9. Using a periodic variation of nonlinearity along the propagation direction, it is possible to achieve high conversion efficiency. This technique is referred to as *quasi phase matching*. The quasi phase matching spatial period required for second harmonic generation is given by

$$\Lambda = \frac{\pi c}{\omega|n(2\omega) - n(\omega)|} = \frac{\lambda_0}{2|n(2\omega) - n(\omega)|} \tag{9}$$

where $n(\omega)$ and $n(2\omega)$ are the refractive indices of the material at the fundamental and second harmonic frequencies respectively and λ_0 is the fundamental wavelength in free space.

10. The second harmonic generation can be considered to be a process in which two photons at frequency ω merge to form a single photon at frequency 2ω. The phase matching condition is nothing but momentum conservation condition for this process.

11. In sum frequency generation one photon at frequency ω_1 and one photon at frequency ω_2 merge to form a single photon at frequency ω_3 given by

$$\omega_3 = \omega_1 + \omega_2 \tag{10}$$

Phase matching needs to be satisfied in this process also. If k_1, k_2 and k_3 are the propagation constants of the waves at the frequencies ω_1, ω_2 and ω_3 then for efficient sum frequency generation we must satisfy the following condition:

$$k_3 = k_1 + k_2 \tag{11}$$

12. An optical fiber does not possess the second order nonlinearity and the nonlinear polarisation is given by

$$P_{nl} = \varepsilon_0 \chi^{(3)} E^3$$

13. The refractive index of a medium possessing third order nonlinearity is given by

$$n = n_0 + n_2 I \tag{12}$$

where,
$$n_2 = \frac{3}{4}\frac{\chi^{(3)}}{c\varepsilon_0 n_0} \tag{13}$$

14. In the presence of nonlinearity, the propagation constant of the mode is given by

$$\beta_{NL} = \beta + \gamma P \tag{14}$$

where,
$$\gamma = \frac{k_0 n_2}{A_{\text{eff}}}; \quad \tilde{A}_{\text{eff}} = 2\pi \frac{\left(\int \psi^2(r) r dr\right)^2}{\int \psi^4(r) r dr} \tag{15}$$

15. The phase shift suffered by an optical beam in propagating through a length L of the optical fiber is given by

$$\Phi = \int_0^L \beta_{NL} dz = \beta L + \gamma P_0 L_{\text{eff}} \tag{16}$$

where,
$$L_{\text{eff}} = \frac{(1 - e^{-\alpha L})}{\alpha} \tag{17}$$

is called the *effective length for nonlinear effects*. Here α is the attenuation co-efficient of the fiber.

PROBLEMS

24.1 Two plane waves at frequencies ω_1 and ω_2 are incident on a nonlinear medium. What decides whether the nonlinear effect will lead to the generation of sum frequency $(\omega_1 + \omega_2)$ or difference frequency $(\omega_1 - \omega_2)$?

24.2 The refractive index of a medium at wavelengths of 1000 nm and 500 nm are given by 2.16 and 2.27 respectively. What is the velocity of the nonlinear polarisation generated at the second harmonic frequency?

24.3 Lithium niobate is a crystal with $n_0 > n_e$ and in which phase matching for *SHG* is possible for propagation along the *X* direction (*XYZ* represents the principal axis system). From phase matching considerations, state what should be the state of polarisation of the fundamental and that of the second harmonic. Write down the corresponding phase matching condition.

24.4 Consider *SHG* in quartz for which corresponding to a fundamental wavelength of 694 nm,

$$n_o^{(\omega)} = 1.541; \quad n_e^{(\omega)} = 1.550; \quad n_o^{(2\omega)} = 1.566; \quad n_e^{(2\omega)} = 1.577.$$

Can one obtain birefringent phase matching? Give brief reasons.

24.5 Consider sum frequency generation as shown below. Obtain the maximum power at the sum frequency that can be generated.

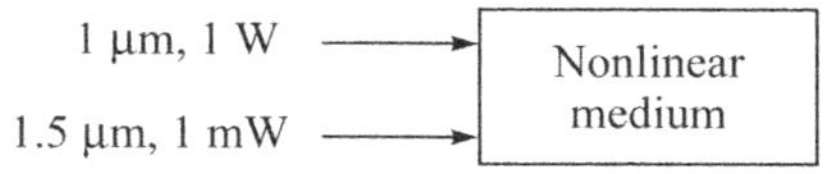

24.6 Consider sum frequency generation with wavelengths of 2 μm and 1 μm.

(a) What is the wavelength of the generated wave?

(b) Write the corresponding phase matching condition in terms of refractive indices at different wavelengths.

24.7 Consider second harmonic generation in lithium niobate and assume that the fundamental is an *e*-wave at 1 μm leading to an *e*-wave at 0.5 μm. Given that the refractive indices at these two wavelengths are 2.15 and 2.25 respectively, obtain the maximum efficiency of second harmonic generation if the input power is 1 W and the area of the beam is 1 mm^2. Assume $d = 30 \times 10^{-12}$ m/V.

24.8 A parametric amplifier operates with a pump wavelength of 1 μm and a signal wavelength of 1.5 μm.

(a) Obtain the wavelength of the idler.

(b) If the input pump power is 1 W and an input signal power of 1 mW is amplified to 1.5 mW at the output, obtain the output power at the idler frequency.

24.9 The ordinary and extraordinary refractive indices of lithium niobate for 1.06 μm and 0.53 μm are given below:

$$n_o(1.06\ \mu\text{m}) = 2.2323; \qquad n_e(1.06\ \mu\text{m}) = 2.1561$$
$$n_o(0.53\ \mu\text{m}) = 2.3247; \qquad n_e(0.53\ \mu\text{m}) = 2.2355$$

(a) If both ω and 2ω waves are extraordinary waves propagating along the *x*-direction, what period Λ_0 would you choose for *QPM* to generate *SHG* most efficiently?

(b) If I choose the 1.06 μm to be an ordinary wave and that at 0.53 μm to be an extraordinary wave what would be the corresponding *QPM* period?

24.10 $LiNbO_3$ is a uniaxial medium with $n_0 > n_e$ with a d tensor given by Eq. (8). It is found that one can achieve noncritical phase matching for *SHG* while propagating along the *X*-axis of the crystal. *XYZ* are principal axes of the crystal. Given that

$$P_i^{(2\omega)} = d_{ijk} E_j^{(\omega)} E_k^{(\omega)} \tag{18}$$

and that the input at ω is *Y* polarised, obtain the components of the nonlinear polarisation at 2ω. Which component of the nonlinear polarisation will be responsible for *SHG*?

24.11 The nonlinear polarisation in an optical fiber is given by

$$P = \varepsilon_0 \chi^{(3)} E^3 \tag{19}$$

Assuming the incidence of waves at frequencies ω_1 and ω_2 (with propagation constants β_1 and β_2) obtain an expression for the nonlinear polarisation generated at frequency ω_1.

24.12 (a) Consider an optical fiber with $n_2 = 3.2 \times 10^{-20}$ m^2/W and an effective mode area of 50 μm^2. If we couple 100 mW into the fiber, obtain the change in the refractive index in the fiber due to nonlinearity, (b) If we propagate a distance of 20 km in the fiber obtain the change of phase due to nonlinearity. Assume a wavelength of 1550 nm and neglect attenuation of the fiber.

24.13 Waves corresponding to frequencies ω_1 and ω_2 are input into an optical fiber. (a) Obtain an expression for the nonlinear polarisation at a frequency $\omega_3 = 2\omega_1 - \omega_2$, and (b) What is the velocity of the nonlinear polarisation wave at ω_3?

24.14 Light waves at frequencies ω_0 and $\omega_0 + \Delta\omega (\Delta\omega << \omega_0)$ are incident in an optical fiber. Under what condition will the generation of the frequency $\omega_0 - \Delta\omega$ be efficient?

24.15 Consider *SHG* for a fundamental wavelength of 1 μm over a crystal of length 2.5 cm. Estimate the maximum allowed value of Δk so that the reduction in peak efficiency due to nonphase matched operation is not less than 81% of the phase matched case.

24.16 The threshold condition of a parametric oscillator is given by

$$\cosh g_{th} L = \frac{1 + R_s R_i}{R_s + R_i} \tag{20}$$

where symbols have their usual meaning. Show that the threshold gain required for a singly resonant *OPO* is much higher than that of a doubly resonant *OPO*.

24.17 Give all possible wavelengths that can be generated using inputs at 800 nm and 1200 nm in a $\chi^{(2)}$ nonlinear medium.

24.18 The phase matched *SHG* efficiency of a 5 cm long *KDP* crystal is 1%. For what value of Δk will the *SHG* efficiency become zero?

24.19 In an *SHG* experiment the second harmonic conversion efficiency is 2% when the input power is 1 W. If the input wavelength is 1 μm, how many photons at the second harmonic frequency are exiting per unit time from the medium?

SOLUTIONS

24.1 The phase matching condition would determine which of the two processes will take place. If the phase matching condition for the sum frequency generation is satisfied, then it will lead to sum frequency generation. Similarly, if the phase matching condition for the difference frequency generation is satisfied then it will lead to difference frequency generation.

24.2 The nonlinear polarisation at the second harmonic frequency travels at the same velocity as the electromagnetic wave at the fundamental frequency. Hence, the velocity of the nonlinear polarisation at the second harmonic frequency is $c/2.16$.

24.3 Since the refractive index increases with increase in frequency and for the given crystal $n_o > n_e$, the fundamental should have ordinary polarisation and the second harmonic should have extraordinary polarisation so that the phase matching condition of $n_o(\omega) = n_e(2\omega)$ can be satisfied.

24.4 Since $n_o^{(2\omega)}, n_e^{(2\omega)} > n_o^{(\omega)}, n_e^{(\omega)}$ it is not possible to achieve birefringent phase matching in this case.

24.5 The wavelength λ_s corresponding to the sum frequency is given by

$$\frac{1}{\lambda_s} = \frac{1}{1} + \frac{1}{1.5}$$

which gives us $\lambda_s = 0.6$ μm.

For complete power conversion by sum frequency generation, the entire power at 1.5 μm would get converted to the sum frequency. In sum frequency generation, one photon at 1.5 μm and 1 photon at 1 μm fuse to form one photon at the sum frequency. Hence, if all the power at 1.5 μm is converted to the sum frequency, then the number of photons exiting at the sum frequency will be equal to the number of photons incident at the wavelength of 1.5 μm. The number of photons corresponding to 1.5 μm (referred to here as the frequency ω_1) entering the crystal per unit time is

$$n = \frac{P}{\hbar\omega_1}$$

which should equal the number of photons exiting the crystal at 0.6 μm. Hence, the power exiting the sum frequency would be

$$P_s = n\hbar\omega_s = \frac{\omega_s}{\omega_1}P = \frac{\lambda_1}{\lambda_s}P = \frac{1.5}{0.6}1 = 2.5 \text{ mW}$$

24.6 (a) Wavelength of the generated wave will be 0.666 μm.

(b) The phase matching condition is given by

$$\frac{n(\lambda_3)}{\lambda_3} = \frac{n(\lambda_1)}{\lambda_1} + \frac{n(\lambda_2)}{\lambda_2} \tag{21}$$

where $n(\lambda)$ is the refractive index of the medium at the wavelength λ and λ's represent free space wavelengths.

24.7 The efficiency for second harmonic generation is given by Eq. (4). In the present case the two waves are not phase matched and hence Δk is nonzero. Substituting values of various parameters we find the efficiency to be approximately 8×10^{-10}. This extremely small value of efficiency is due to nonphase matched interaction.

24.8 (a) Idler wavelength is given by

$$\frac{1}{\lambda_i} = \frac{1}{\lambda_p} - \frac{1}{\lambda_s} \tag{22}$$

Substituting values of the pump and signal wavelengths we obtain the idler wavelength to be 3 μm.

(b) The signal power increases by 0.5 mW. Since the number of photons generated at the idler must be equal to the number of signal photons added to the signal, the exiting power at the idler must be equal to the number of additional signal photons multiplied by the idler photon energy:

$$P_i = \frac{\Delta P_s}{\hbar\omega_s}\hbar\omega_i = \frac{\lambda_s}{\lambda_i}\Delta P_s = 0.25 \text{ mW}$$

24.9 (a) The *QPM* period required is given by Eq. (9). Substituting the values of the extraordinary indices at ω and 2ω frequencies we obtain the required *QPM* period to be 6.67 μm.

(b) In this case, we need to use the ordinary index at 1.06 μm and the extraordinary index at 0.53 μm to obtain the *QPM* period which comes out to be 165.6 μm.

24.10 Since the fundamental wave is *Y*-polarised, we have

$$E_x^{(\omega)} = 0, \quad E_y^{(\omega)} \neq 0, \quad E_z^{(\omega)} = 0 \tag{23}$$

Hence, the nonlinear polarisation generated is given by the following:

$$\begin{pmatrix} \tilde{P}_x \\ \tilde{P}_y \\ \tilde{P}_z \end{pmatrix} = \varepsilon_0 \begin{pmatrix} 0 & 0 & 0 & 0 & d_{15} & -d_{22} \\ -d_{22} & d_{22} & 0 & d_{15} & 0 & 0 \\ d_{31} & d_{31} & d_{33} & 0 & 0 & 0 \end{pmatrix} \begin{pmatrix} 0 \\ E_y^{(\omega)}E_y^{(\omega)} \\ 0 \\ 0 \\ 0 \\ 0 \end{pmatrix}$$

$$= \begin{pmatrix} 0 \\ d_{22}E_y^{(\omega)}E_y^{(\omega)} \\ d_{31}E_y^{(\omega)}E_y^{(\omega)} \end{pmatrix} \tag{24}$$

Since the phase matching is achieved using birefringence phase matching, the generated second harmonic would be *z*-polarised (extraordinary polarisation). Hence, the *z* component of the nonlinear polarisation will be responsible for the generation of the second harmonic.

24.11 The total incident electric field will be given by

$$E = E_1\cos(\omega_1 t - \beta_1 z) + E_2\cos(\omega_2 t - \beta_2 z)$$
$$= \frac{1}{2}(E_1 e^{i(\omega_1 t - \beta_1 z)} + c.c.) + \frac{1}{2}(E_2 e^{i(\omega_2 t - \beta_2 z)} + c.c.) \quad (25)$$

Substituting in the expression for the nonlinear polarisation we obtain the nonlinear polarisation at the frequency ω_1 to be given by

$$P_{NL}^{(\omega_1)} = \frac{\varepsilon_0 \chi^{(3)}}{8}[(3|E_1|^2 + 6|E_2|^2)E_1 e^{i(\omega_1 t - \beta_1 z)} + c.c.] \quad (26)$$

The first term within the brackets represents self phase modulation and the second term the cross phase modulation.

24.12 (a) Using the expression for the change in index we obtain

$$\Delta n = n_2 I = n_2 \frac{P}{A_{\text{eff}}} \approx 6 \times 10^{-11}$$

(b) The nonlinear change in phase is given by

$$\Delta\phi_{NL} = \frac{2\pi}{\lambda_0}\Delta n L = \frac{2\pi}{1.55 \times 10^{-6}} 6 \times 10^{-11} \times 20 \times 10^3 \approx 1.55\pi$$

which represents a large change in index. The change in phase is large inspite of a very small index change since the length of propagation is very large compared to the wavelength.

24.13 Using a similar procedure as in Problem 24.11 we obtain the nonlinear polarisation at ω_3 to be

$$P_{NL}^{(\omega_1)} = \frac{\varepsilon_0 \chi^{(3)}}{8}[3E_1^2 E_2^* \exp[i(\omega_3 t - (2\beta_1 - \beta_2)z)] + c.c.] \quad (27)$$

The velocity of the nonlinear polarisation wave is given by

$$v_{NL} = \frac{\omega_3}{2\beta_1 - \beta_2} \quad (28)$$

24.14 From Problem 24.12, we see that this situation corresponds to $\omega_1 = \omega_0$, and $\omega_2 = \omega_0 - \Delta\omega$. For efficient generation of the new frequency, we need to satisfy the condition that the velocity of the nonlinear polarisation and the electromagnetic wave at $\omega_0 - \Delta\omega$ be the same. This would happen if

$$\Delta\beta = \beta(\omega_0 + \Delta\omega) - \{2\beta(\omega_0) - \beta(\omega_0 - \Delta\omega)\} = 0 \quad (29)$$

Since $\Delta\omega << \omega_0$ we can make Taylor series expansion of the propagation constants about the frequency ω_0 and obtain the following condition:

$$\Delta\beta = \beta(\omega_0) + \Delta\omega \left.\frac{d\beta}{d\omega}\right|_{\omega=\omega_0} + \frac{(\Delta\omega)^2}{2}\left.\frac{d^2\beta}{d\omega^2}\right|_{\omega=\omega_0}$$
$$-\left\{2\beta(\omega_0) - \beta(\omega_0) + \Delta\omega \left.\frac{d\beta}{d\omega}\right|_{\omega=\omega_0} - \frac{(\Delta\omega)^2}{2}\left.\frac{d^2\beta}{d\omega^2}\right|_{\omega=\omega_0}\right\}$$
$$= (\Delta\omega)^2 \left.\frac{d^2\beta}{d\omega^2}\right|_{\omega=\omega_0} = 0 \quad (30)$$

Thus, the required condition is that the fiber should have zero dispersion at $\omega = \omega_0$.

24.15 The second harmonic efficiency decreases as $\text{sinc}^2\left(\frac{\Delta k L}{2}\right)$. Thus, if the efficiency has to be higher than 81% for the case with $\Delta k = 0$, then we must have $\Delta k L/2 < 1.1$ or $\Delta k < 0.88\ \text{cm}^{-1}$.

24.16 In a singly resonant oscillator $R_i = 0$. Thus, the threshold condition is given by

$$\cosh g_{th,s} L = \frac{1}{R_s} \tag{31}$$

For high reflectivity, $R_s \sim 1$ and we can write $R_s = 1 - \delta_s$ with $\delta_s << 1$. Simplifying the above equation we get,

$$g_{th,s} L = \sqrt{2(1 - R_s)} \tag{32}$$

For the doubly resonant oscillator assuming $R_i \sim 1$, we get

$$g_{th,d} L = \sqrt{(1 - R_s)(1 - R_i)} \tag{33}$$

Thus, the ratio of the threshold gain coefficients is given by

$$\frac{g_{th,s}}{g_{th,d}} = \sqrt{\frac{2}{(1 - R_i)}} \tag{34}$$

which is very large for $R_i \sim 1$ showing that the threshold gain coefficient for singly resonant oscillator is much higher than the threshold gain coefficient for a doubly resonant oscillator.

24.17 The new wavelengths would correspond to second harmonics of the two waves, the sum and difference frequencies. These correspond to 400 nm, 600 nm, 480 nm, and 2400 nm.

24.18 The efficiency would be zero when $\Delta k\ L/2 = \pi$. This corresponds to $\Delta k = 0.4\pi\ \text{cm}^{-1}$.

24.19 Since the efficiency is 2%, the output power at the second harmonic is given by 20 mW. For a wavelength of 500 nm, this corresponds to about 5×10^{16} photons.

Table 24.1 Values of second order nonlinear co-efficients of KDP and Lithium Niobate

Material	d_{ij} *(m/V)*
KDP	$d_{36} = 0.42 \times 10^{-12}$
	$d_{14} = 0.42 \times 10^{-12}$
$LiNbO_3$	$d_{31} = d_{51} = 5.95 \times 10^{-12}$
	$d_{33} = 34.4 \times 10^{-12}$
	$d_{22} = 3.07 \times 10^{-12}$

APPENDIX

Multiple Choice Questions

1. Spherical aberration of a thin lens can be reduced by
   (a) Using monochromatic light
   (b) Using a doublet combination
   (c) Using a circular annular mask over the lens
   (d) Increasing the size of the lens
2. A short linear object of length b lies along the axis of a thin convex lens of focal length f at a distance u from the center of the lens. The size of the image is approximately equal to

   (a) $b\left(\frac{u-f}{f}\right)^{1/2}$ (b) $b\left(\frac{u-f}{f}\right)$

   (c) $b\left(\frac{f}{u-f}\right)^{1/2}$ (d) $b\left(\frac{f}{u-f}\right)^{2}$
3. A converging lens is used to form an image on a screen. When the upper half of the lens is covered by an opaque screen, then
   (a) The upper half of the image will disappear
   (b) The lower half of the image will disappear
   (c) The complete image will be formed
   (d) The image will become smaller
4. **Consider a thin lens (of refractive index 1.5) placed in air. The radii of the first** and second surfaces are $R_1 = +20$ cm and $R_2 = -20$ cm. The focal length of the lens will be

   (a) +20 cm (b) –20 cm

   (c) $+\frac{1}{20}$ cm **(d) Infinite**
5. If the radius of curvature of a spherical mirror is 20 cm, the focal length of the mirror is approximately
   (a) 5 cm (b) 10 cm (c) 20 cm (d) 30 cm
6. A plane wave given by $\psi(x, y, z, t) = Ae^{i(\omega t + kz)}$
   (a) Propagates along the $+z$ direction
   (b) Propagates along the $-z$ direction
   (c) Propagates in the x-y plane
   (d) Represents a standing wave

7. A light wave with a free space wavelength of 1000 nm and propagating in a medium of refractive index $\sqrt{3}$ is incident at 60° at an interface with free space. The speed of propagation of the wave in the rarer medium is
   (a) 2×10^8 m/s (b) 3×10^8 m/s
   (c) $2/\sqrt{3} \times 10^8$ m/s (d) $\sqrt{3} \times 10^8$ m/s
8. In the case of total internal reflection
   (a) There is no energy in the rarer medium.
   (b) Energy is present in the rarer medium and it propagates parallel to the interface.
   (c) Energy is present in the rarer medium and it propagates normal to the interface.
   (d) Energy is present in the rarer medium and propagates towards the interface.
9. When a light wave propagates from one medium to another, which of the following associated quantity does not change
   (a) Velocity (b) Frequency
   (c) Wavelength (d) Intensity
10. A wave of frequency ω and wave vector $(\hat{\mathbf{x}} + \hat{\mathbf{y}} - \hat{\mathbf{z}})\omega/c$ is propagating through a medium. The magnitude of the phase difference between the points $A(0, 1, 2)$ and $B(2, 1, 0)$ is
    (a) $\dfrac{6\omega}{c}$ (b) 0 (c) $\dfrac{\omega}{c}$ (d) $\dfrac{4\omega}{c}$
11. Which of the following represents a wave propagating along the negative z-direction (here t is measured in seconds and ψ and z are measured in centimeters)
    (a) $\psi = 2\cos\pi(z - 2t)$ (b) $\psi = 3\sin\pi(z + 4t)$
    (c) $\psi = 3\sin\pi z\cos 2\pi t$ (d) $\psi = 3\sin 2\pi t e^{-2z}$
12. The displacement represented by the following equation
    $$y(x, t) = a\cos(kx + \omega t)$$
    represents a
    (a) Transverse wave propagating in the $+x$ direction
    (b) Longitudinal wave propagating in the $+x$ direction
    (c) Transverse wave propagating in the $-x$ direction
    (d) Longitudinal wave propagating in the $-x$ direction
13. A wave is represented by the following equation
    $$y(x, t) = 5\sin(2x + 3t)$$
    where, x and y are measured in meters and t is seconds. The velocity of the wave is
    (a) 3 m/s (b) $\dfrac{2}{3}$ m/s (c) $\dfrac{3}{2}$ m/s (d) $\dfrac{3}{5}$ m/s

14. The displacement associated with a wave is described by the equation

$$\psi(x, y, z, t) = A\cos(3y - 4z - 5t)$$

where, x, y and z are measured in centimeters and t is seconds. The wave length of the wave is given by

(a) $4\pi\ \text{cm}^{-1}$ (b) $0.4\pi\ \text{cm}^{-1}$

(c) $4\ \text{cm}^{-1}$ (d) $0.4\ \text{cm}^{-1}$

15. The displacement associated with a wave is described by the equation

$$\psi(x, y, z, t) = A\cos(3y - 4z - 5t)$$

where x, y and z are measured in centimeters and t is seconds. The unit vector along the propagating of the wave is given by

(a) $0.6\hat{\mathbf{y}} - 0.8\hat{\mathbf{z}}$ (b) $-0.6\hat{\mathbf{y}} + 0.8\hat{\mathbf{z}}$

(c) $0.6\hat{\mathbf{y}} + 0.8\hat{\mathbf{z}}$ (d) $-0.6\hat{\mathbf{y}} - 0.8\hat{\mathbf{z}}$

16. In a medium characterised by the refractive index variation $n^2(x) = n_1^2 - \gamma^2 x^2$, the ray paths are given by

(a) $x(z) = A + Bz$

(b) $x(z) = Ae^{\alpha z} + Be^{-\alpha z}$

(c) $x(z) = A\sin(\Gamma z + \phi)$

(d) $x(z) = A + Bz^2$

where A, B, α, ϕ and Γ are constants.

17. Wavelength of gamma rays is of the order of

(a) 5000 Å (b) 1 cm (c) 10^{-13} cm (d) 1 Å

18. The **k** vector for the wave described by the equation $E_1 = A\cos\left[\omega t - \dfrac{\omega(x+z)}{c\sqrt{2}}\right]$ is given by

(a) $\mathbf{k} = \dfrac{\omega}{c}\hat{\mathbf{x}} + \dfrac{\omega}{c}\hat{\mathbf{z}}$ (b) $\mathbf{k} = \dfrac{\omega}{c\sqrt{2}}\hat{\mathbf{x}} + \dfrac{\omega}{c\sqrt{2}}\hat{\mathbf{z}}$

(c) $\mathbf{k} = \dfrac{\omega}{c}\hat{\mathbf{x}} - \dfrac{\omega}{c}\hat{\mathbf{z}}$ (d) $\mathbf{k} = \dfrac{\omega}{c\sqrt{2}}\hat{\mathbf{x}} - \dfrac{\omega}{c\sqrt{2}}\hat{\mathbf{z}}$

19. Two sources are said to be coherent when

(a) Their phase difference is π

(b) They are in phase

(c) Their phase difference remains constant with time

(d) Their phase difference depends on time

20. A medium characterised by

$$n(z) = 1.5 + 0.001\sin(2\pi\alpha z)$$

(z is in μm) has a maximum reflectivity at a wavelength of 1500 nm under normal illumination. The value of α (in μm^{-1}) is

(a) 2 (b) 1 (c) π (d) 0.5

21. In an interference pattern between two coherent sources, the maximum of intensity is $9I_0$ and the minimum is I_0. The intensities of the individual waves are

(a) I_0 and I_0 (b) $5I_0$ and $4I_0$

(c) $4I_0$ and I_0 (d) $9I_0$ and I_0

22. The entire setup of Young's double hole experiment is immersed in water. The fringe spacing will

(a) Increase (b) Decrease

(c) Remain the same (d) No fringes will be observed

23. A water film (refractive index 4/3) in air is illuminated by light of wavelength 600 nm at normal incidence. In the reflected light, an interference minimum is observed. If instead, an interference maximum is to be observed, the minimum change required in the thickness of the film (in nm) is

(a) 112.5 (b) 225.0 (c) 300.0 (d) 600.0

24. In a Young's double hole experiment, both holes are covered by thin sheets of refractive indices n_1 and n_2 and thicknesses t_1 and t_2 respectively. If it is required that the fringes should occupy the same position as in the absence of the two sheets, the ratio t_1/t_2 should be equal to

(a) $\frac{n_2}{n_1}$ (b) $\frac{n_2-1}{n_1-1}$ (c) $\frac{n_1-1}{n_2-1}$ (d) $\frac{n_1}{n_2}$

25. Two waves $E_1 = A\cos\left[\omega t - \frac{\omega(x+z)}{c\sqrt{2}}\right]$ and $E_2 = A\cos\left[\omega t - \frac{z\omega}{c}\right]$ are simultaneously propagating through a medium. The planes describing intensity maximum would be given by

(a) $x + z = m\lambda$

(b) $x - z(\sqrt{2} - 1) = m\sqrt{2}\lambda$

(c) $x - z = m\lambda$

(d) $x + z(\sqrt{2} - 1) = m\sqrt{2}\lambda$

26. S_1 and S_2 are two coherent sources (see Fig. 1). The interference fringes formed on the screen LL' will be

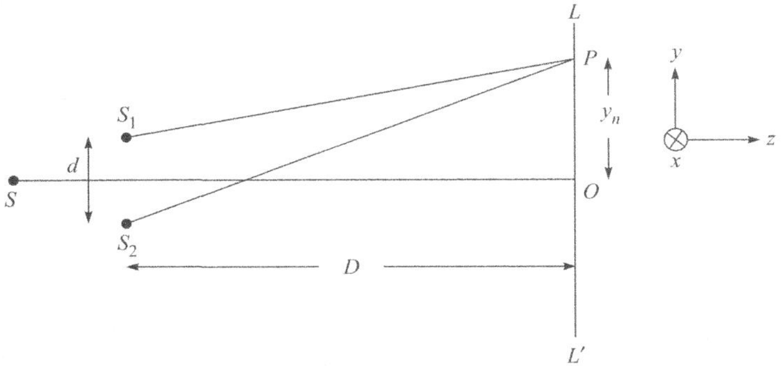

Fig. 1

(a) Strictly circular
(b) Strictly hyperbolae but approximately straight lines
(c) Strictly straight lines
(d) Strictly elliptical with major axis not equal to minor axis

27. S_1 and S_2 are two coherent sources (see Fig. 2). The interference fringes formed on the screen PP' will be
(a) Strictly circular
(b) Strictly hyperbolae but approximately straight lines
(c) Strictly straight lines
(d) Strictly elliptical with major axis not equal to minor axis

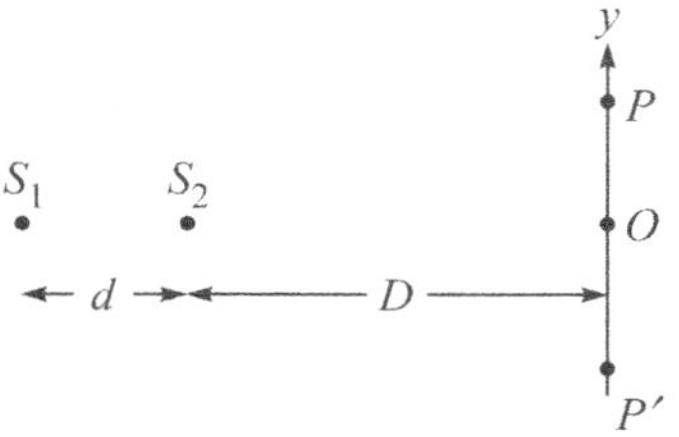

Fig. 2

28. Two coherent plane waves, moving at angle θ with respect to each other, are incident on a screen, placed normal to one of them. They form bright-and-dark fringes on the screen. The separation between two bright fringes is 1 mm and the wavelength of the waves is 6.33×10^{-5} cm. The angle θ will be approximately
(a) 0.0036° (b) 0.036° (c) 0.36° (d) 3.6°

29. A patch of oil on the surface of water produces beautiful colours. This is due to
(a) Diffraction (b) Interference
(c) Total internal reflection (d) Dispersion

30. A microscope lens of refractive index 1.55 is to be coated with a MgF_2 film (n = 1.38) to increase transmission of normally incident yellow light (λ = 5500 Å). The minimum thickness of the film deposited on the lens will be about
(a) 10^{-7} cm (b) 10^{-5} cm
(c) 10^{-3} cm (d) 10^{-1} cm

31. Newton's rings are formed by reflection in the air film between a plane surface and a spherical surface of radius 100 cm. If the radius of third dark ring is 0.09 cm and of twenty eighth 0.25 cm, the wavelength of light used is
(a) 1.038×10^{-5} cm (b) 2.176×10^{-5} cm
(c) 4.352×10^{-5} cm (d) 8.704×10^{-5} cm

32. A Gaussian beam is incident on a converging lens of focal length f with its waist at the front focal plane of the lens. The intensity distribution on the back

focal plane of the lens (assumed to have a transverse dimension much larger than the Gaussian beam width) would be

(a) An Airy pattern
(b) A Gaussian distribution of the same width as at input.
(c) A Gaussian distribution with a width that is inversely proportional to the width of the incident Gaussian beam.
(d) A Gaussian distribution with a width that is directly proportional to the width of the incident Gaussian beam.

33. Diffraction of light
(a) Always leads to divergence of the beam
(b) Can lead to convergence or divergence
(c) Always leads to convergence
(d) Does not take place if the slit size is large

34. Resolution of a telescope
(a) Depends only on the objective lens
(b) Depends both on the objective and the eye piece
(c) Depends on the eye piece only
(d) Depends on the spatial separation between the objective and the eyepiece

35. Consider a grating with $d = 3b$. The ratio of the intensity of the first order to that in the zero order is

(a) 1 (b) $27/4\pi^2$ (c) $9/4\pi^2$ (d) 3/4

36. A laser beam of diameter $2b$ is incident on a convex lens of diameter $2a$ with $a > b$. The radius of the spot on the focal plane of the lens would be about

(a) $\lambda f/a$ (b) $\lambda f/b$ (c) $\lambda a/f$ (d) $\lambda b/f$

37. A plane light wave of wavelength λ_0 is incident on a converging lens. The intensity at the focus is I_0. If the wavelength is increased to $2\lambda_0$ and the incident intensity remains the same, the intensity at the focus will be

(a) I_0 (b) $I_0/2$ (c) $I_0/4$ (d) $2I_0$

38. In a single slit Fraunhoffer diffraction pattern, the intensity of the central maximum is I_0. If the slit width is doubled, the intensity of the central maximum would be

(a) I_0 (b) $I_0/2$ (c) $2I_0$ (d) $4I_0$

39. A beam of light of wavelength 600 nm from a distant source falls on a single slit 1 mm wide and the resulting diffraction pattern is observed on a screen placed 2 m away. The distance between the first dark fringes on either side of the central bright fringe is

(a) 1.2 cm (b) 1.2 mm (c) 2.4 cm (d) 2.4 mm

40. The Fraunhoffer diffraction pattern of a circular aperture is observed on a screen placed at the focal plane of the lens. If the aperture is shifted upwards along its plane
(a) The diffraction pattern will shift downwards.
(b) The diffraction pattern will shift upwards.

(c) The diffraction pattern will not shift.
(d) The diffraction pattern will disappear.

41. A single slit is illuminated by light of wavelengths λ_1 and λ_2 so chosen that the first diffraction minimum of λ_1 coincides with the second minimum of λ_2. This implies that λ_1/λ_2 is equal to
(a) 3/2 (b) 2/3 (c) 1/2 (d) 2

42. As the $f^{\#}$ of a camera lens increases
(a) The image resolution becomes better.
(b) The image resolution worsens.
(c) The image resolution does not change.
(d) The image becomes brighter.

43. For a converging lens forming an image on a screen,
(a) The resolution is determined only by the diameter of the lens.
(b) The resolution is determined by both the diameter and the focal length of the lens.
(c) The resolution is determined only by the focal length of the lens.
(d) The resolution does not depend on the focal length of the lens.

44. A parallel beam of light ($\lambda = 6 \times 10^{-7}$ m) passes through a circular aperture of radius r. A good geometrical shadow will be formed on the screen when
(a) $r = 1$ cm (b) $r = 10^{-3}$ cm
(c) $r = 6 \times 10^{-5}$ cm (d) $r = 10^{-8}$ cm

45. The resolution by a microscope will be better if
(a) Wavelength of light is increased
(b) Wavelength of light is decreased
(c) Focal length of the eyepiece is increased
(d) Focal length of the eyepiece is decreased

46. The spatial frequencies in an object distribution given by

$$g(x) = A + B\cos 6\pi x; \quad (x \text{ in mm})$$

are
(a) 0 and 3 mm^{-1} (b) 0 and 6 mm^{-1}
(c) 3 mm^{-1} (d) 6 mm^{-1}

47. The field distribution on the front focal plane of a lens is given by

$$g(x, y) = 2 + \cos^3\left(\frac{2\pi x}{a}\right)$$

On the back focal plane of the lens we would observe
(a) 5 spots along the x-axis
(b) 5 spots along the y-axis
(c) 2 spots along the x-axis and 2 spots along the y-axis with one spot at the center
(d) 2 spots along the x-axis

48. When a linearly polarised light wave passes through a quarter wave plate, the output state of polarisation is
    (a) Always circularly polarised
    (b) Can be only circularly or linearly polarised
    (c) Always linearly polarised
    (d) Can be linearly or circularly or elliptically polarised
49. When a light wave propagates in a uniaxial medium
    (a) SOP changes if the propagation direction is along the optic axis.
    (b) SOP changes for any direction of propagation.
    (c) SOP changes if the propagation direction is other than along the optic axis.
    (d) SOP does not change for any propagation direction.
50. A right circularly polarised wave is incident normally on a block of uniaxial medium of thickness 4 μm and having $n_o = 1.66$ and $n_e = 1.49$; the optic axis of the medium is parallel to the surface of the block. If the wavelength of the wave is 680 nm, the SOP of the emerging light will be
    (a) Same as at input
    (b) Linearly polarised
    (c) Left circularly polarised
    (d) Linearly polarised along the optic axis
51. An elliptically polarised wave can always be converted to a linearly polarised light wave with the help of a
    (a) $\lambda/4$ plate (b) $\lambda/2$ plate
    (c) λ plate (d) 2λ pate
52. The state of polarisation of a wave with electric field $\mathbf{E} = 0.5(\hat{\mathbf{x}} + \hat{\mathbf{y}}) \cos(\omega t - kz)$ is
    (a) Right circular (b) Left circular
    (c) Elliptical (d) Linear
53. A left circularly polarised beam is incident normally on a polaroid. The intensity of the emergent beam
    (a) Will be almost zero
    (b) Will almost remain the same
    (c) Will decrease by about half
    (d) Will increase slightly
54. A right circularly polarised beam is incident normally on a quarter wave plate. The emergent beam will be
    (a) Unpolarised
    (b) Left Circularly polarised
    (c) Linearly polarised
    (d) Elliptically polarised but not circularly or plane plane polarised

55. A right circularly polarised beam is incident normally on a half wave plate. The emergent beam will be
    (a) Unpolarised
    (b) Left Circularly polarised
    (c) Linearly polarised
    (d) Elliptically polarised but not circularly or plane plane polarised
56. A polarised beam is described by the following equations

$$E_x = A\cos(\omega t - kz) \quad \text{and} \quad E_y = A\sin(\omega t - kz).$$

    The wave is
    (a) Right Circularly polarised
    (b) Left Circularly polarised
    (c) Linearly polarised
    (d) Elliptically polarised but not circularly or plane plane polarised
57. A plane wave propagates along a direction with the unit vector $\hat{\kappa} = \frac{\sqrt{3}}{2}\hat{\mathbf{y}} + \frac{1}{2}\hat{\mathbf{z}}$ in a uniaxial medium. The unit vector along the direction of the **D** of the extraordinary wave is
    (a) $\hat{\mathbf{n}} = -\frac{1}{2}\hat{\mathbf{y}} + \frac{\sqrt{3}}{2}\hat{\mathbf{z}}$
    (b) $\hat{\mathbf{n}} = \frac{1}{2}\hat{\mathbf{y}} + \frac{\sqrt{3}}{2}\hat{\mathbf{z}}$
    (c) $\hat{\mathbf{n}} = -\frac{1}{2}\hat{\mathbf{y}} - \frac{\sqrt{3}}{2}\hat{\mathbf{z}}$
    (d) $\hat{\mathbf{n}} = \hat{\mathbf{x}}$
58. Suppose you wish to make a resonator in which one of the mirrors is a convex mirror of radius of curvature 1 m. If the length of the resonator is to be 1 m, for the resonator to be stable the second mirror should be
    (a) A plane mirror
    (b) A concave mirror of radius of curvature between 1 m and 2 m
    (c) A concave mirror of radius of curvature greater than 2 m
    (d) A concave mirror of radius of curvature less than 1 m
59. A gas laser of length 15 cm oscillates simultaneously in two adjacent longitudinal modes around a wavelength of 600 nm. The wavelength spacing between two longitudinal modes is
    (a) 1.2 pm (b) 1.2 nm (c) 2.4 pm (d) 2.4 nm
60. The SI unit of Einstein's B co-efficient is
    (a) J s^{-1} (b) $\text{m}^3\ \text{s}^{-1}$
    (c) $\text{J}^{-1}\ \text{s}^{-2}\ \text{m}^3$ (d) $\text{J s}^{-1}\ \text{m}^{-2}$
61. If A and B represent the Einstein co-efficients corresponding to a pair of nondegenerate energy levels separated by energy $h\nu$, then the ratio A/B

(a) Does not depend on ν
(b) Is proportional to ν
(c) Is proportional to ν^3
(d) Is proportional to ν^{-3}

62. A laser can be made to operate in a single longitudinal mode
(a) By increasing the length of the cavity
(b) By increasing the diameter of the cavity
(c) By decreasing the length of the cavity
(d) By increasing the pump power

63. Consider an optical fiber having a loss coefficient of 0.25 dB/km. If a power of 0.5 mW is launched at the input, the power exiting the fiber after 68 km of propagation would be
(a) 10 μW (b) 0.125 mW
(c) 0.05 mW (d) 0.1 mW

64. Consider a parabolic index medium with the following refractive index variation:

$$n^2(x) = n_1^2\left[1 - 2\Delta\left(\frac{x}{a}\right)^2\right]$$

where Δ is positive. A ray is launched at $x = a/10$, $z = 0$ parallel to the z-axis. The
(a) Ray will propagate straight without any bending.
(b) Ray will bend away from the axis.
(c) Ray will bend towards the axis.
(d) The ray cannot propagate in the medium.

65. When a ray propagates in the x-z plane in a medium with a refractive index depending only on x-co-ordinate, if $\theta(x)$ is the angle made by the ray with the z-axis, then as the ray propagates
(a) $n(x)\sin\theta(x)$ remains constant
(b) $n(x)\cos\theta(x)$ remains constant
(c) $n(x)\tan\theta(x)$ remains constant
(d) $\sin\theta(x)/n(x)$ remains constant

66. In a symmetric step index planar waveguide
(a) All modes have a finite cutoff V-value
(b) Only the fundamental TE mode has no cutoff
(c) Only the fundamental TM mode has no cutoff
(d) Both the fundamental TE and the fundamental TM modes have no cutoff

67. If the intensity pattern corresponding to an LP_{lm} mode has one zero along the radial direction and four zeros the azimuthal direction, then the values of l and m are
(a) 1 and 1 respectively (b) 0 and 1 respectively
(c) 2 and 1 respectively (d) 2 and 2 respectively

68. Two perfectly aligned single mode fibers will have no loss
    (a) Only if both fibers are identical.
    (b) If both fibers have the same mode field diameter.
    (c) If both fibers have the same V-number.
    (d) If both fibers have the same core radius.
69. Consider a multimode optical fiber having a numerical aperture of 0.2. When all the modes are equally excited, then at the output
    (a) The maximum angle of emergence of the cone of rays is less in water than in air
    (b) The maximum angle of emergence of the cone of rays is more in water than in air
    (c) The maximum angle of emergence of the cone of rays is the same in water as in air
    (d) The change in the maximum angle of emergence of the cone of rays would depend on the core diameter.
70. If the output power is 0.001 mW for an input power of 1 mW, the attenuation in the fiber is
    (a) 10 dB (b) 20 dB (c) 30 dB (d) 0.001 dB
71. Intermodal dispersion is highest in
    (a) Step index single mode fiber
    (b) Step index multimode fiber
    (c) Parabolic index multimode fiber
    (d) Graded index single mode fiber
72. Consider a pulse propagating in the $+x$ direction. At $x = 0$, the time variation is given by

$$\psi(x=0,t) = E_0 e^{-i\omega_0 t} \qquad |t| < \frac{1}{2}\tau$$
$$= 0 \qquad |t| > \frac{1}{2}\tau$$

The spectral widths $\Delta\omega$ of the pulse is given by

(a) $\sim \dfrac{\omega_0}{\tau^2}$ (b) $\sim \omega_0^2 \tau$ (c) $\sim \dfrac{1}{\tau}$ (d) $\sim \dfrac{1}{\tau^2}$

73. Consider a Gaussian pulse given by

$$\psi(x=0,t) = E_0 \exp\left[-\frac{t^2}{2\tau^2}\right] e^{-i\omega_0 t}$$

The spectral widths $\Delta\omega$ of the pulse is given by

(a) $\sim \dfrac{\omega_0}{\tau^2}$ (b) $\sim \omega_0^2 \tau$ (c) $\sim \dfrac{1}{\tau}$ (d) $\sim \dfrac{1}{\tau^2}$

74. The power of a 2 mW laser beam decreases to 15 μW after traversing through 25 km of a single mode option fiber. The attenuation of the fiber is
    (a) 0.085 dB/km (b) 0.85 dB/km
    (c) 8.5 dB/km (d) 85 dB/km

75. A 5 mW laser beam through a 26 km fiber of loss 0.2 dB/km. The power at the output end is
 (a) 3 mW (b) 1.5 mW
 (c) 0.3 mW (d) 0.15 mW

76. Consider a step index fiber with $n_1 = 1.5$, $a = 40$ μm and $\Delta = \dfrac{n_1 - n_2}{n_1} = 0.01$ operating at 850 nm with a spectral width of 20 nm. The value of $\dfrac{d^2n}{d\lambda_0^2} \approx 0.0297\,(\mu\text{m})^{-2}$. The material dispersion is
 (a) 0.17 ns/km (b) 1.7 ns/km
 (c) 17 ns/km (d) 170 ns/km

77. For pure silica, $n(\lambda_0) \approx 1.451 - 0.003\left(\lambda_0^2 - \dfrac{1}{\lambda_0^2}\right)$ where λ_0 is measured in μm.
 The zero material dispersion wavelength is approximately given by
 (a) 0.8 μm (b) 1.32 μm
 (c) 1.55 μm (d) 2.5 μm

78. In a phase matched difference frequency generation setup an incident wave at 1000 nm and having a power of 1 W interacts with a wave at 1500 nm and having a power of 1 mW. If the power exiting at 1500 nm is 1.1 mW, the power exiting at the difference frequency would be
 (a) 0.1 mW (b) 0.05 mW
 (c) 0.5 mW (d) 1 mW

79. Consider a medium with $n_o = 2.26$ and $n_e = 2.20$. A light wave having ordinary polarisation and propagating along the $-x$ direction interacts with an acoustic wave propagating along $+x$ direction and gets diffracted to a wave propagating along $-x$ direction. If Bragg condition is satisfied, the diffracted wave will
 (a) Have the same frequency as the incident wave
 (b) Have a higher frequency than the incident wave
 (c) Have a lower frequency than the incident wave
 (d) Will contain both higher and lower frequencies

80. For the extra-ordinary wave propagating in an uniaxial crystal (with the optic axis along the z-direction)
 (a) $\mathbf{D} \cdot \mathbf{k}$ is always zero.
 (b) D_z is always zero.
 (c) $\mathbf{D}$ is always at right angles to $\mathbf{E}$.
 (d) $\mathbf{D} \times \mathbf{k}$ is always zero.

81. For a Gaussian beam (whose phase front is plane at $z = 0$) and whose spot size at $z = 0$ is w_0, the spot size at large values of z is approximately given by
 (a) $w(z) \approx \dfrac{\lambda w_0}{\pi z}$
 (b) $w(z) \approx \dfrac{\lambda z}{\pi w_0}$

(c) $w(z) \approx \dfrac{w_0 z}{\pi\lambda}$

(d) $w(z) \approx \dfrac{w_0^2 z}{\pi\lambda^2}$

82. A nonlinear medium has the following refractive indices: at 1000 nm, $n_o = 4.80$, $n_e = 6.25$ and at 500 nm, $n_o = 4.86$ and $n_e = 6.32$. If this medium is used for birefringence phase matching, then
   (a) The fundamental will be ordinary (o) and second harmonic will be extraordinarily (e) polarised
   (b) The fundamental will be e-polarised and the second harmonic will be o-polarised
   (c) Both fundamental and second harmonic will be o-polarised
   (d) Both the fundamental and second harmonic will be e-polarised

Answers to Multiple Choice Questions

1. (b)	2. (d)	3. (c)	4. (a)	5. (b)
6. (b)	7. (a)	8. (b)	9. (b)	10. (d)
11. (b)	12. (c)	13. (c)	14. (b)	15. (a)
16. (c)	17. (c)	18. (b)	19. (c)	20. (a)
21. (c)	22. (b)	23. (a)	24. (b)	25. (b)
26. (b)	27. (a)	28. (b)	29. (b)	30. (b)
31. (b)	32. (c)	33. (b)	34. (a)	35. (c)
36. (b)	37. (c)	38. (d)	39. (d)	40. (c)
41. (d)	42. (b)	43. (b)	44. (a)	45. (b)
46. (a)	47. (a)	48. (d)	49. (c)	50. (a)
51. (a)	52. (d)	53. (c)	54. (c)	55. (b)
56. (a)	57. (a)	58. (b)	59. (a)	60. (c)
61. (c)	62. (c)	63. (a)	64. (c)	65. (b)
66. (d)	67. (c)	68. (b)	69. (a)	70. (c)
71. (b)	72. (c)	73. (c)	74. (b)	75. (b)
76. (b)	77. (b)	78. (b)	79. (b)	80. (a)
81. (b)	82. (b)			

References and Suggested Readings

Ab1. Abramowitz M and I Stegun, *Handbook of Mathematical Functions*, National Bureau of Standards, Washington DC, (1965).

An1. Ankiewicz A and C Pask, *Geometric Optics Approach to Light Acceptance and Propagation Ingraded Index Fibers*, *Opt. Quant. Electr.* **9**, 87, (1977).

Bo1. Born M and E Wolf, *Principles of Optics*, Seventh Ed., Cambridge University Press, New York (1999).

Co1. Collier R J, C B Burckhardt and L H Lin, *Optical Holography*, Academic Press, New York (1971).

Da1. Davis C C, *Lasers and Electro-Optics*, Cambridge University Press, Cambridge (1996).

Di1. Ditchburn R W, *Light*, Academic Press, London, (1976).

Fe1. Feynman R P, R B Leighton and M Sands, *The Feynman Lectures on Physics*, Vol. I, Addison Wesley Publishing Co., Reading, Mass. (1965).

Fe2. Feynman R P, R B Leighton and M Sands, *The Feynman Lectures on Physics*, Vol. II, Addison Wesley Publishing Co., Reading, Mass. (1965).

Fr1. Francon M, *Optical Interferometry*, Academic Press, New York, (1966).

Gh1. Ghatak A, *Optics*, Fifth edition, Tata McGraw-Hill, New Delhi (2009); reprinted by McGraw-Hill, New York (2009).

Gh2. Ghatak A and K Thyagarajan, *Contemporary Optics*, Plenum Press, New York (1978); reprinted by Macmillan India, New Delhi..

Gh3. Ghatak A, I C Goyal and R Varshney, *FIBER OPTICA : A Software for Characterising Fiber and Integrated-Optic Waveguides*, Viva Books, New Delhi (1999).

Gh4. Ghatak A, A Sharma and R Tewari, *Understanding Fiber Optics on a PC*, Viva Books, New Delhi (1999).

Gh5. Ghatak A and K Thyagarajan, *Introduction to Fiber Optics*, Cambridge University Press, New York (1998).

Gh6. Ghatak A and K Thyagarajan, *Optical Electronics*, Cambridge University Press, Cambridge (1988).

Gh7. Ghatak A and K Thyagarajan, *Optical Waveguides and Fibers* in *Fundamentals of Photonics* (Editors: A Guenther, L Pedrotti and C Roychoudhuri), Materials developed under project STEP (Scientific and Technology Education in Photonics) by University of Connecticut and CORD, National Science Foundation, USA.

Go1. Goodman J W, *Introduction to Fourier Optics*, Third Edition, Roberts & Co., Englewood, Co. USA (2005).

He1. Hecht E, *Optics*, Pearson Education, Singapore (2002).

Ir1. Irving J and N Mullineux, *Mathematics in Physics and Engineering*, Academic Press, New York (1959).

Je1. Jenkins F A and H E White, *Fundamentals of Optics*, McGraw-Hill, New York, USA (1976).

Kl1. Klein M V and T E Furtak, *Optics*, John Wiley, New York (1986).

Ku1. Kumar A and A Ghatak, *Polarization of Light with Applications in Optical Fibers*, SPIE Tutorial Text (#TT90) in Optical Engineering, SPIE Press, Bellingham, Wa., USA (2011).

Li1. Lin C, *Optical Communications: Single Mode Optical Fiber Transmission Systems,* in *Optoelectronic Technology and Lightwave Communication Systems* (Ed: C Lin), Van Nostrand Reinhold, New York (1989).

Ma1. Marcuse D, *Gaussian Approximation of the Fundamental Modes of a Graded Index Fibers*, J Opt. Soc. Am. **Vol. 68**, 103 (1978).

Ma2. Marcuse D, *Interdependence of Waveguide and Material Dispersion*, App. Opt., **Vol. 18**, pp. 2930-2932, (1979).

My1. Mynbaev D K and L L Scheiner, *Fiber-Optic Communications Technology*, Prentice Hall, USA (2001).

Pa1. Paek U C, G E Peterson and A Carnevale, *Dispersionless Single Mode Light Guides with α* ***Index Profiles,*** *Bell Syst. Tech. J.* **60**, 583 (1981).

Pa2. Pal B P (Ed.), *Guided Wave Optical Components and Devices: Basics, Technology and Applications,* Academic Press (2006).

Ri1. Rinard P M, *Large Scale Diffraction patterns from Circular Objects*, American Journal of Physics, Vol. **44**, p.70, 1976.

Sa1. Saleh B E A and M C Teich, *Fundamentals of Photonics*, John Wiley, New York (1991).

Si1. Siegman A E, *Lasers*, Oxford University Press, Oxford, UK (1986).

Sl1. Sladkova J, *Interference of Light*, Iliffe Books Ltd., London, (1968).

Sm1. Smith F G and T A King, *Optics and Photonics*, John Wiley, Chicester, (2000)

Sm2. Smith H M, *Principles of Holography*, Wiley Interscience, New York (1975).

Sn1. Snyder A W and J D Love, *Optical Waveguide Theory*, Chapman and Hall, London (1983).

St1. Steel W H, *Interferometry*, Cambridge University Press, Cambridge, London, (1967).

St2. Steward E G, *Fourier Optics: An Introduction*, Second Edition, Dover Publications, New York (2004).

Th1. Thyagarajan K and A Ghatak, *LASERS: Fundamentals and Applications*, Springer, New York (2010).

Th2. Thyagarajan K and A Ghatak, *Nonlinear Optics*, in *Encyclopedia of Modern Optics* (Editors: Bob Guenther, Alan Miller, Leopold Bayvel and John Midwinter), Elsevier Ltd. (2005).

To1. Tolansky S, *Multiple Beam Interferometry of Surfaces and Films*, Oxford University Press, London, (1948).

To2. Tolansky S, *An Introduction to Interferometry*, Longmans Green and Co., London, (1955).

Ve1. Verdeyen J T, *Laser Electronics*, Prentice Hall, Englewoodcliffs, N. J (1989).

Made in United States
North Haven, CT
04 November 2026

21730347R10196